Current Topics in Bioenergetics

Volume 8
Photosynthesis: Part B

Current Topics in Bioenergetics

Edited by
D. RAO SANADI

Boston Biomedical Research Institute
Boston, Massachusetts

and
LEO P. VERNON

Brigham Young University
Provo, Utah

VOLUME 8
PHOTOSYNTHESIS: PART B

1978

ACADEMIC PRESS
NEW YORK SAN FRANCISCO LONDON

A Subsidiary of Harcourt Brace Jovanovich, Publishers

ACADEMIC PRESS, INC.
111 Fifth Avenue, New York, New York 10003

United Kingdom Edition published by
ACADEMIC PRESS, INC. (LONDON) LTD.
24/28 Oval Road, London NW1 7DX

LIBRARY OF CONGRESS CATALOG CARD NUMBER: 66–28678

ISBN 0–12–152508–2

PRINTED IN THE UNITED STATES OF AMERICA

Contents

ELECTRON TRANSPORT AND PHOTOPHOSPHORYLATION

Alternate Fates of the Photochemical Reducing Power Generated in Photosynthesis: Hydrogen Production and Nitrogen Fixation

NORMAN I. BISHOP AND LARRY W. JONES

THE PHOTOSYNTHETIC MEMBRANE

Chlorophyll–Protein Complexes and Structure of Mature and Developing Chloroplasts

N. K. BOARDMAN, JAN M. ANDERSON, AND D. J. GOODCHILD

Dynamic Structural Features of Chloroplast Lamellae

CHARLES J. ARNTZEN

Structure and Development of the Membrane System of Photosynthetic Bacteria

GERHART DREWS

GENETIC CONTROL OF THE PHOTOSYNTHETIC MEMBRANE

Genetic Control of Chloroplast Proteins

N. W. GILLHAM, J. E. BOYNTON, AND N.-H. CHUA

Mutations and Genetic Manipulations as Probes of Bacterial Photosynthesis

BARRY L. MARRS

List of Contributors

Numbers in parentheses indicate the pages on which the authors' contributions begin.

JAN M. ANDERSON (35), *Commonwealth Scientific and Industrial Research Organization, Division of Plant Industry, Canberra, Australia*

CHARLES J. ARNTZEN (111), *USDA/ARS: Department of Botany, University of Illinois, Urbana, Illinois*

NORMAN I. BISHOP (3), *Department of Botany and Plant Pathology, Oregon State University, Corvallis, Oregon*

N. K. BOARDMAN (35), *Commonwealth Scientific and Industrial Research Organization, Division of Plant Industry, Canberra, Australia*

J. E. BOYNTON (211), *Departments of Zoology and Botany, Duke University, Durham, North Carolina*

N.-H. CHUA (211), *The Rockefeller University, New York, New York*

GERHART DREWS (161), *Lehrstuhl für Mikrobiologie, Institut für Biologie 2 der Albert Ludwigs-Universität, Freiburg, West Germany*

N. W. GILLHAM (211), *Departments of Zoology and Botany, Duke University, Durham, North Carolina*

D. J. GOODCHILD (35), *Commonwealth Scientific and Industrial Research Organization, Division of Plant Industry, Canberra, Australia*

LARRY W. JONES (3), *Department of Botany, University of Tennessee, Knoxville, Tennessee*

BARRY L. MARRS (261), *Edward A. Doisy Department of Biochemistry, St. Louis University School of Medicine, St. Louis, Missouri*

Preface

Photosynthesis has always been a fruitful field for investigation, not only because of its inherent importance, but also because it is an integrated biological process which has traditionally attracted the interest of the physicist, the chemist, and the biologist. The major advances in our understanding of the process include contributions from each of the disciplines: the fundamental study by Priestley of the gases involved, the quantum requirement experiments of Warburg and Emerson, Arnold's research into the size of the photosynthetic unit, the pathway of carbon dioxide fixation by Calvin's group, Van Niel's elegant work on comparative photosynthesis, and so on.

The articles contained in this volume and the preceding Volume 7 reflect this same broad approach to the modern study of photosynthesis. The experimental sophistication has extended markedly to permit picosecond measurements of the primary photophysical and photochemical reactions, the study of individual polypeptide components of the chlorophyll–protein reaction center complexes, detection of the individual complexes and their arrangement in the photosynthetic membrane, and the measurement of proton and other ion movement across the membrane as a function of the photosynthetic electron transfer reactions.

In reviewing the manuscripts for Volumes 7 and 8, it became evident that we were no longer dealing with black boxes with alphabets in them, but with specific molecules, their function, and their interaction with other specific molecules. One gains the impression that the major areas of investigation of photosynthesis have been defined and we are in a "mopping up" phase in which we supply those final, critical pieces of the puzzle that tie everything together. It is obvious that we are moving quickly on all fronts, just as in most areas of biology, and the pieces are quickly falling into place. We feel, however, that there are still a few surprises left before the picture is complete.

D. Rao Sanadi

Contents of Previous Volumes

xiii

Electron Transport and Photophosphorylation

Alternate Fates of the Photochemical Reducing Power Generated in Photosynthesis: Hydrogen Production and Nitrogen Fixation

NORMAN I. BISHOP
Department of Botany and Plant Pathology
Oregon State University
Corvallis, Oregon

LARRY W. JONES
Department of Botany
University of Tennessee
Knoxville, Tennessee

I. Introduction

The impact of the successful oil embargo experienced by the United States in 1973 was sufficiently severe to cause an increased awareness of and interest in alternate sources of energy, including biosolar energy conversion devices (NSF/NASA Report, 1972; NSF/RANN Report, 1972, 1973). Nature's primary process for the conversion of light energy to a biologically useful chemical energy form is photosynthesis. It is this process that provided energy for the production of the fossil fuel supply of the earth and continues to supply the yearly organic energy supply and reserves for the world's population. With the ever-increasing global population combined with the insatiable appetite for all energy forms in countries throughout the world, it has become paramount (1) to increase the productivity of agriculture (the Green Revolution) by manipulation of a variety of extrinsic and intrinsic factors known to influence photosynthesis, (2) to explore the possibilities of increasing the inherent efficiency of the basic mechanism of photosynthesis, and (3) to investigate the possibilities of utilizing the primary stable photoreductant generated in photosynthesis in a way more direct than that represented by the mechanism for carbon dioxide assimilation in green plant photosynthesis. It is the primary purpose of this article to examine item (3) above in terms of the *in vivo* capacity of three major groupings of photosynthetic organisms—the green algae, the blue-green algae, and the photosynthetic bacteria—to perform the two light-dependent processes of hydrogen production and nitrogen fixation; both reactions represent natural processes for utilization of the energy provided by either the generation of the primary photoreductant of photosynthesis or ATP, or both. Furthermore, the rate-limiting aspects of the nitrogen-fixing reaction sequence represent potential points of attack for improving the yield of photosynthesis under conditions where the availability of reduced nitrogen is a major limiting factor for plant productivity.

II. Nature of the Primary Photoreductant

In the process of photosynthesis the existence of two photosystems allows for one of them to generate a strong "photooxidant," which is responsible for the photolysis of water, and the other to generate a strong "photoreductant," which reduces the primary electron acceptor (Clayton, 1965). In one sense the primary photoreductant is equivalent to the reduced state of some component of the reaction center of PS I, i.e., P700 (see the chapter by Ke in Volume 7 of this series). Because of the short lifetime of such intermediates, it is biochemically more pertinent to describe the primary photoreductant in terms of the first stable reductant generated by the photoactivation of P700. The chemical candidates for this particular function have changed continually during the past 20 years as new components of the chloroplast have been discovered. As concerns reactions in which hydrogen gas is formed or the photochemistry is linked to nitrogen fixation, two or three chloroplast constituents require attention: these are NADP, ferredoxin, and a "bound" ferredoxin. Through observations on the light-induced (EPR) signal at 77°K it was deduced that a bound ferredoxin functions as this primary electron acceptor in PS I (Bearden and Malkin, 1975; see also the chapter by Ke in Volume 7 of this series for a discussion of P430 and bound ferredoxin). Bolton and colleagues (see Bolton and Warden, 1976) are of the opinion that some substance other than bound ferredoxin fulfills the role of the primary acceptor of PS I; this substance is recognized only by its g components of 1.75 and 2.07. Since Ke (1973) observed that the kinetics of appearance and disappearance of the oxidized forms P700 and of the bound ferredoxin were identical at low temperature, more credence is currently afforded to the interpretation that bound ferredoxin is the primary acceptor in PS I of green plants and algae. It is unlikely, however, that bound ferredoxin reacts directly in photohydrogen production or in the light-driven nitrogen fixation of blue-green algae and photosynthetic bacteria. In the case of the photosynthetic bacteria, the primary acceptor is viewed as being both iron and ubiquinone (Bolton and Warden, 1976). From an energetic viewpoint it is assumed that if it is not the bound form then it is certainly the free form of ferredoxin that provides the reduction potential sufficient to produce hydrogen gas in concert with the conventional hydrogenase of the green algae. For hydrogen evolution and/or nitrogen fixation by the blue-green algae and the photosynthetic bacteria, there is the additional requirement, perhaps exclusive, for the provision of ATP by the photosynthetic machinery.

Although it has been recognized for some time that reduced ferredoxin and the $NADP^+$ oxidoreductase system cooperate in the formation of reduced NADP, and that the reduced NADP is then utilized in specific reductive steps in the Calvin cycle, it is questionable whether this system participates directly in either hydrogen evolution or nitrogen fixation (Arnon and Yoch, 1974).

III. Hydrogen Production by Green Algae

In continuation of studies on the anaerobic metabolism of the green algae *Scenedesmus obliquus*, Gaffron and Rubin (1942) observed that thoroughly adapted cells when illuminated in the absence of both hydrogen and carbon dioxide produced hydrogen gas. This phenomenon, since termed photohydrogen production, was separable from a dark fermentative hydrogen metabolism through the action of dinitrophenol

(DNP); this substance caused complete inhibition of the dark hydrogen production and apparent stimulation of photohydrogen evolution (Gaffron and Rubin, 1942; Gaffron, 1944). They recognized that this form of metabolism for a normal aerobic-type organism was unique, that it was dependent upon the adaptable hydrogenase of *Scenedesmus*, and that it was most likely representative of an anaerobic photooxidation of some unknown intermediate formed in fermentation. Because of the esoteric nature of the problem, of the apparent miniscule rate of hydrogen evolution, and of the inability of anaerobically adapted cells to show a sustained production of hydrogen, only limited additional studies have been conducted, and these principally by subsequent students of Gaffron.

It is now known that the reactions catalyzed by anaerobically adapted algal cells that possess an adaptable hydrogenase include the following:

Light-dependent reactions

1. Photosynthesis: $\quad CO_2 + 2H_2O + light \longrightarrow (CH_2O) + O_2 + H_2O$
2. Photoreduction: $\quad CO_2 + 2H_2 + light \xrightarrow{H_{2ase}} (CH_2O) + H_2O$
3. H_2 photoproduction: $\quad XH_2 + light \xrightarrow{H_{2ase}} X + H_2$

Dark reactions

4. Oxy-hydrogen reaction: $\quad 2H_2 + O_2 \xrightarrow{H_{2ase}} 2H_2O$
5. Dark CO_2 fixation: $\quad CO_2 + 2H_2 + energy \xrightarrow{H_{2ase}} (CH_2O) + H_2O$
6. H_2 production: $\quad RH_2 \xrightarrow{H_{2ase}} R + H_2$
7. H_2 uptake: $\quad R + H_2 \xrightarrow{H_{2ase}} RH_2$
8. Respiration: $\quad RH_2 + \frac{1}{2}O_2 \xrightarrow{H_{2ase}} R + H_2O$

The increased complexity of the metabolism associated with activation of the hydrogenase system, as summarized above, must be evaluated and understood in order to explain the many factors that influence photohydrogen production. The prevention of reactions 1, 2, and 6, by dinitrophenol inhibition, for example, would greatly simplify the reaction sequences.

Although the presence of an adaptable hydrogenase and the ability of algal cells to perform photoreduction and the several associated reactions listed above was originally studied in a restricted few species, notably *Scenedesmus obliquus*, strain D_3, it is now known (Kessler, 1974) that numerous additional species of algae have similar anaerobic physiology. Recently, we have examined over 100 species of algae from ten classes of algae for their capacity for photoreduction and photohydrogen evolution. Practically all the species possessing an adaptable hydrogenase were members of the class Chlorophyceae and within the orders Volvacales or Chlorococcales. The majority of species having the potential for hydrogen photoproduction were found in the latter order and included a number of *Chlorella* species. None of the new species examined had a capacity for photohydrogen production greater than that observed for *Scenedesmus obliquus* (Bishop *et al.*, 1977).

A. THE ROLE OF PHOTOSYSTEM I IN PHOTOHYDROGEN PRODUCTION

Since the early 1960s photosynthesis has been recognized to involve two photosystems, photosystem I (PS I) and photosystem II (PS II), which are functionally connected by an electron transport system. This formulation, in which water serves as the electron donor and $NADP^+$ as the terminal electron acceptor for these two cooperative photosystems, represents the generally accepted mechanism for photosynthesis of green plants. The experimental findings culminating in the elaboration of this proposed mechanism have been extensively reviewed (Boardman, 1970; Bishop, 1966, 1971, 1973; Avron, 1967; Brown, 1973; Trebst, 1974; Cheniae, 1970) and formulations representing this mechanism, i.e., the so-called Z-scheme of photosynthesis, occur in practically all general biology, botany, and biochemistry textbooks.

The anaerobic, DCMU insensitive, assimilation of carbon dioxide and hydrogen gas by adapted cells of *Scenedesmus*, the process termed photoreduction (Gaffron, 1940), was shown by Bishop and Gaffron (1962) to be an exclusive PS I type of reaction. In this reaction H_2 gas, through the intervention of the alga's hydrogenase system, serves as the source of electrons for the photoreduction of carbon dioxide. This reaction is, in a qualified sense, similar to the process of carbon dioxide assimilation in bacterial photosynthesis and to the ascorbate-dichloro-phenolindophenol (DCPIP), DCMU-insensitive, photoreduction of $NADP^+$ by isolated chloroplasts (Vernon and Zaugg, 1960). The initial attempts to determine whether photohydrogen production was also a PS I-catalyzed reaction revealed that both photosystems were apparently involved (Bishop and Gaffron, 1963). Since both the photoreduction process and photohydrogen production require similar circumstances for their activation and function and, furthermore, appear to compete for the primary photoreductant generated by PS I, the apparent participation of PS II activity for hydrogen photoproduction seemed anomalous. However, if water serves as the electron donor for this reaction, as was originally suggested by Gaffron and Rubin, then the participation of both photosystems would be mandatory. The possibility of water serving as the substrate for hydrogen formation has been, and continues to be, a point of controversy. Additional lines of evidence confirming the requirement of at least PS II, if not water photolysis, in hydrogen evolution will be discussed in a later section of this review.

The clearest lines of evidence demonstrating the absolute requirement for PS I in photohydrogen evolution was obtained by mutational studies with *Scenedesmus*. Data summarizing the consequence of the loss of either P700, cytochrome f, or cytochrome b-563 upon photohydrogen evolution and a variety of other partial reactions of whole cells and isolated chloroplast particles are shown in Table I. Originally Bishop and Gaffron (1963) noted that mutant PS-8 of *Scenedesmus* was deficient in the capacity for photohydrogen evolution; concurrently, it was observed that this mutant lacked detectable P700 (Weaver and Bishop, 1963; Butler and Bishop, 1963; Beinert and Kok, 1963). As indicated in Table I, this mutant also showed decreased capacity for photohydrogen evolution. A number of additional mutants have been isolated that also lack P700 [determined on the basis of light or chemically induced absorbancy changes at 697 nm and by sodium dodecyl sulfate (SDS)–polyacrylamide gel electrophoresis (PAGE)]; like mutant 8, these strains also do not evolve hydrogen in the light but, in general, show appreciable PS II activity either in whole-cell or chloroplast reactions (Table I).

TABLE I

WHOLE-CELL AND CHLOROPLAST REACTIONS OF P700- AND CYTOCHROME-DEFICIENT MUTANTS OF *Scenedesmus Obliquus*[a]

Mutant strain	Photosynthesis	Photoreduction	Photohydrogen	H_2O-MV
		P700-deficient types		
PS-21	0	0	0	5.6
PS-24	5.6	4.5	2	9.1
PS-30	0	0	0	0
PS-46	5.7	2.8	3	6.3
PS-8	8.3	3.4	1	8.2
		Cytochrome-deficient types		
PS-34	5.3	2.3	5.1	5.7
PS-50	6.8	3.5	1	7.1
PS-123	5.1	2.1	1	5.2
PS-134	1.4	tr	0	2.5
PS-141	10.4	3.5	8.9	11.6

[a] All values given are in percentages of the respective rates of comparable measurements performed on whole cells or chloroplasts of the wild-type strain. These values are: photosynthesis, 670 μl of O_2 per hour per milligram chlorophyll; photoreduction, 200 μl of CO_2 per hour milligram chlorophyll; photohydrogen, 70 μl of H_2/hour per 10 μl packed-cell volume (PCV); H_2O-methylviologen, 100 μl of O_2/hour per mg chlorophyll.

The P700 deficient mutants all possess an active ferredoxin and hydrogenase.

Also from the data of Table I, it is clear that the genetic deletion of cytochrome f-553 (Bishop, 1972) inhibits photosynthesis, photoreduction, and photohydrogen evolution without major inhibition of PS II reactions. $HgCl_2$, which is a known inhibitor of the activity of plastocyanin (Katoh, 1962; Kimimura and Katoh, 1972, 1973), prevents hydrogen evolution (N. I. Bishop and M. Frick, unpublished results). Dibromomethylisopropylbenzoquinone (DBMIB), a synthetic antagonist of the electron transport potential of plastoquinone, inhibits photohydrogen evolution in addition to its inhibitory effects upon PS II-type reactions (Ben-Amotz and Gibbs, 1975). Confirmatory evidence for the action of DBMIB upon the anaerobic hydrogen metabolism in *Scenedesmus* is shown in Fig. 1; concentrations less than 5 μM cause greater than 50% inhibition. The efficiency of this compound as an inhibitor is comparable to that of DCMU (see data of Fig. 4).

From the results obtained with the various mutation types and inhibitors examined, it is apparent that an intact and functional PS I is an absolute prerequisite for the light-dependent hydrogen metabolism both in algae and in coupled chloroplast reactions leading to hydrogen formation. That the photoreduction of P700, in combination with an appropriate electron donor, drives the formation of a stable photoreductant (in this case either reduced ferredoxin or reduced NADP), which then interacts through the hydrogenase to form hydrogen gas, appears to be a logical formulation. The requirement for an intact photosynthetic

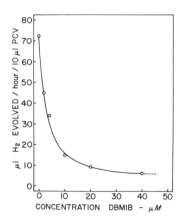

FIG. 1. Inhibition of photohydrogen evolution in heterotrophic *Scenedesmus oblique,* strain D_3, by the plastoquinone antagonist dibromomethylisopropylbenzoquinone (DBMIB). Temperature = 25°C. Light intensity = 3×10^5 erg cm^{-2}sec^{-1}. PCV = 10 μl.

electron transport system to at least the level of plastoquinone indicates that the natural electron donor for hydrogen formation must enter into the *in vivo* mechanism at a point prior to plastoquinone.

B. Independence of Photohydrogen Evolution from an ATP-Dependent Reverse Electron Flow

While it is clear that light mediates an electron flow from either an *in situ* organic hydrogen donor and/or water, it is also equally apparent that the ATP-dependent hydrogenase, which is involved in the hydrogen and nitrogen metabolism of the blue-green algae and photosynthetic bacteria, is not required for photohydrogen evolution in green algae. The early observations that dinitrophenol stimulates photohydrogen production but inhibits dark hydrogen fermentation reactions has been interpreted by many as an additional classic example of uncoupling of electron transport from phosphorylation. However, Gaffron and Rubin (1942) offered a more appropriate interpretation in suggesting that the observed augmentation of rates of hydrogen evolution by DNP was due to the elimination of the two competitive reactions, photosynthesis and photoreduction. Subsequent studies on the action of CCCP on photosynthesis, photoreduction, and photohydrogen evolution (Bishop and Gaffron, 1963; Kaltwasser *et al.*, 1969; Stuart and Kaltwasser, 1970; Stuart and Gaffron, 1971, 1972a,c) revealed that the alternative interpretation for the action of phosphorylation uncouplers was the more appropriate one. By inhibition of cyclic photophosphorylation, those reactions requiring ATP (and consuming hydrogen) are prevented and, consequently, more of the primary photoreductant is directed toward hydrogen production (see Section III, reactions 1–8). Consideration of the various reactions that can be performed by anaerobically adapted cells of *Scenedesmus* reveals that the loss of photosynthesis and photoreduction removes not only systems that compete strongly for the primary photoreductant, but also two systems that consume hydrogen gas (reactions 2 and 4) as well as one that produces oxygen gas (reaction 1), a classic inhibitor of hydrogenase activity.

The failure of uncouplers of phosphorylation to inhibit photohydrogen evolution in the green algae underlines the differences in mechanisms for hydrogen evolution by this system and that of the blue-green algae and photosynthetic bacteria. Earlier Bishop and Gaffron (1963) stressed that the two then-known systems for producing hydrogen photochemically utilized different mechanisms. At that time it was not apparent from the work of Gest and his colleagues (Gest *et al.*, 1962; Ormerod and Gest, 1962) that the ATP-dependent nature of hydrogen evolution in the photosynthetic bacteria was associated with their nitrogen-fixing capacity. Progress in this area (see Section V,A) now allows for a clearer distinction to be made between systems utilizing the ATP-dependent and ATP-independent hydrogenase systems.

C. PARTICIPATION OF PHOTOSYSTEM II IN PHOTOHYDROGEN PRODUCTION

In their original publication Gaffron and Rubin proposed that the source of hydrogen was either an organic hydrogen donor or water. Their predilection was toward the former interpretation because of their observations on the dark-hydrogen fermentation, which appeared to result from a dehydrogenation of an organic donor system. In an elegant series of experiments, Spruit (1954, 1958) found that under the conditions required for photohydrogen evolution, a simultaneous production of hydrogen and oxygen could be measured in a suspension of *Chlorella* cells. From this observation, plus his measurements on the stoichiometry of the hydrogen and oxygen produced, Spruit favored the concept that the primary source of electrons for photohydrogen production was water. He recognized that, because of the anaerobic circumstances of the algal cells, any oxygen produced by them would be rapidly scavenged by respiration and by the oxy-hydrogen reaction. He noted also the rapid dark uptake of hydrogen gas following a period of photohydrogen production (see Section III, reactions 4, 7, and 8). Combinations of these reactions would easily result in erroneous values for the stoichiometry, and only under idealized conditions would it be possible to measure a value of 2, for the ratio of hydrogen : oxygen, i.e.,

$$2H_2O + light \; \frac{PS \; I \; and \; PS \; II}{H_2ase} \; 2H_2 + O_2$$

Recently we have repeated and extended the observations of Spruit with a two-electrode system designed to provide simultaneous measurements on hydrogen and oxygen within a liquid-phase system (Jones and Bishop, 1976). When intermittent illumination (30 seconds light, 30 seconds dark) was employed and the intensity of the "flash" was adjusted so that oxygen concentrations sufficient to inactivate the hydrogenase did not accumulate, it was possible to obtain stoichiometry values approaching 2 with relative ease (Fig. 2A). Increasing the intensity or the length of the "flash" gave aberrant and low values (Fig. 2B) whereas a decrease in intensity to values that allowed limited net production of oxygen resulted in $H_2 : O_2$ ratios as high as 9.

Under the experimental conditions employed for the amperometric determination of hydrogen and oxygen, the normal kinetics for produc-

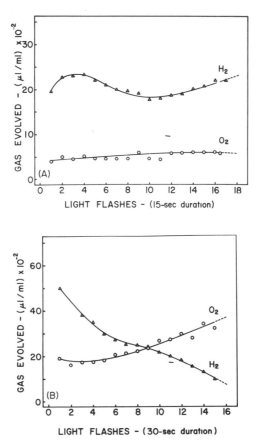

FIG. 2. (A) Measurements on the simultaneous photoproduction of hydrogen and oxygen with 15-second light and 30-second dark periods with cultures of *Scenedesmus* previously adapted in the dark for 4 hours. White light intensity $= 8.3 \times 10^4$ erg cm^{-2}sec^{-1}. (B) Measurements as in (A), but with a longer illumination period at lower light intensity; 4.8×10^4 erg cm^{-2}sec^{-1}.

tion of oxygen are quite different from those observed in the flashing-light regime. When suspensions of cells are exposed to saturating white light, a pattern like that shown in Fig. 3 is seen. Hydrogen gas appears immediately, and after about a 7-second delay oxygen begins to be detected. This delay is important and indicates, we believe, an internal consumption of oxygen by the reactions represented in Section III. The biphasic nature of the oxygen curve and the peaking of the hydrogen curve represent both the activity of the oxy-hydrogen reaction and the inactivation of the hydrogenase as a result of the increased amount of oxygen produced. This inactivation is noted by the failure of the cells to generate hydrogen during the second period of illumination.

The pattern shown in Fig. 3 for the production and consumption of the two gases can be drastically changed by manipulation of parameters influencing the availability of oxygen. For example, the addition of dithionite to adapted algal cells prevents the accumulation of oxygen and stimulates the rate of hydrogen evolution by removing the competition introduced by the presence of oxygen. Dithionite does not appear to act as an additional electron donor, since only the rate of hydrogen evolution is affected, not the final yield. Other electron acceptors, such as nitrite and nitrate, inhibit hydrogen production and stimulate oxygen production and, necessarily, the inactivation of the hydrogenase (Bishop *et al.*, 1977).

The obvious complexity of a system (a) in which the primary photoreductant can be utilized either to form hydrogen gas through the mediation of the hydrogenase or to react directly with oxygen; (b) where both products of the light reaction, H_2 and O_2, can recombine through the oxy-hydrogen reaction and thus both gases can be reabsorbed directly by the algae (reactions 7 and 8, Section III); and (c) where the essential enzyme, hydrogenase, is inactivated by the oxygen produced by the system—all serve to explain the variability of the stoichiometry. However, when these problems are recognized and the appropriate

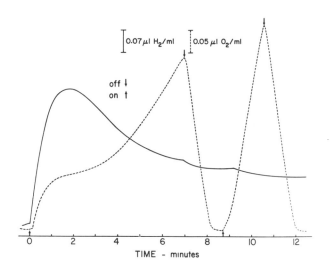

FIG. 3. Simultaneous production of hydrogen and oxygen by anaerobically adapted cultures of heterotrophic *Scenedesmus obliquus*, strain D_3, measured in continuous light. Temperature = 25°C. Light intensity = 3×10^5 erg cm^{-2}sec^{-1}. —, hydrogen; - - - -, oxygen.

experiment is performed, the predicted stoichiometry can be measured. But are such values fortuitous? Might they not arise from restraints placed upon the measurements and result from a simultaneous dehydrogenation of an organic hydrogen donor and normal photosynthetic generation of oxygen? These questions, plus many other unposed ones that cannot be considered because of space limitations, are pertinent and need extra evaluation before an absolutely definitive answer can be provided.

However, it is relevant to ask whether the machinery of PS II is required in photohydrogen production. We have indicated earlier that the light-dependent evolution of hydrogen appeared to have a requirement for both photosystems (Bishop and Gaffron, 1963). Although this observation has not been repeated or extended, the findings remain pertinent, since it has also been shown that inhibitors of PS II activity, i.e., DCMU and simazine, are equally effective against photohydrogen production (Bishop and Gaffron, 1963). Although the ability of DCMU to inhibit this reaction has been a controversial subject (Healey, 1970a; Stuart and Kaltwasser, 1970; Stuart and Gaffron, 1972b), it is now clear that its effectiveness is variable. This variability is traceable, in part, to the algal species employed, whether the culture type utilized was obtained by photoautotrophic, heterotrophic, or mixotrophic growth, and whether inhibitory effects upon the initial rates of hydrogen production (the so-called fast phase) or upon the prolonged but slowed phase of hydrogen production (Stuart and Gaffron, 1971) are being described.

A typical response to DCMU concentration of photohydrogen production by cells of *Scenedesmus obliquus*, obtained by either photoautotrophic or heterotrophic growth, is shown in Fig. 4. Clearly the DCMU exhibits greater inhibition with the heterotrophic cultures, but the concentration range within which this compound is effective is approximately the same for both culture types. Nearly identical results have been obtained with comparable culture types of *Kirchneriella lunaris*, *Chlorella fusca, Selenastrum* sp., and numerous additional species of green algae (Bishop *et al.*, 1977). With either photoautotrophic or mixotrophic cultures it is also of importance to evaluate the stage of the cell cycle at which cultures are selected for measurements. For example, synchronous cultures of *Scenedesmus* show a differential sensitivity of their photosynthesis to DCMU, which is apparently dependent upon permeability characteristics of the different cell types formed during the algal cell cycle (Senger and Frickel-Faulstitch, 1975).

As part of the initial application of photosynthetic mutants of *Scenedesmus* to studies on the mechanism of photohydrogen evolution Bishop and Gaffron (1963) presented evidence that mutants blocked within PS II, although having normal photoreductive capacities, were unable to

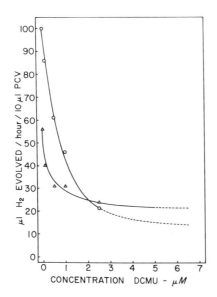

FIG. 4. Inhibition of photohydrogen evolution by DCMU in photoautotrophic (○) and heterotrophic (△) cultures of *Scendesmus*. Rates determined from initial slopes as shown in Fig. 3.

sustain photohydrogen production. For reasons yet not clear, it was later reported that one of the PS II mutants employed in the earlier studies (mutant strain PS 11) had a normal capacity for photohydrogen evolution (Stuart and Kaltwasser, 1970; Kaltwasser *et al.*, 1969). Consequently, this aspect of the study has been reinvestigated utilizing the amperometric technique for H_2 determination (Jones and Bishop, 1976) rather than the classic manometric technique. The results of our observations on ten independently isolated PS II-type mutants of *Scenedemus* are summarized in Table II. In no instance was a rate of photohydrogen evolution detected that was greater than the rate of photosynthesis in any of these mutants; according to one of the criteria utilized in selecting this mutant type, all strains showed an unimpaired rate of photoreduction. It is now recognized that this mutant type represents the manifestations of a single gene mutation whose effects are pleiotropic. Numerous factors of PS II, including part of the plastoquinone pool, substance Q, C-550, cytochrome *b*-559 (H.P.) and the reaction center of PS II, P680, are affected by the mutation; stated otherwise, these mutants have lost not only the primary reaction center of PS II but also most of the known components of the reducing side of this photosystem (Bishop and Wong, 1971, 1974). If it is true that PS II is essential for photohydrogen evolution, then the only potential for this

TABLE II
WHOLE CELL AND CHLOROPLAST REACTIONS OF MUTANTS OF *Scenedesmus*
Obliquus DEFICIENT IN PHOTOSYSTEM II ACTIVITY[a]

Mutant strain	Photosynthesis	Photoreduction	Photohydrogen	H_2O-MV
PS-4	0	104	0	0
PS-5	0	107	0	0
PS-11	0	98	2	2
PS-79	13	100	10	7
PS-84	12	100	7	5
PS-96	0	104	0	0
PS-102	0	101	0	0
PS-110	2	100	tr	1
PS-112	3	104	tr	2
LF-1	0	102	tr	2

[a] All values given are in percentages of the respective rates of comparable measurements performed on whole cells or chloroplast of the wild-type strain. Actual values are as presented in Table I. MV = methylviologen.

reaction in these mutations would be derived from an exclusive PS I-mediated hydrogen evolution; on occasion minute bursts of hydrogen have been observed.

An alternative, and obvious, interpretation of the inability of the PS II type mutants to perform photohydrogen evolution is that they have lost preferentially the site for electron donation on the reducing side of PS II, e.g., substance Q, which would be identical mechanistically to DCMU inhibition. Hence the loss of oxygen-evolving capacity in these mutants would not be directly related to their inability to produce hydrogen in the light. Evidence that contradicts this interpretation has recently been obtained with a newly isolated mutant of *Scenedesmus* (Bishop *et al.*, 1977) which lacks PS II activity, has a normal low fluorescence enhanced by DCMU, shows no major modifications of any of the known components of PS II, and performs DCMU-sensitive PS II-dependent electron transport with alternate electron donor–acceptor systems. This low fluorescent mutant strain is apparently identical to the mutant strains of *Chlamydomonas reinhardtii* described earlier (Epel and Levine, 1971; Epel *et al.*, 1972; Epel and Butler, 1972). The *Scenedesmus* mutant, LF-1, has neither photosynthetic activity nor photohydrogen production, but it retains a normal photoreduction (Table II). Although the low-fluorescence strains of *Chlamydomonas* and *Scenedesmus* show a decreased content of cytochrome *b*-559 (H.P.), but a normal total cytochrome *b*-559 content, it is believed that that is not the underlying cause of the loss of oxygen-evolving capacity, but rather is an indicator of a fundamental change within the ultrastructure of the chloroplast membranes. Our preliminary results suggest that both the reduction in content of the high-potential form of cytochrome b-559 and the loss of oxygen-evolving capacity are manifestations of modifications in the chloroplast lipids of LF-1. This alteration would affect, consequently, only the water photolysis step and, by inference, photohydrogen evolution.

Additional lines of evidence from other mutant types support the interpretation drawn above. Specifically, mutations in which carotenoid biosynthesis is blocked develop a functional PS I but neither PS II nor photohydrogen evolution activities. Light-sensitive mutants that preferentially lose PS II activity when exposed to high intensity irradiation also lose the potential for hydrogen evolution. Temperature-sensitive mutants of *Scenedesmus* which lack PS II activity when grown at a nonpermissive temperature also lack photohydrogen evolution capacity (Bishop *et al.*, 1977).

D. NATURE OF THE ELECTRON DONOR IN PHOTOHYDROGEN
    EVOLUTION

As indicated earlier in this review, two schools of thought prevail concerning both the nature of the substance dehydrogenated during photohydrogen production and the photosystems utilized. In the original research of Gaffron and Rubin, an organic hydrogen donor was favored as the substrate since the addition of glucose caused an increase in the amount of hydrogen evolved and because a period of active photosynthesis prior to adaptation stimulated hydrogen photoproduction. Kaltwasser *et al.* (1969), Stuart (1971), and Stuart and Gaffron (1971) attempted to provide additional specifics about the mechanism of glucose stimulation in *Scenedesmus*. Stuart and Gaffron (1971) showed that, of thirteen potential substrates for glycolysis and intermediary metabolism of algae, only glucose was preferentially utilized. Kaltwasser *et al.* (1969) observed that the increased production of carbon dioxide, and supposedly also of hydrogen, resulted primarily from the breakdown of glucose by the Embden–Meyerhof pathway since most of the carbon dioxide evolved originated from positions 3 and 4 of labeled glucose. Although this deduction is logical as concerns the source of carbon dioxide evolved either in the dark or light anaerobic metabolism of adapted algal cells, it cannot predict any specifics about what portion of the glucose molecule is utilized for the production of hydrogen. Actual determination of the stoichiometry of hydrogen produced and glucose utilized showed that 1 mole of hydrogen gas was formed per 2 moles of glucose metabolized (Kaltwasser *et al.*, 1969). If photohydrogen production in anaerobic cells were to serve as an alternative mechanism for reoxidizing the reduced NAD formed in normal glycolysis (Kok, 1973), then a stoichiometry of 2, not 0.5, would be expected. Since the green algae possess a mixed fermentative mechanism producing both lactic acid and alcohol, not all the reduced NAD would need to be reoxidized through the hydrogenase system. Perhaps because of these difficulties in evaluating mechanisms of photoanaerobic metabolism of algae, Stuart and Gaffron (1971) concluded: "It is thus likely that H_2 photoproduction removes electrons from a 'side path' of glucose (anaerobic) catabolism."

The glucose stimulation of photohydrogen production is restricted to the slow phase of photohydrogen production, is more apparent in starved or older cells, is maximally stimulated by CCCP and is not inhibited by DCMU. Because these measurements of photohydrogen metabolism are performed over long periods of time, i.e., 6–8 hours of manometric measurements, it is difficult to correlate the observations on the slow evolution of hydrogen to those that occur in illumination periods of less than 5–10 minutes. This latter, rapid, phase requires the participation of both photosystems, is inhibited by DCMU, is not stimulated by glucose, and appears to be correlated with a functional PS II, as we have discussed previously. Whether an organic hydrogen donor functions in both phases, where electrons derived from this source can be donated either to PS I or to both photosystems operating in a normal sequence, must remain an open question at this time.

IV. Nitrogen Fixation and Hydrogen Evolution in Blue-Green Algae

In several genera of blue-green algae a significant proportion of the reducing power and ATP generated by photosynthesis can ultimately be used to reduce nitrogen to ammonia via the nitrogenase enzyme. Several excellent and extensive literature reviews and books have appeared recently, so only the more recent literature will be treated here. The general physiology and cytology of the blue-green algae is extensively covered in "The Biology of the Blue-Green Algae" edited by Carr and Whitton (1973), "The Blue-Green Algae" by Fogg *et al.* (1973), and in the article by Wolk (1973). Reviews and monographs concerned with various aspects of nitrogen fixation, which cover the blue-green algae, include the following: Stewart (1974) in "The Biology of Nitrogen Fixation"; "Nitrogen Fixation in Bacteria and Higher Plants" by Burns and Hardy (1975); Nitrogen Fixation by Free-Living Micro-organisms" by Stewart (1975); "Molecular Biology of Nitrogen Fixation" by Shanmugan and Valentine (1975); "Biological Nitrogen Fixation" by Postgate

(1976); and "Proceedings of the 1st International Symposium on Nitro-gen Fixation" edited by Newton and Nyman (1976).

A. METABOLIC REQUIREMENTS FOR NITROGEN FIXATION IN THE BLUE-GREEN ALGAE

The ability to fix atmospheric nitrogen appears to be a widely distributed characteristic of the blue-green algae. Environmental require-ments seem primarily to include the lack of a readily available source of fixed nitrogen and an increased requirement of molybdenum (Wolfe, 1954) and perhaps calcium (Allen, 1956) and sodium (Allen and Arnon, 1955; Brownell and Nicholas, 1967; Ward and Wetzel, 1975). Oxygen, which inactivates the nitrogenase enzyme, must either be excluded from the site of nitrogen fixation or be present in small concentrations in the environment (microaerophilic).

Considerable effort in many laboratories during the late 1960s and early 1970s was directed toward proving that the heterocyst was the major, if not the sole, site of nitrogen fixation in the blue-green algae. Despite the many diverse and ingenious methods employed, the net result as summarized by Stewart (1974, p. 216) is that "only more experimentation will resolve the problem." The heterocyst is a unique cell that differentiates in both morphology and function to resemble, at least functionally, its cousin the photosynthetic bacterium—it lacks PS II (Bradley and Carr, 1971), but retains an active photosystem I and an active oxidative metabolism. It therefore appears to be ideally specialized for nitrogen fixation in that it (1) maintains a low oxygen tension necessary for nitrogenase activity because of an impervious cell wall, a highly developed internal membrane structure, and an active oxidative metabolism; and (2) provides large amounts of ATP and reduced ferredoxin in the light (via oxidative metabolism, PS I, and cyclic phosphorylation), which are necessary for nitrogenase activity.

It is well established that heterocyst frequency is correlated with the nitrogen source in the growth medium—the presence of ammonium blocks heterocyst formation entirely, while nitrate suppresses produc-tion to varying degrees over that observed under atmospheric nitrogen alone (Fogg, 1949; Mickelson et al., 1967). During logarithmic growth in nitrate medium, cells of Anabaena cylindrica obtain more than 50% of their cellular nitrogen via nitrogen fixation (Ohmori and Hattori, 1972; see also Bone, 1971). That a finely tuned metabolic control correlates the rates of nitrogen fixation and carbon assimilation was postulated in 1960 by Fogg and Than-Tun. Cobb and Myers (1964) found a linear but inverse correlation between the ratio of carbon:nitrogen assimilated and the C:N ratio of the cells—the higher the cellular C:N ratio, the greater the proportion of photosynthetic reducing power used for nitrogen fixation. Kulasooriya *et al.* (1972) described experiments in which *Anabaena cylindrica* cells grown on ammonium (with no heterocysts) were switched to nitrogen-free medium. No heterocysts were produced until the C:N ratio of the cells increased to between 5 and 6, and maximum activity required a C:N ratio of about 8. Similar correlations of heterocyst formation and nitrogenase activity were reported for an *Anabaena* sp. when switched from nitrate medium to nitrogen-free medium under N_2 or argon (Neilson *et al.*, 1971). Cells grown without a nitrogen source (under argon and CO_2), conditions sufficient to produce extreme C:N ratios, had rates of nitrogen fixation 2–3 times greater and increased heterocyst frequency above those switched to N_2 and CO_2. Thus nitrogen starvation (or high C:N ratios) seems to be the internal stimulus which induces heterocyst formation and nitrogenase activity in this specialized cell.

Nitrogen fixation in cultures of nonheterocystous blue-green algae, beginning with the first convincing evidence for unicellular *Gloeocapsa alpicola* by Wyatt and Silvey (1969), has been reported in increasingly diverse species, such as *Plectonema boryanum* 549 (Stewart and Lex, 1970); *Lyngbya, Oscillatoria,* and *Phormidium* (Kenyon *et al.*, 1972); *Aphanothece* (Singh, 1973); and the marine oscillatorian *Trichodesmium* (Taylor *et al.*, 1973). Of these, *Gloeocapsa* and *Trichodesmium* have maximum nitrogen-fixation rates under aerobic conditions—all others are apparently active only under microaerophilic conditions. Nonheterocystous species react to nitrogen starvation (increasing C:N ratios) in a manner similar to the heterocystous species, except for differentiating heterocysts. This reaction has been identified as "nitrogen chlorosis" of aging cultures, phycocyanin being viewed as one of the emergency nitrogen sources (Fogg *et al.*, 1973).

Katoh and Ohki (1975, 1976) have shown that *Anabaena variablis* (which closely resembles other *Anabaena* species but does not differentiate heterocysts or produce nitrogenase) reacts to nitrogen deficiency (increasing C:N ratios) by losing only PS II activity and phycocyanin, but maintains the other parts of photosynthetic electron transport, PS I, and photosynthetic pigments. Katoh and Okhi (1975) concluded that the nitrogen-depleted cells of *Anabaena variablis* closely resemble those of heterocysts differentiated from normal vegetative cells in (1) their deficiency of phycobilin pigments, (2) their loss of PS II activity, (3) their photoorganotrophic mode of cell growth, depending upon externally provided organic substances, and (4) their induction by nitrogen deficiency (increasing C:N ratio). These same conditions of nitrogen depletion in other species of *Anabaena* result in the apparent irreversible production of heterocysts; these too lack PS II activity and phycocyanin but retain a functional PS I. It is interesting that measurements by Fay (1969) found the nitrogen content of vegetative cells, heterocysts, and spores of old *Anabaena cylindrica* cultures to be 7.1, 2.5, and 4.8% of dry weight, respectively. The heterocysts must maintain their high $C:N$ ratio, and thus their activity, by actively transporting the fixed nitrogen to the neighboring vegetative cells.

A similar situation has now been reported in the nonheterocystous blue-green algae *Oscillatoria* (*Trichodesmium*), which had previously been noted as an efficient aerobic nitrogen fixer (Taylor *et al.*, 1973). Carpenter and Price (1976) reported that the internal cells of the trichomes of this species appear to have lost their phycocyanin content (by visual inspection) and do not fix labeled CO_2. They hypothesized that these cells maintain the low oxygen tensions necessary for nitrogenase activity because of the loss of oxygen production activity (and phycocyanin). That this method of oxygen exclusion would be much more susceptible to disruption by currents and turbulence than heterocysts correlates with abundance of these marine organisms only in very calm waters.

The heterocyst then seems to be only a physiological extension of the typical response of blue-green algal cells to nitrogen depletion and apparently has no metabolic capabilities not found in vegetative cells of other nitrogen fixing blue-green algae exposed to similar conditions. This agrees with Van Gorkom and Donze's (1971) inference from observations of phycocyanin synthesis that under aerobic conditions only heterocysts of *Anabaena* fix nitrogen, but under anaerobic conditions apparently all cells of the filament fix nitrogen.

B.   NATURAL SOURCES OF REDUCTANT AND ATP FOR NITROGENASE ACTIVITY

The requirement for both a low potential reductant and ATP is common to preparations of nitrogenase from all sources studied (Ljones, 1974). Ferredoxin seems to be the immediate physiological electron donor to nitrogenase in the blue-green algae, since it is required for the activity of all the *in vitro* systems studied so far (Bothe, 1972; Smith *et al.*, 1971; Haystead and Stewart, 1972; and Smith and Evans, 1971). An alternate donor—phytoflavin (flavodoxin) as purified from *Anacystis* (Bothe, 1972)—is also active but is thought to be important only in iron-depleted algae. However, the possible role of the phytoflavins as an alternate low-potential intermediate generated by PS I and required for nitrogen fixation has had only limited evaluation. The increase in concentrations of the flavodoxins observed when the blue-green algae, or other nitrogen-fixing organisms, are grown under iron-deficient conditions has been looked upon as an alternative mechanism for the organism to tolerate low ferredoxin levels, i.e., that the flavodoxins (phytoflavins) replace ferredoxin. This interpretation may be completely unjustified and has hindered, perhaps, a more thorough examination of the normal function of the phytoflavins.

In the heterocystous algae both the reduced ferredoxin and ATP could come either from oxidative metabolism or PS I activity, both of which are present in high levels in the heterocyst. That many species of blue-green algae can grow heterotrophically in the dark (albeit slowly) while fixing nitrogen (Fay, 1976; Hoare et al., 1971; Khoja and Whitton, 1971) shows that light as such is not required for nitrogen fixation. However, since maximum rates of nitrogen fixation (and growth) occur only in the light, even with exogenous carbon sources, and since the rate of nitrogen fixation varies directly with light intensity (Fay, 1968), photosynthesis must be the major and immediate source of an essential substrate or cofactor. Several lines of evidence are available which show that heterocystous algae grown under aerobic, nitrogen-fixing conditions do not produce reduced ferredoxin directly from the photolysis of water (see Stewart, 1974, for a summary). It is generally held that some form of reduced carbon transports the electrons from the vegetative cells, where photolysis of water and oxygen production occur, to the heterocyst, where they are used for nitrogen fixation. The possibility that light produces reduced ferredoxin via PS I (utilizing organic substrates as hydrogen donors), which is active in the heterocysts, has not been ruled out.

Dark, oxidative metabolism in the blue-green algae proceeds almost exclusively via the pentose phosphate pathway; less than 1% is believed to occur via the tricarboxylic acid cycle [see Hoare et al. (1972) for a summary of their own and earlier work]. Winkenbach and Wolk (1973) found higher levels (7 to 8 times) of pentose phosphate enzymes activities in heterocyst preparations than in vegetative cells and suggested therefore that the major source of electrons for reductive processes in heterocysts is sugars produced photosynthetically in the vegetative cells. Peterson and Burris (1976) found that key intermediates in the pentose phosphate pathway (glucose 6-phosphate, fructose 6-phosphate, and 6-phosphogluconate) enhanced oxygen uptake in isolated heterocysts; Krebs cycle intermediates were largely inactive. They suggested that the oxidation of pentose phosphate intermediates may provide the reductant for nitrogenase.

Pyruvate was also implicated in early *in vivo* experiments as the molecule that carries the reducing power from the vegetative cells to the heterocyst (Cox, 1966; Cox and Fay, 1969). However, studies on *in vitro* preparations from heterocysts gave mixed support for this role of pyruvate, since minimal stimulation was obtained by exogenous pyruvate additions and only low activities of pyruvate dehydrogenase were found (Cox and Fay, 1967; Smith *et al.*, 1971; Haystead and Stewart, 1972). The discovery of a ferredoxin- and coenzyme A-dependent stimulation of pyruvate metabolism in *Anabaena cylindrica* (Leach and Carr, 1971) circumvented the lack of pyruvate dehydrogenase activity in heterocyst preparations. This increased activity, as found also in the non-nitrogen-fixing blue-green *Anabaena variablis*, is caused by pyruvate–ferredoxin oxidoreductase (Bothe and Falkenberg, 1973, Codd and Stewart, 1973). Preparations of heterocysts have been obtained in which the activity of this enzyme was about 30 times greater than the pyruvate-dependent rates of acetylene reduction observed in the same preparations (Codd *et al.*, 1974). However, Bennett *et al.* (1975) found from *in vivo* studies that pyruvate stimulates nitrogenase activity only when the supply of pyruvate is severely limited, i.e., under low oxygen tensions in the dark. They suggested that either this enzyme and respiratory electron transport are not major contributors of substrates in the light or that endogenous levels of pyruvate are sufficiently high to saturate any exogenous requirement.

If the ferredoxin active in nitrogen fixation is reduced directly through this enzyme system (pyruvate:ferredoxin oxidoreductase, EC 1.2.7.1.), the stimulation of nitrogen fixation by light would be strictly due to the ATP supplied through cyclic photophosphorylation via PS I.

C. PHOTOPRODUCTION OF HYDROGEN IN THE BLUE-GREEN ALGAE

One of the classic features of the nitrogenase system, regardless of its source, is that H_2 gas is a competitive inhibitor of nitrogen fixation and that nitrogen-fixing organisms, in the absence of nitrogen gas, can produce significant amounts of hydrogen. It is now recognized that nitrogenase catalyzes several different reactions depending upon the substrates present and also upon external conditions (see Ljones, 1974; Hwang and Burris, 1972). Through recent studies on the nitrogen-fixing potential of blue-green algae and also because of the recent interest in potential biosolar energy conversion systems, it is now recognized that blue-green algae containing a functional nitrogenase produce hydrogen photochemically when nitrogen gas is limited (Benemann and Weare, 1974; Benemann *et al.*, 1973; and Jones and Bishop, 1976). Except for the role of light in producing organic substrates to serve as hydrogen donors, this photohydrogen metabolism is independent of a direct photolysis of water, i.e., no PS II requirement, requires PS I activity at least for the cyclic photophosphorylation capacity, and perhaps requires this photosystem for the generation of reduced ferredoxin (Bothe and Loos, 1972). Recent studies have shown that the plastoquinone antagonist, DBMIB, and the ferredoxin inhibitor, DSPD, interfere with photohydrogen production in *Anabaena cylindrica* (Bothe *et al.*, 1976). These authors have also demonstrated that the strain of *Anabaena* employed also contains an "uptake" hydrogenase, since maximum rates of photohydrogen production were obtained only when both carbon monoxide and acetylene were present. Since CO inhibits nitrogen fixation, but not hydrogen evolution, in a nitrogenase-dependent system, and both acetylene and CO inhibit the conventional hydrogenase, all reducing potential of the ATP-activated nitrogenase is drained off as hydrogen gas. In the normal situation cells containing the uptake-hydrogenase might act to prevent the loss of reduction potential expressed as hydrogen evolution.

Attempts to show the presence of an aerobic assimilation of carbon dioxide with hydrogen gas acting as the electron donor have been largely unsuccessful in the blue-green algae. Frenkel *et al.* (1950) reported that *Synechococcus elongatus* and *Chroococcus* sp. possessed a hydrogenase that could be coupled to carbon dioxide assimilation. Nakamura reported in 1938 that a strain of *Oscillatoria* grew in the presence of hydrogen sulfide. Both cases have not been fully substantiated by further investigation. In our own work on a variety of *Anabaena* species, no evidence for photoreduction has been noted. Thus, if a normal hydrogenase is present in the blue-green algae, it may be formed only under special conditions, for example, perhaps microaerophilic, and when present would be associated only with the functioning of the nitrogenase and nitrogen fixation.

V. Nitrogen Fixation and Hydrogen Evolution in the Photosynthetic Bacteria

Biological nitrogen fixation occurs in numerous photosynthetic and nonphotosynthetic organisms and employs the same basic mechanism. With increased recognition of the importance of this natural phenomenon, in step with the awareness of the high cost in energy for producing ammonia by nonbiological processes, there has occurred a proliferation of monographs, proceedings, and reviews containing an inclusive coverage of all aspects of nitrogen fixation, including that of the photosynthetic bacteria. Some of these have been listed in an earlier section of this article; for the interested reader additional specific publications concerned with nitrogen and hydrogen metabolism of photosynthetic bacteria include the following: Frenkel (1970) in *Biological Reviews*; Keister and Fleischman (1973) in *Photophysiology*; Clayton (1973) in *Annual Review of Bioenergetics and Biophysics*; Arnon and Yoch (1974) and Yoch and Arnon (1974) in "The Biology of Nitrogen Fixation"; and Winter and Burris (1976) in *Annual Review of Biochemistry*. Because of the intensive and recent reviewing of the area of photosynthetically linked nitrogen fixation, only limited additional space need be given to it.

A. NATURE OF THE REDUCTANT AND THE SOURCE OF ATP FOR NITROGEN FIXATION

Both photosynthesis and biological nitrogen fixation require an appreciable input of energy in the form of reducing potential (ferredoxin?) and ATP. In organisms such as the photosynthetic bacteria and the blue-green algae, the coupling of nitrogen fixation to the photosynthetic process potentially relieves some of the immediate energetic problems normally encountered by the nonphotosynthetic aerobic nitrogen fixers. In this latter group it is recognized that a strong oxidative metabolism is required for both the formation of the low-potential reductant and for the photoreduction of the ATP required for the reduced iron–Mg–ATP complex. The generation of the low-potential reductant is envisaged to occur through an ATP-dependent reverse electron flow (Winter and Burris, 1976). The nature of the reductants utilized for the nitrogenase activity has been extensively surveyed and reviewed (Evans and Phillips, 1975). As Yoch and Arnon (1974) described, the capacity of the photosynthetic baceria to generate ATP through a cyclic-type photophosphorylation potentially relieves some of the energetic problems faced by the classic organotrophic nitrogen-fixing organisms. But is this the only requirement for light in nitrogen fixation in these organisms? Previously, Bennett *et al.* (1975) proposed that the provision of extra ATP through a light-dependent mechanism would suffice energetically in the photosynthetic bacteria. However, the question remains whether the additional photoreduction of a low potential reductant would not favor an increased capacity for nitrogen fixation. Earlier studies on chromatophores of *Rhodospirillum rubrum* (see Evans and Phillips, 1975) implicated a role of bacterial ferredoxin in the light-dependent nitrogen fixation. Although ferredoxin has been directly implicated in mechanisms of carbon dioxide reduction in certain photosynthetic bacteria (Arnon and Yoch, 1974), its role as a reductant for the nitrogenase in these organisms is questionable. More recent studies on the green sulfur bacterium *Chloropseudomonas ethylicum* (Evans and Smith, 1971) have revealed a definitive role for reduced bacterial ferredoxin in nitrogen fixation. Whether this is a general characteristic of all nitrogen-fixing photosynthetic bacteria remains to be clarified. Yates (1972) observed that the hydroquinone form of flavodoxin also would serve as an electron donor for the nitrogenase of *Azotobacter chroococcum*. Since substances similar to flavodoxin are found in photosynthetic bacteria and blue-green algae (Evans and Phillips, 1975), it would be appropriate to examine their potential function as primary reductants in nitrogen fixation.

As we indicated earlier in this review, the primary photoreductant generated by photosynthetic bacteria involves both an iron–sulfur component (referred to as phytoredoxin by Bearden and Malkin, 1975) and ubiquinone (Clayton, 1973; Bearden and Malkin, 1975; Bolton and Warden, 1976). The estimated reduction potential for the iron–sulfur component has been measured by a number of laboratories, and the value ranges from about -100 mV to -135 mV. A reduction potential of about 0 mV to 50 mV has been estimated for the midpoint potential of ubiquinone (Clayton, 1973; Bearden and Malkin, 1975). The ferredoxins isolated from the photosynthetic bacteria all have a strongly negative midpoint potential (about -420 mV). This apparent contradiction between the reduction potentials of the primary acceptors in the photosynthetic bacterial photochemical reaction centers and of bacterial ferredoxin most likely is the primary cause for the general inability of bacterial chromatophores (except perhaps for the green sulfur bacteria) to reduce ferredoxins photochemically. Obviously, further studies are needed in this area.

B. PHOTOPRODUCTION OF HYDROGEN IN THE PHOTOSYNTHETIC
   BACTERIA

Numerous photosynthetic and nonphotosynthetic bacteria possess a hydrogen metabolism. Characteristically this metabolism is mediated by one or more types of hydrogenases; these include (1) the classic hydrogenase, which either can utilize hydrogen gas to generate reducing potential or can evolve hydrogen gas as a means of dissipating excessive reduction potential; and (2) the ATP-dependent hydrogenase activity associated with the nitrogenase–enzyme complex. As we have discussed earlier, the photoevolution of hydrogen by green algae is mediated through the first type of hydrogenase, does not require ATP, is sensitive to carbon monoxide, and requires the photochemical generation of a strong photoreductant, probably reduced ferredoxin. In contrast to this type of hydrogen evolution, the photosynthetic bacteria form hydrogen photochemically through a system which is ATP dependent, inhibited by nitrogen gas, and carbon monoxide insensitive. Clearly this is characteristic of the nitrogenase–enzyme complex, which has been shown by many investigators to produce hydrogen gas when supplied with an external hydrogen donor and a source of ATP. Thus the mechanism for generating hydrogen photochemically in either the blue-green algae or the photosynthetic bacteria requires principally the participation of the nitrogen-fixing apparatus. Because of the strong negative reduction potential generated by this system, it is not necessarily surprising that proton reduction occurs. This is viewed as an inherent inefficiency or "leakage" in the system, which is perhaps compensated for through the presence of an uptake of hydrogen. The association of this type of hydrogenase to the nitrogen-fixing apparatus might serve to increase its efficiency.

Recent studies on photohydrogen evolution by the photosynthetic bacteria have been reviewed by Arnon and Yoch (1974), and it would serve no useful purpose to restate their excellent coverage and interpretation. It is now recognized that photohydrogen evolution by photosynthetic organisms is not a single phenomenon, but each type of organism—the green algae, the blue-green, etc.—has either a separate or a modified mechanism in which different "hydrogenases" function. We

TABLE III

GENERAL CHARACTERISTICS OF THE LIGHT-DEPENDENT HYDROGEN METABOLISM IN
GREEN AND BLUE-GREEN ALGAE AND IN THE PHOTOSYNTHETIC BACTERIA

Factor[a]	Green algae	Blue-green algae	Bacteria
Photosystem requirement	PS I and PS II	Only PS I	Not pertinent
Action of PS II inhibitors:			
DCMU, 5 μM	90% Inhibition	No immediate inhibition	No inhibition
Simazine, 10 μM	Total inhibition	No inhibition	Not tested
NH$_2$OH, 10 μM	Total inhibition	No inhibition	Not tested
Action of phosphorylation uncouplers (DNP and CCCP)	None (sometimes stimulatory)	Inhibitory	Inhibitory
Nature of electron donor	Water?	Organic substrate	Organic substrate
Nitrogenase requirement	Not involved	Essential	Essential
Classic hydrogenase requirement	Absolute	Not required	Not required
ATP-dependent "hydrogenase"	Not required	Essential	Essential
Effects of tricarboxylic cycle intermediates	None	Stimulatory	Stimulatory
Effect of glycolytic pathway intermediates	Slight	Not tested	Stimulatory
Inhibition by carbon monoxide	Complete	Slight	None

[a] DCMU, 3(3,4 dichlorophenyl) 1,1-dimethyl urea; DNP, 2,4 dinitrophenol; CCCP, *m*-chloro-carbonyl-cyanide-phenylhydrazone.

include, as a partial summary for the review, a comparison of the various characteristics of the different types of hydrogen evolution by photosynthetic organisms (Table III). Future research, regardless of its motivation, must take these general features into account for any additional interpretations of mechanisms.

ACKNOWLEDGMENTS

The authors wish to acknowledge the expert technical assistance of Ms. Marianne Frick and Mr. James Wong; their efforts in the reported studies on hydrogen metabolism of the green algae were of paramount importance. Financial support for the research reported herein and for the preparation of the manuscript was provided through a grant from the National Science Foundation (BMS-7518023).

REFERENCES

Allen, M. B. (1956). *Sci. Mon.* **83,** 100.
Allen, M. B., and Arnon, D. I. (1955). *Plant Physiol.* **8,** 653.
Arnon, D. I., and Yoch, D. C. (1974). *In* "The Biology of Nitrogen Fixation" (A. Quispel, ed.), p. 168. Am. Elsevier, New York.
Avron, M. (1967). *Curr. Top. Bioenerg.* **2,** 1.
Bearden, A. J., and Malkin, R. (1975). *Q. Rev. Biophys.* **7,** 131.
Beinert, H., and Kok, B. (1963). *N.A.S.—N.R.C. Publ.* **1145,** 131.
Ben-Amotz, A., and Gibbs, M. (1975). *Biochem. Biophys. Res. Commun.* **64,** 355.
Benemann, J. R., and Weare, N. M. (1974). *Science* **184,** 174.
Benemann, J. R., Berneson, J. A., Kaplan, N. O., and Kamen, M. S. (1973). *Proc. Natl. Acad. Sci. U.S.A.* **70,** 2317.
Bennett, K., Silvester, W. B., and Brown, J. M. A. (1975). *Arch. Microbiol.* **105,** 61.
Bishop, N. I. (1966). *Annu. Rev. Plant Physiol.* **17,** 185.
Bishop, N. I. (1971). *Annu. Rev. Biochem.* **40,** 197.
Bishop, N. I. (1972). *Photosynth., Two Centuries Its Discovery Joseph Priestley, Proc. Int. Congr. Photosynth. Res., 2nd, 1971* p. 459.
Bishop, N. I. (1973). *Photophysiology* **8,** 65.
Bishop, N. I., and Gaffron, H. (1962). *Biochem. Biophys. Res. Commun.* **6,** 471.
Bishop, N. I., and Gaffron, H. (1963). *N.A.S.—N.R.C., Publ.* **1145,** 441.
Bishop, N. I., and Wong, J. (1971). *Biochim. Biophys. Acta* **234,** 433.
Bishop, N. I., and Wong, J. (1974). *Ber. Dtsch. Bot. Ges.* **87,** 359.
Bishop, N. I., Frick, M., and Jones, L. W. (1977). *In* "Biological Solar Energy Conversion" (A. Mitsui, S. Miyachi, A. San Pietro, and S. Tamura, eds.), p. 1. Academic Press, New York.
Boardman, N. K. (1970). *Annu. Rev. Plant Physiol.* **21,** 115.
Bolton, J. R., and Warden, J. T. (1976). *Annu. Rev. Plant Physiol.* **27,** 375.
Bone, D. H. (1971). *Arch. Mikrobiol.* **80,** 234.
Bothe, H. (1972). *Photosynth., Two Centuries Its Discovery Joseph Prestley, Proc. Res., 2nd, 1971* p. 2169.
Bothe, H., and Loos, E. (1972). *Arch. Mikrobiol.* **86,** 241.
Bothe, H., and Falkenberg, B. (1973). *Plant Sci. Lett.* **1,** 151.
Bothe, H., Tennigkeit, J., and Eisbrenner, G. (1976). Personal communication.
Bradley, S., and Carr, N. G. (1971). *J. Gen. Microbiol.* **68,** 13.

Brown, J. B. (1973). *Photophysiology* **8**, 97.
Brownell, P. F., and Nicholas, D. J. D. (1967). *Plant Physiol.* **42**, 915.
Burns, R. C., and Hardy, R. W. F. (1975). "Nitrogen Fixation in Bacteria and Higher Plants." Springer-Verlag, Berlin and New York.
Butler, W. L., and Bishop, N. I. (1963). *N.A.S.—N.R.C., Publ.* **1145**, 91.
Carpenter, E. J., and Price, C. C. (1976). *Science* **191**, 1278.
Carr, N. G., and Whitton, B. A. (1973). "The Biology of the Blue-Green Algae." Univ. of California Press, Berkeley and Los Angeles.
Cheniae, G. M. (1970). *Annu. Rev. Plant Physiol.* **21**, 467.
Clayton, R. K. (1965). "Molecular Physics in Photosynthesis." *Ginn* (Blaisdell), New York.
Clayton, R. K. (1973). *Annu. Rev. Biophys. Bioeng.* **2**, 131.
Cobb, H. D., and Myers, J. (1964). *Am. J. Bot.* **51**, 753.
Codd, G. A., and Stewart, W. D. P. (1973). *Arch. Mikrobiol.* **94**, 11.
Codd, G. A., Rowell, P., and Stewart, W. D. P. (1974). *Biochem. Biophys. Res. Commun.* **61**, 424.
Cox, R. M. (1966). *Arch. Mikrobiol.* **53**, 263.
Cox, R. M., and Fay, P. (1967). *Arch. Mikrobiol.* **58**, 357.
Cox, R. M., and Fay, P. (1969). *Proc. R. Soc. London, Ser. B* **172**, 357.
Epel, B. L., and Butler, W. L. (1972). *Biophys. J.* **12**, 922.
Epel, B. L., and Levine, R. P. (1971). *Biochim. Biophys. Acta* **226**, 154.
Epel, B. L., Butler, W. L., and Levine, R. P. (1972). *Biochim. Biophys. Acta* **275**, 395.
Evans, H. J., and Phillips, D. A. (1975). *In* "Nitrogen Fixation by Free-Living Micro-organisms" (W. D. P. Stewart, ed.), p. 289. Cambridge Univ. Press, London and New York.
Evans, M. C. W., and Smith, R. V. (1971). *J. Gen. Microbiol.* **65**, 95.
Fay, P. (1965). *J. Gen. Microbiol.* **39**, 11.
Fay, P. (1968). *Biochim. Biophys. Acta* **216**, 353.
Fay, P. (1969). *Arch. Mikrobiol.* **67**, 62.
Fogg, G. E. (1949). *Ann. Bot. (London)* [N.S.] **13**, 241.
Fogg, G. E., and Than-Tun. (1960). *Proc. R. Soc. London, Ser. B* **153**, 111.
Fogg, G. E., Stewart, W. D. P., Fay, P., and Walsby, A. E. (1973). "The Blue-Green Algae." Academic Press, New York.
Frenkel, A. W. (1970). *Biol. Rev. Cambridge Phil. Soc.* **76**, 5568.
Frenkel, A. W., Gaffron, H., and Battley, E. H. (1950). *Biol. Bull.* **99**, 157.
Gaffron, H. (1940). *Am. J. Bot.* **27**, 273.
Gaffron, H. (1944). *Bacteriol. Rev.* **19**, 1.
Gaffron, H., and Bishop, N. I. (1963). *Colloq. Int. C. N. R. S.* **119**, 645.
Gaffron, H., and Rubin, J. (1942). *J. Gen. Physiol.* **26**, 209.
Gest, H., Ormerod, J. G., and Ormerod, K. S. (1962). *Arch. Biochem. Biophys.* **97**, 21.
Haystead, A., and Stewart, W. D. P. (1972). *Arch. Mikrobiol.* **82**, 325.
Healey, F. P. (1970a). *Plant Physiol.* **45**, 153.
Healey, F. P. (1970b). *Planta* **91**, 220.
Hoar, D. S., Hoare, S. L., and Smith, A. H. (1970). *In* "Taxonomy and Biology and Blue-Green Algae" (T. V. Desikachary, ed.), p. 27. University of Madras, Madras.
Hoare, D. S., Ingram, L. O., Thurston, E. L., and Walkup, R. (1971). *Arch. Mikrobiol.* **78**, 310.
Hwang, J. C., and Burris, R. H. (1972). *Biochim. Biophys. Acta* **238**, 339.
Jones, L. W., and Bishop, N. I. (1976). *Plant Physiol.* **57**, 659.
Kaltwasser, H., Stuart, T. S., and Gaffron, H. (1969). *Planta* **89**, 309.
Katoh, S. (1972). *Plant Cell Physiol.* **13**, 273.

Katoh, T., and Ohki, K. (1975). *Plant Cell Physiol.* **16**, 815.
Katoh, T., and Ohki, K. (1976). *Plant Cell Physiol.* **17**, 525.
Ke, B. (1973). *Biochim. Biophys. Acta* **301**, 1.
Keister, D. L., and Fleischman, D. E. (1973). *Photophysiology* **8**, 157.
Kenyon, C. H., Rippka, R., and Stanier, R. Y. (1972). *Arch. Mikrobiol.* **83**, 216.
Kessler, E. (1974). *In* "Algal Physiology and Biochemistry" (W. D. P. Stewart, ed.), p. 456. Blackwell, Oxford.
Khoja, T., and Whitton, B. A. (1971). *Arch. Mikrobiol.* **79**, 280.
Kimimura, M., and Katoh, S. (1972). *Biochim. Biophys. Acta* **283**, 279.
Kimimura, M., and Katoh, S. (1973). *Biochim. Biophys. Acta* **325**, 167.
Kok, B. (1973). "NSF/RANN Report," Proc. Workshop Bio-Solar Conversion, p. 22.
Kulasooriya, S. A., Langs, N. J., and Fay, P. (1972). *Proc. R. Soc. London, Ser. B* **181**, 199.
Leach, C. K., and Carr, N. G. (1971). *Biochim. Biophys. Acta* **245**, 165.
Lindeman, W., and Spruit, C. J. P. (1963). *Recl. Trav. Chim. Pays-Bas* **82**, 671.
Ljones, T. (1974). *In* "The Biology of Nitrogen Fixation" (A. Quispel, ed.), p. 617. Am. Elsevier, New York.
Mickelson, J. C., David, E. B., and Tischer, R. C. (1967). *J. Exp. Bot.* **18**, 397.
Nakamura, H. (1938). *Acta Phytochim.* **10**, 271.
Neilson, A., Rippka, R., and Kunisawa, R. (1971). *Arch. Mikrobiol.* **76**, 139.
NSF/NASA Report. (1972). "An Essessment of Solar Energy as a National Energy Resource" (P. Donovan, W. Woodward, F. H. Morse, and L. O. Herwig, eds.).
NSF/RANN Report. (1972). "An Inquiry into Biological Energy Conversion" (W. T. Synder and E. Volkin, eds.). Gatlinburg.
NSF/RANN Report. (1973). "Proceedings of the Workshop on Bio-Solar Conversion" (A. San Pietro and S. Lien, eds.).
Newton, W. E., and Nyman, C. J., eds. (1976). "Proceedings of the 1st International Symposium on Nitrogen Fixation," Vols. I and II. Washington State Univ. Press, Pullman.
Nutman, P. S. (1975). "Symbiotic Nitrogen Fixation in Plants." Cambridge Univ. Press, London and New York.
Ohmori, M., and Hattori, A. (1972). *Plant Cell Physiol.* **13**, 589.
Ormerod, J. G., and Gest, H. (1962). *Bacteriol. Rev.* **26**, 51.
Peterson, R. B., and Burris, R. H. (1976). *Arch. Microbiol.* **108**, 35.
Postgate, J. R. (1972). "Biological Nitrogen Fixation." Merrow Publ. Co. Ltd., Watsford, England.
Senger, H., and Frickel-Faulstitch, G. (1975). *Proc. Int. Congr. Photosynth. Res., 3rd,* Vol. 1, p. 715.
Shanmugan, K. T., and Valentine, R. C. (1975). *Science* **187**, 919.
Singh, P. K. (1973). *Arch. Mikrobiol.* **93**, 59.
Smith, R. V., and Evans, M. C. W. (1971). *J. Bacteriol.* **105**, 913.
Smith, R. V., Boy, R. J., and Evans, M. S. W. (1971). *Biochim. Biophys. Acta* **253**, 104.
Spruit, C. J. P. (1954). *Proc. Int. Photobiol. Congr., 1st, 1954* p. 323.
Spruit, C. J. P. (1958). *Meded. Landbouwhogesch. Wageningen* **58**, 1.
Stewart, W. D. P. (1974). *In* "The Biology of Nitrogen Fixation" (A. Quispel, ed.), p. 202. Am. Elsevier, New York.
Stewart, W. D. P., ed. (1975). "Nitrogen Fixation by Free-living Microorganisms." Cambridge Univ. Press, London and New York.
Stewart, W. D. P., and Lex, M. (1970). *Arch. Mikrobiol.* **73**, 250.
Stuart, T. S. (1971). *Planta* **96**, 81.
Stuart, T. S., and Gaffron, H. (1971). *Planta* **100**, 228.

Stuart, T. S., and Gaffron, H. (1972a). *Planta* **106**, 91.
Stuart, T. S., and Gaffron, H. (1972b). *Planta* **106**, 101.
Stuart, T. S., and Gaffron, H. (1972c). *Plant Physiol.* **50**, 136.
Stuart, T. S., and Kaltwasser, H. (1970). *Planta* **91**, 220.
Taylor, B. G., Lee, C. D., and Bunt, J. S. (1973). *Arch. Mikrobiol.* **88**, 205.
Trebst, A. (1974). *Annu. Rev. Plant Physiol.* **25**, 423.
Van Gorkom, H. J., and Donze, M. (1971). *Nature (London)* **234**, 321.
Vernon, L. P., and Zaugg, W. S. (1960). *J. Biol. Chem.* **235**, 2728.
Ward, A. K., and Wetzel, R. G. (1975). *J. Phycol.* **11**, 357.
Weare, N. M., and Benemann, J. R. (1974). *J. Bacteriol.* **119**, 258.
Weaver, E. C., and Bishop, N. I. (1963). *Science* **140**, 1095.
Winkenbach, F., and Wolk, C. P. (1973). *Plant Physiol.* **52**, 480.
Winter, H. C., and Burris, R. H. (1976). *Annu. Rev. Biochem.* **45**, 409.
Wolfe, M. (1954). *Ann. Bot. (London)* [N.S.] **18**, 309.
Wolk, C. P. (1973). *Bacteriol. Rev.* **37**, 32.
Wyatt, J. T., and Silvey, J. K. G. (1969). *Science* **165**, 908.
Yates, M. G. (1972). *FEBS Lett.* **27**, 63.
Yoch, D. C., and Arnon, D. I. (1974). *In* "The Biology of Nitrogen Fixation" (A. Quispel, ed.), p. 87. Am. Elsevier, New York.

The Photosynthetic Membrane

Chlorophyll–Protein Complexes and Structure of Mature and Developing Chloroplasts

N. K. BOARDMAN, JAN M. ANDERSON, and D. J. GOODCHILD
Commonwealth Scientific and Industrial Research Organization,
Division of Plant Industry, Canberra, Australia

I. Introduction

The function of most of the chlorophyll *a* molecules and all the chlorophyll *b* molecules of the chloroplast thylakoid membrane is to absorb light quanta and transfer the excitation energy to a small fraction of special chlorophyll *a* molecules in the photochemical reaction centers. The reaction-center chlorophylls are closely associated with electron donor and acceptor molecules in the thylakoid membrane, and their function is to convert the excitation energy into chemical free energy in the form of a charge separation between the oxidized donor molecule and reduced acceptor molecule. Photosynthesis in plants and algae involves the cooperation of two photochemical reactions, photosystem I (PS I) and photosystem II (PS II) at two different types of reaction-center chlorophylls (Govindjee, 1975). A photosynthetic unit, defined as the minimum number of chlorophyll molecules associated with one reaction center of PS I and one of PS II consists of some 400 chlorophyll molecules. Chlorophyll *a in vivo* is composed of a number of spectral forms (French *et al.*, 1972), which presumably reflect the differing interactions of the pigment molecules, with themselves and with protein molecules.

Studies over the past decade have shown that at least two-thirds of the chlorophyll of the thylakoid membrane is isolated as two chlorophyll–protein complexes by extraction with sodium dodecyl sulfate (SDS) (Thornber, 1975). Anderson (1975a) and Thornber (1975) have advanced arguments for the conclusion that the two major chlorophyll complexes are not artifacts of the solubilization procedures, and chlorophylls are complexed to the proteins *in vivo*. The isolated chlorophyll–protein complexes account for some 70% of the intrinsic proteins of the thylakoid membranes. Except in the case of a water-soluble bacterio-chlorophyll–protein complex (Fenna and Matthews, 1975, 1977), nothing is known about the arrangement of chlorophyll molecules in the complexes or the mode of interaction of the pigment and protein.

The etioplasts of dark-grown seedlings contain a protochlorophyllide–protein complex, which has been isolated in its photoactive state either by extraction into aqueous buffer (Boardman, 1966) or by detergent treatment (Henningsen and Kahn, 1971). During greening of the dark-grown seedlings, the protochlorophyllide is converted to chlorophyll *a* via a number of intermediate spectral forms of chlorophyllide *a* and chlorophyll *a*.

Freeze-fracture electron microscopy has indicated that the thylakoid membrane is composed of particles, embedded in what is assumed to be the fluid lipid matrix of the membrane. This chapter reviews current knowledge of the protochlorophyllide–, chlorophyllide–, and chlorophyll–protein complexes, and their possible significance to the substructure of the etioplast tubular membranes and chloroplast thylakoid membranes. The differentiation of the thylakoid system into stacked (grana) and unstacked (stroma) regions is also considered in relation to the chlorophyll–protein complexes.

II. Chloroplast Ultrastructure

Chloroplast ultrastructure has been the subject of several recent reviews (Park and Sane, 1971; Kirk, 1971; Anderson, 1975a; Arntzen and Briantais, 1975), and it is the intention here to describe a membrane model that satisfies the data available. The discussion is mainly restricted to the internal chloroplast membrane network, since this is location of the chlorophyll–protein complexes.

A. CHLOROPLAST ENVELOPE AND PERIPHERAL RETICULUM

The chloroplast envelope, which delineates the internal membrane network and stroma from host cytoplasm in most plant species, is made up of two unit membranes, the inner and outer envelope membranes, separated by a space whose width varies with different preparations but is about 20 nm. Isolation and characterization of the envelope membrane suggest that it contains carotenoids but no chlorophyll and is of different lipid and polypeptide compositions than the internal chloroplast membranes (Douce *et al.*, 1973; Poincelot, 1973; Mackender and Leech, 1974; Jeffrey *et al.*, 1974; Joy and Ellis, 1975).

Sprey and Laetsch (1976), using the freeze-etch technique as an ultrastructural marker for spinach chloroplast envelopes, suggested, however, that the membrane envelope fractions thus far studied are predominantly of outer envelopes. Their evidence indicates the outer envelope membrane to have particles 9.0 nm in diameter on both fracture faces with densities of about 150 particles/μm^2. The inner envelope membrane differs markedly from this in having about 975 particles/μm^2 of diameter 9.0 nm on the outer fracture face and 1820 particles/μ^2 of diameter 7.0 nm on the inner fracture face. These particle numbers for the fracture faces of the inner envelope membrane are similar to those reported by Sprey and Laetsch (1976) for the fracture faces of thylakoid membranes. This observation lends support to the view discussed by Arntzen and Briantais (1975), that the internal chloroplast membranes arise by invagination of the inner envelope membrane.

Other evidence (Armond *et al.*, 1977) suggests that in developing thylakoids the addition of light-harvesting components during development results in an increase in size of a basic number of 8.0-nm particles to form the familiar large particles of the thylakoid fracture face of mature chloroplasts. The size distribution of the inner envelope fracture-face particles supports the contention that thylakoid membranes arise by invagination of the inner envelope membrane. Alternatively, the asymmetry and relatively high particle number per square micrometer in the inner envelope membrane may reflect a regulatory function, as suggested by Sprey and Laetsch (1976), since the capacity for the biosynthesis of some membrane components, e.g., fatty acids, resides in chloroplast stroma (Stumpf, 1975) and there is, as yet, no conclusive evidence to implicate the inner chloroplast envelope in new membrane biosynthesis.

In some plant species, the inner chloroplast envelope membrane forms an anastomosing system of tubules known as the peripheral reticulum. It is particularly prominent in plants that form C_4 dicarboxylic acids as the primary products of photosynthesis. Its structure and possible role have been reviewed by Laetsch (1974). There is no evidence for an association of chlorophyll with peripheral reticulum, although the fragile membrane network has not been isolated to establish unequivocally the presence or the absence of chlorophyll. Peripheral reticulum is occasionally seen in plants that form 3-phosphoglyceric acid as the primary product of photosynthesis, but is poorly developed (Laetsch, 1974).

B. INTERNAL MEMBRANE NETWORK

By electron microscopy of fixed and sectioned material from higher plants, the internal membrane network of the chloroplast is seen to consist of two distinctly different regions. In one region, the grana region, the membranes are arranged as stacks of small saclike disks, termed thylakoids by Menke (1962). A grana stack consists of 2 or more thylakoids, and the size and number of grana in a chloroplast may vary greatly. The region where two thylakoids are in contact is termed the partition, or appressed, region, and this region appears as a heavy dense line in material fixed in potassium permanganate or glutaraldehyde followed by osmium tetroxide. With osmium alone the partition region is seen as a 4.0-nm gap (Nir and Pease, 1973). The interior space of each thylakoid, enclosed by its membranes, has been termed the loculus. Grana stacks are connected by membranes (frets, stroma thylakoids, or stroma lamellae) that are not appressed, one to another, but whose loculi are continuous with the loculi of grana thylakoids. Many studies have attempted to define the structural relationship between stroma thylakoids and grana thylakoids, and the model shown in Fig. 1 (Gunning and Steer, 1975) was largely derived from the work of Paolillo (1970). In this model each stroma thylakoid is connected by short tubes to several grana thylakoids in an ascending helical pattern with respect to each granum, thus supporting the view of Heslop-Harrison (1963): "The entire lamellar system of the chloroplast, including all the grana,

FIG. 1. Model of the structural organization of the internal membranes of the higher plant chloroplast, showing the relationship between grana thylakoids and stroma thylakoids. From Gunning and Steer (1975).

constitutes a single, enormously complex, membrane-bounded entity, separate and distinct from the stroma.'' Such a model does not entirely account for the complexity of the membrane network within a chloroplast. The number of thylakoids per granum varies from one granum to another, and there is variation in the ratio of stroma thylakoids to grana thylakoids.

C. THYLAKOID SUBSTRCTURE

Our knowledge of the thylakoid membrane substructure is now largely due to freeze-etch studies. With the recognition that, in this technique, membranes fracture internally to produce matching faces from a single membrane break (Branton, 1973; Bullivant, 1974), it has been possible to interpret the freeze-fracture images in relation to the position of the particles observed on the fracture faces. The interpretation has also been aided by deep-etching to reveal the membrane surfaces adjacent to fracture faces.

The information derived from many studies is represented diagrammatically in Fig. 2. Two adjoining thylakoids in a granum are shown with the membranes extending into the chloroplast stroma as stroma thylakoids. The model is largely derived from the data of Park and Pfeifhofer (1969) and reviewed by Park and Sane (1971), but their convention for designating the fracture faces has been altered to conform with agreed convention (Branton et al., 1975). According to this convention, when a membrane is split, the half closest to the chloroplast stroma (protoplasmic half) is designated P and the half closest to the loculus (endoplasmic half) designated E. PF and EF are thus used to designate the fracture faces; and PS and ES, the true surfaces exposed by deep etching. For the chloroplast it is also convenient to use the subscripts s to denote the grana region where the thylakoids are stacked and u (unstacked) to denote the stroma thylakoid region where there is no membrane stacking (Armond et al., 1977). The chloroplast fracture faces which originally were termed B and C by Branton and Park (1967) are now EF_s and PF_s. By recovery of matching fracture faces, using the double-replica technique, the EF_s and PF_s faces have been shown to be complementary, as depicted in the model (Wehrli et al., 1970; Mühlethaler, 1972).

Deep-etching has revealed the A' and D surfaces, now PS_s and ES_s (Park and Pfeifhofer, 1969). An A surface was also described by Branton and Park (1967), who thought it to be due to an unusual fracture along the partition between two thylakoids, as this region was considered to be hydrophobic and they had postulated the splitting of membranes to be due in part to hydrophobic bonding. However, the partition region is

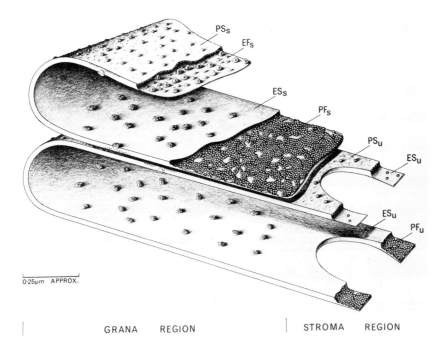

0·25µm   APPROX.

GRANA   REGION   STROMA   REGION

FIG. 2. Model of the substructure of the thylakoid membrane, derived from freeze-fracture electron microscopy. Two adjoining thylakoids in a granum are shown with the membranes extending into the chloroplast stroma as stroma thylakoids. See text for explanation of nomenclature.

probably not hydrophobic (Kirk, 1971; Nir and Pease, 1973; Anderson, 1975a), and from the number and size of the particules seen by Branton and Park (1967) on this fracture face, it is probable that it was actually a C fracture face (now PF_u) in a stroma thylakoid region. Certainly the pictures of Goodenough and Staehelin (1971) show that in *Chlamydomonas* there are minor differences in the same particle fracture face within a granal stack (designated by them CS) and in the stroma thylakoid region (CU). These pictures are similar to Fig. 11 of Branton and Park (1967), where the A (PF_s) and C (PF_u) faces of spinach chloroplasts appear to be differentiated along the same fracture face with respect to particle numbers. This observation is supported by the data of Sprey and Laetch (1976) for spinach chloroplasts.

From deep-etching experiments Park and Pfeifhofer (1969) concluded that their A′ surface (PS_s) is the exterior thylakoid surface. This surface was shown to have on it particles 10–12 nm in diameter, and as these were removed by washing with EDTA, it was concluded that they represent chloroplast coupling factor (CF_1). In other experiments, using both negative staining, and freeze-fracturing and deep-etching, particles have been shown to reappear on PS_s surfaces when purified CF_1 was added to spinach thylakoids denuded of particles by silicotungstate treatment (Garber and Steponkus, 1974) or with 300 mM sucrose, 2 mM Tris, pH 7.5 (Miller and Staehelin, 1976). CF_1 has also been visualized in sectioned and stained chloroplasts, where it has been shown to occur on both stroma thylakoids and end-grana thylakoids, but not in partition regions (Oleszko and Moudrianakis, 1974). This observation has been confirmed by Miller and Staehelin (1976), who, by a combination of deep-etching and grana unstacking and restacking experiments, demonstrated a conservation of CF_1 particles, leading to the conclusion that coupling factor is excluded from grana regions but is capable of movement during grana unstacking and restacking. By using deep-etching, and antibody-labeling techniques that aggregate CF_1 on the PS surface, Berzborn et al. (1974) concluded that there were two particle size classes on the surface: 14-nm CF_1 and a 10nm particle that was probably ribulose-1,5-diphosphate (RuDP) carboxylase. However, when sodium pyrophosphate was used to remove RuDP carboxylase (Strotmann et al., 1973), 30% of the 14–15-nm particles were removed (Miller and Staehelin, 1976) without affecting CF_1 activity. Thus, by the deep-etching technique both CF_1 and RuDP carboxylase are visualized as 14–15-nm particles on PS surfaces, but both are within the 9–10 nm size range by negative staining, and CF_1 also by positive staining. Deep-etching of purified CF_1 supports a size of 14–15 nm for this procedure (Garber and Steponkus, 1974; Oleszko and Moudrianakis, 1974; Miller and Staehelin, 1976). RuDP carboxylase is the major soluble protein of chloroplast stroma and its appearance on PS surfaces may be due to nonspecific absorption.

The first fracture face seen if a thylakoid is fractured in the manner shown in the model is the EF_s fracture face, which is characterized by particles 17.5 nm in diameter. From the shape of their shadow, it is estimated that the particles are 9 nm high and probably somewhat conical (Branton and Park, 1967). In some images these particles are seen to be arranged in regular arrays, sometimes of a paracrystalline nature, but neither the significance of these arrays nor the conditions for their production are known (Park and Pfeifhofer, 1969). Particle height and diameter are greatly influenced by the roughness of the surface being shadowed, the amount of shadow deposited, and the fracturing process itself, which may lead to "plastic" deformation of the particles (Clark and Branton, 1968). For these reasons, workers have been reluctant to place great reliance on the absolute values obtained by such measurements. However, it is probably not unreasonable to assume that when adequate precautions are taken to ensure reproducible shadowing conditions, particle diameters can be compared on a relative basis. The number of particles on fracture faces varies, but in general there are between 200 and 1300 particles/μm^2 on the EF$_s$ fracture face (Park and Sane, 1971). Also, when particle size distributions are constructed, it is seen that the EF$_s$ particles cover a size range between 7 nm and 20 nm with 2 or 3 peaks (Goodenough and Staehelin, 1971; Ojakian and Satir, 1974; Garber and Steponkus, 1974; Armond et al., 1977). The possible functional role for these particles is discussed elsewhere.

The ES$_s$ surface, revealed by deep-etching, is seen to have particles of very low profile that may have 4 subunits. Their distribution, size (15.5 × 18.5 nm), and number suggest that they are related to the particles on the EF$_s$ face, which may also have 4 subunits (Branton and Park, 1967). The ES$_s$ surface probably provided the shadowed image from which the quantasome concept developed (Park and Biggins, 1964; Park and Sane, 1971).

The PF$_s$ face is complementary to the EF$_s$ face (Wehrli et al., 1970) but has many more particles, up to 5000/μm^2. These particles are described by Branton and Park (1967) as being about 11 nm in diameter, although the particle size distributions of Goodenough and Staehelin (1971), Ojakian and Satir (1974), and Armond et al. (1977) show peaks at about 7–8 nm with an almost normal distribution. The double-replica technique has revealed that the small areas without particles sometimes seen on the PF$_s$ face do frequently correspond to the presence of large particles on the EF$_s$ complementary face, suggesting that the particles on the two complementary faces interdigitate, and if the height of the EF$_s$ face particles really approaches 9 nm, they could well span the membrane. In some cases, Mühlethaler (1972) has observed what appear to be broken particles in the double replicas suggesting that the tops of the large EF$_s$ particles may be bound in the PF$_s$ face and break off during fracturing. Miller (1976) has also provided evidence for membrane spanning by particles in spinach, by showing that where paracrystalline arrays occur on the EF$_s$ face and ES$_s$ surface (Park and Pfeifhofer, 1969), a regular pattern can also be seen on the PS$_s$ surface with a similar repeat spacing and tetrameric substructure. This finding and that of Garber and Steponkus (1974) suggest that at least the 16-nm particles on the EF$_s$ face may span the membrane. Garber and Steponkus (1974) observed that removal of CF$_1$ from the PS$_s$ surface resulted in an almost complete absence of 16-nm particles from the ES$_s$ face. This finding, however, is puzzling since CF$_1$ particles have been shown to be absent in grana regions, so that this observation would have to be restricted to end grana membranes.

These differences in the EF$_s$ and PF$_s$ fracture faces demonstrate membrane asymmetry in thylakoids. Further evidence is provided by Park and Pfeifhofer (1969), who observed, in deep-etching, that the EF_s fracture face is partially or wholly destroyed by etching times that do not affect the PF_s fracture face, suggesting that the EF_s fracture face is more fragile than the PF_s. This finding is supported by Garber and Steponkus (1974).

In the stroma thylakoid region, the EF_u face that runs out from the same EF_s face in the grana has been shown to have few particles, and these are of smaller size than the EF_s face particles. Pictures first suggesting this possibility were published by Remy (1969). Further evidence for a differentiation in structure and function along chloroplast membranes was obtained by detergent and mechanical fractionation of the chloroplast lamellar system into separate grana and stroma thylakoid fractions. In these studies (reviewed by Park and Sane, 1971; Arntzen and Briantais, 1975) the grana fraction has been shown to have both EF_s and PF_s faces and PS_s and ES_s surfaces, while the stroma lamellar fraction only has a PF_s type face and a PS_s type surface. Since these studies, other freeze-fracture images from whole-chloroplast preparations have been published that show continuity between EF_s and EF_u fracture faces that supports differentiation along the membrane. In two studies with *Chlamydomonas* (Goodenough and Staehelin, 1971; Ojakian and Satir, 1974) the EF_u fracture face has fewer particles, 600 compared with 1900, but seems to have a unique particle size distribution with few particles 16 nm in diameter. In the green alga *Oocystis,* the EF_u fracture face has no particles at all (Penland and Aldrich, 1973).

In spinach chloroplasts, Sprey and Laetsch (1976) reported an EF_u face with fewer particles, of 13 nm average diameter, than the 16 nm average diameter particles on the EF_s face. Their PF_u face also had particles of average diameter 13 nm compared with 10 nm on the PF_s face, with possibly a few more on the PF_u face. For pea, Armond *et al.* (1977) presented particle size distributions that reflect similar differences, although particle numbers per square micrometer and sizes are slightly different from those of Sprey and Laetsch (1976). Thus, it appears that the fracture faces in stacked and unstacked thylakoids have particle numbers per square micrometer and size distributions that are unique to their position. This suggests a possible functional differentiation along the thylakoid membranes, since stroma and grana thylakoids are continuous.

1. The Role of Large Particles in Membrane Stacking

In the model (Fig. 2) particles do not cross the partition region. However, Goodenough and Staehelin (1971) accounted for their observations on *Chlamydomonas* by suggesting that the large particles (16 nm) observed on the EF_s face form in response to the stacking process. In the model of Goodenough and Staehelin (1971), the large particles traverse the partition from one thylakoid to another and are shared by the two adjacent thylakoids. The evidence for this suggestion was based on studies with two mutants of *Chlamydomonas*. One mutant could be grown under conditions where membrane stacking did not occur and no large particles could be found on the EF_s fracture face. The other grew naturally without stacking, but when the chloroplasts were isolated in a high-salt buffer, grana stacks were formed and large particles appeared in the particle size distributions. If a low-salt Tricine buffer was used for isolation, the chloroplasts remained unstacked and there were no large particles. Ojakian and Satir (1974) have confirmed Goodenough and Staehelins' work on the mutant that stacks in the presence of high salt.

These views on the role of large particles in the EF_s face are not in agreement with the data of Park and Pfeifhofer (1968, 1969) and Arntzen *et al.* (1969), which showed that the large particles in the EF_s fracture face did not disappear in unstacked chloroplasts of higher plant. Further evidence to suggest that the large particles may not play a role in membrane stacking has been obtained with the barley mutant that lacks chlorophyll *b*. Stacking does occur in this mutant, but to a lesser degree than in the wild-type barley. A freeze-etch study shows that in the grana region the wild-type barley has a typical spinach-type EF_s face, but in the mutant the distribution on this face is unique. The larger particles seen in the particle size distribution of the wild type (16.0 nm) are missing, but are replaced by 12.5–13.0-nm particles and an increase in 11.0-nm particles (Goodchild, unpublished observations). Thus the role of large particles in promoting membrane stacking remains unresolved.

2. Particle Movement

Current membrane models feature the concept of fluidity wherein the membrane is a fluid mosaic structure in which proteins are free to move in the lipid matrix. Although no unequivocal evidence has been presented to prove that the freeze-fracture particles seen in most membranes are protein, or lipoprotein, circumstantial evidence suggests that they probably are protein (Bullivant, 1974). In the case of chloroplasts there is some correlation between the type of fracture face and the chlorophyll–protein complexes seen on SDS gel electrophoresis (Anderson, 1975a) (see Sections V and VII).

Chloroplast membranes *in vitro* can be induced to swell or shrink, stack or unstack, and thin by a variety of procedures involving changes in buffer systems, salt concentration, pH, and light. Wang and Packer (1973), using weak-acid anions, reported that the packed volume of isolated spinach chloroplasts (a measure of shrinkage and swelling) decreased 42% on illumination, and this was accompanied by a 2.2-fold increase in particle density on the EF_s face and an increase in particle diameter. When a similar chloroplast preparation was induced to increase its packed volume (swell) upon illumination in high salt, there was a 32% decrease in the number of particles per square micrometer and a decrease in average particle diameter. Further evidence for particle movement has been obtained by Ojakian and Satir (1974) with *Chlamydomonas*. They measured the sizes and numbers of particles per square micrometer on the EF and PF faces in both stacked and unstacked regions, and on these faces in another preparation that had been induced to unstack in low-salt buffer. They then calculated what would be the expected numbers of particles per square micrometer if the particles in the stacked regions were assumed to redistribute laterally and evenly along the EF and PF faces. Their data strongly support redistribution on both fracture faces not only with respect to the total number of particles per square micrometer, but also with respect to all size classes seen in their particle size distribution histograms. When the normally unstacked mutant was induced to stack in the presence of high salt, a similar conservation of particle size and numbers was observed, their large (16 nm) EF_s face particles increasing by 20%.

III. Development of Chloroplast Structure

The etiolated tissue of dark-grown plants contains a special type of plastid, the etioplast (Kirk and Tilney-Bassett, 1967), that is devoid of chlorophyll. The etioplasts accumulate a small amount of protochlorophyllide, which is converted to chlorophyllide *a* on illumination of the dark-grown tissue, followed by estification with phytyl alcohol to form chlorophyll *a* (Boardman, 1966). Etioplast structure and the subsequent structural changes leading to mature chloroplasts on greening have been reviewed by Rosinski and Rosen (1972) and Kirk (1971). The description that follows will attempt to present concepts of the ultrastructure of etioplasts and developing chloroplasts that relate to an understanding of the role of the protochlorophyllide and chlorophyll(ide)–protein complexes in membrane structure.

A. ETIOPLAST ULTRASTRUCTURE

The mature etioplast, as seen in the electron microscope, is an organelle bounded by a double-membrane envelope of similar dimensions to the chloroplast envelope and having, as its major internal feature, a prolamellar body or bodies (Gunning and Jagoe, 1967). The prolamellar body is composed of membrane tubules arranged in a three-dimensional lattice sometimes referred to as a paracrystalline, or semicrystalline, lattice (Kirk and Tilney-Bassett, 1967). Single thylakoids are frequently seen radiating from the lattice extremities and short lengths of appressed regions may also be present, although it is unusual to find more than two thylakoids appressed. These thylakoids are frequently perforated by pores (Weier and Brown, 1970). Plastoglobuli may also be seen (Gunning and Steer, 1975).

Etioplast tubules, measured from leaf sections, are 18–20 nm in outside diameter with a lumen diameter of 8 nm, thus making the tubular membrane wall about 5–6 nm. Several lattice types have been described for the paracrystalline arrays seen in prolamellar bodies (Gunning and Jagoe, 1967; Wehrmeyer, 1965a,b; Ikeda, 1968), and well-defined dislocations can often be observed within a single prolamellar body. In one lattice type, the cubic lattice, the six tubular membranes branch outward at right angles in six directions (Gunning and Jagoe, 1967). Other lattice types include a tetrahedral lattice and pentagonal dodecahedra (Gunning and Steer, 1975). All types can be found in one plant species.

Electron microscopy of isolated prolamellar body fragments by negative-staining techniques (Kahn, 1968) indicated an outer diameter of about 28 nm for the tubules and a lumen diameter of 6–6.5 nm. On the basis of these measurements the tubular membrane wall should be between 10 and 11 nm. This study also revealed regularly arranged macromolecules in the tubular walls with a close-packed axial center-to-center distance of 8 nm, but an average diameter of 10 nm in nontubular membrane segments.

Since the protochlorophyllide–protein complex (termed protochlorophyllide holochrome) of bean etioplasts is an approximately isodiametric macromolecule of average diameter 10 nm (Boardman, 1962a), Kahn (1968) suggested that the macromolecules observed in the tubular membrane wall were protochlorophyllide holochrome. The content of protochlorophyllide holochrome in the etiolated bean leaf was consistent with the estimated number of macromolecules in the tubular walls of the prolamellar body membranes.

The freeze-etch technique has proved to be somewhat disappointing for elucidating the structure of the prolamellar body membranes, possibly because of the complexity of the membrane lattices. Bronchart (1970) proposed an egg-crate type of lattice based on freeze etch evidence, but it appears more likely that his pictures are indicative of a complex prolamellar body formed from large tetrahedral units (see Gunning and Steer, 1975, Plate 37g) and represent only one possible type of lattice structure. Some freeze-etch pictures have shown particles 10–11 nm in diameter in the walls of the tubular membranes, which are consistent with Kahn's observations with negative staining. Fluorescence microscopy (Kahn, 1968) supports the view that the protochlorophyllide holochrome is located in the walls of the prolamellar body tubules.

The lamellae that radiate from a prolamellar body into the etioplast stroma were shown by freeze-fracture electron microscopy to contain particles. The EF_s face had 300–1300 particles/μm^2 of diameter 8–12 nm and the PF_s face 4000–5500 particles/μm^2 of diameter 6–12 nm (Phung Nhu Hung et al., 1970). This study showed that the lamellae are perforated, but studies using thin-section techniques are divided on this issue (Weier and Brown, 1970).

The development of the membrane lattice of the prolamellar body during the growth of dark-grown seedlings has been studied by several investigators (von Wettstein, 1958; Gunning and Jagoe, 1967; Henningsen and Boynton, 1969; Weier and Brown, 1970; Bradbeer et al., 1974a; Robertson and Laetsch, 1974). The young tissue of dark-grown seedlings contain relatively undifferentiated proplastids (approximately 1 μm in diameter) that develop into etioplasts. Bradbeer et al. (1974a) observed invaginations of the inner envelope membrane of the proplastids after 6 days' growth of bean seedlings in the dark at 23°C, and these developed into a membrane network of perforated sheets with membrane condensations of the prolamellar body-type in localized areas. The condensations developed into the regular tight lattice structure of a mature prolamellar body after 14 days in the dark. Similar findings were reported by Weier and Brown (1970), but they grew beans at 25°C and the sequence of events was much faster. Robertson and Laetsch (1974) studied etioplast development in barley. Since leaves of monocots develop basipetally, they were able to recognize five stages of etioplast development along a single leaf. Gunning and Steer (1975) reported a diminution in the area of perforated thylakoids surrounding the prolamellar body just before the latter reaches maximum volume and inferred that the two forms of membrane are in a precursor–product relationship throughout the period of development.

B. Ultrastructural Changes during Greening

When etiolated plants are illuminated, a rearrangement of membrane tubules of the prolamellar body takes place. The first stage, as seen by electron microscopy, is a loss of the symmetrical framework of the lattice structure, which has been termed tube transformation, but few of the tubules between the units seem to be broken (Gunning, 1965). In a 14-day-old bean seedling grown at 23°, the half-time for tube transformation in high intensity illumination is 30–45 minutes (Bradbeer et al., 1974b). The process is more rapid in young tissue (Robertson and Laetsch, 1974; Henningsen, 1970), where the prolamellar body tubules are not organized as tightly. In early studies (von Wettstein, 1958; Eriksson *et al.*, 1961; Klein *et al.*, 1964; Gyldenhohm and Whatley, 1968), when potassium permanganate was used as a fixative, tube transformation appeared to take place immediately after the etioplasts received light and to correspond to photoconversion of protochlorophyllide to chlorophyllide *a*. Freeze-etch studies (Goodchild, unpublished observations) support the results obtained with glutaraldehyde fixation and suggest that potassium permanganate gives rise to artifacts. Nevertheless, it seems that short-term illumination of etioplasts potentiates a change, possibly conformational, that makes the tubular lattice sensitive to potassium permanganate. A conformational change in the protochlorophyllide holochrome, which is discussed later, may be related to the permanganate effect.

After tube transformation, the prolamellar body tubules disperse and essentially two-dimensional sheets of double lamellae (thylakoids) are generated from the three-dimensional network of tubes (Wellburn and Wellburn, 1971). The membranes at this stage are perforated, and in section profile they appear as rows of vesicles; this process was termed "vesicle dispersal" (Virgin *et al.*, 1963; Gunning, 1965; Lemoine, 1968). With continued illumination the perforations disappear, and the membranes are then termed primary thylakoids. The disappearance of the performations is possibly due to contraction of the membrane material (Henningsen and Boynton, 1974). Measurements of total membrane area suggests that, up to this stage, all membrane material is derived from the existing prolamellar body (Bradbeer *et al.*, 1974b; Henningsen and Boynton, 1974; Gunning and Steer, 1975). Membrane overlaps then form, pairing begins, and appressed regions appear. Finally, fully developed grana profiles form (Gunning, 1965; Lemoine, 1968; Henningsen and Boynton, 1974; Weier *et al.*, 1974; Robertson and Laetsch, 1974; Bradbeer *et al.*, 1974b).

IV.   Protochlorophyllide– and Chlorophyllide–Protein Complexes

A.   Spectroscopic Forms of Protochlorophyll(ide)* and Chlorophyll(ide)

1. Protochlorophyll(ide)

Absorption spectrometry of etiolated leaves or isolated prolamellar body membranes at the temperature of liquid nitrogen indicates three spectroscopically distinguishable forms of protochlorophyll(ide) with

---

* Protochlorophyll(ide) and chlorophyll(ide) are collective terms to denote both the esterified and nonesterified pigments.

absorption maxima at 628 nm (PChl-628) 637 nm (PChl-637) and 650 nm (PChl-650) (Shibata, 1957; Kahn *et al.*, 1970). Curve analyses of the absorption spectra of etioplasts at room temperature also indicate three gaussián components with maxima at 630, 637, and 650 nm (Horton and Leech, 1975). Two fluorescence forms of protochlorophyll(ide) are observed at 77°K with maxima at 630–633 nm and 655 nm (Litvin and Krasnovsky, 1957; Kahn *et al.*, 1970). An excitation spectrum of the fluorescence band at 655 nm shows two bands at 650 nm and 638 nm, indicating that the 655 nm emission is activated by light absorbed by both PChl-650 and PChl-637. Light energy absorbed by PChl-637 is transferred to PChl-650 with high efficiency, and this accounts for the lack of a fluorescence band corresponding to PChl-637. The fluorescence emission at 630 nm originates from PChl-628.

The relative amounts of the three forms of protochlorophyll(ide) depend on the age of the dark-grown seedlings. For example, PChl-650 and PChl-637 are present in approximately equal amounts in etiolated bean leaves 10–14 days old, and together they account for 85–90% of the total protochlorophyll(ide) of the leaf. Young seedlings contain higher proportions of PChl-628 and PChl-637 (Thorne, 1971a; Klein and Schiff, 1972); in a 3-day-old bean leaves, the proportion of PChl-650 is small and PChl-628 accounts for more than 50% of the protochlorophyll(ide). Between day 3 and day 7 of leaf development, PChl-650 accumulates faster than PChl-637 and PChl-628. The formation of PChl-650 correlates with the formation and enlargement of the prolamellar bodies.

2. Chlorophyll(ide) a

A number of spectroscopic forms of chlorophyll(ide) are observed when dark-grown leaves are illuminated (Fig. 3). PChl-637 and PChl-650 are photoconverted to a form of chlorophyllide absorbing at 678 nm (Chl-678), which is convered within 30 sec at room temperature to Chl-682, followed by a slower spectral shift (Shibata shift) to 672 nm (Chl-672) (see review in Kirk, 1970). During several hours of greening, the absorption maximum shifts gradually from 672 nm to 678 nm, which corresponds with the absorption maximum of chlorophyll in the fully

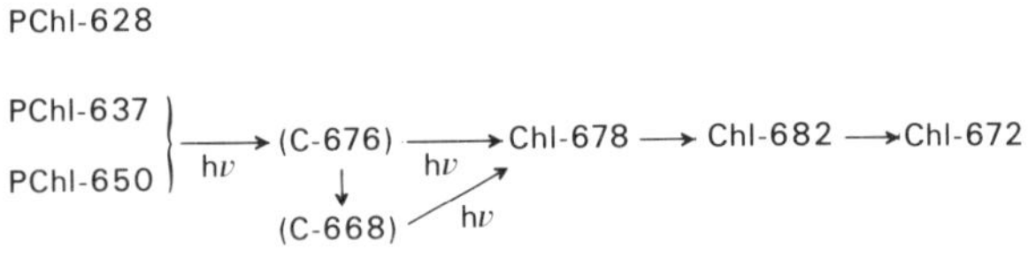

FIG. 3. Spectroscopic forms of protochlorophyll(ide) and chlorophyll(ide). See text for explanation.

greened leaf. Intermediate pigment forms absorbing at 676 nm (C-676) and 668 nm (C-668) are observed after very short flashes of high-intensity light or illumination of dark-grown leaves with low-intensity light (Litvin and Belyaeva, 1968, 1971; Thorne, 1971a). C-668 appears to be derived from C-676 and is stable in the dark. C-668 is not observed if the leaves are illuminated at 0° instead of room temperature (Litvin and Belyaeva, 1971). PChl-628 is not transformed into chlorophyllide by light.

The various spectroscopic intermediates exhibit different fluorescence emission maxima. The rapid spectral shift from 678 nm to 682 nm is accompanied by a lowering of the fluorescence intensity, the quantum yield of fluorescence of Chl-682 being about one-half that of Chl-678. There is no change in the quantum yield of fluorescence during the Shibata shift from 682 to 672 nm, but on completion of the shift there is a doubling of the quantum yield.

The photoconvertible forms (PChl-637 and PChl-650) are protochloro-phyllide whereas PChl-628 is esterified protochlorophyllide (Wolff and Price, 1957; Virgin, 1960; Bovey *et al.*, 1974; Ogawa *et al.*, 1975). Recent work suggests that the esterifying alcohol is geranylgeraniol, rather than phytol (Liljenberg, 1974). Large amounts of inactive proto-chlorophyllide absorbing about 630 nm are accumulated on feeding leaves with δ-aminolevulinic acid. The pool of inactive pigment can serve as a precursor for photoconvertible protochlorophyllide (Sund-quist, 1970, 1973; Murray and Klein, 1971; Gassman, 1973).

3. Energy Transfer

If dark-grown leaves or isolated prolamellar bodies are illuminated for a short period so as to convert only a fraction of the convertible protochlorophyllide and then cooled to 77°K, resonance energy transfer (Förster, 1959) is observed between the remaining photoconvertible protochlorophyllide and the various spectroscopic forms of chlorophyl-lide *a*, including Chl-672 (Thorne, 1971a). Resonance transfer is not observed after the doubling of the fluorescence yield, following the Shibata shift. Figure 4 shows the fluorescence emission spectrum of a partly converted bean leaf (ca. 10% photoconversion). A high proportion of the fluorescence is emitted by chlorophyllide *a* and a smaller proportion from protochlorophyllide. It is estimated that 40% of the energy absorbed by protochlorophyllide at 77° is transferred to chloro-phyllide *a* (Kahn *et al.*, 1970). Energy transfer from protochlorophyllide to chlorophyllide *a* also has been demonstrated at room temperature (Vaughan and Sauer, 1974; Thorne, unpublished observations).

For chlorophyll molecules in monolayers, resonance energy transfer by the Förster mechanism occurs over distances of up to 3–7 nm (Tweet

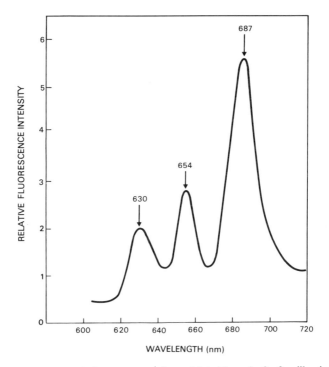

FIG. 4. Fluorescence emission spectrum of an etiolated bean leaf, after illumination for a brief period to convert 10% of the protochlorophyllide to chlorophyllide *a*. This demonstrates energy transfer from protochlorophyllide to chlorophyllide *a* and indicates that the pigment molecules are organized into units. From Kahn *et al*. (1970).

et al., 1964; Colbow, 1973). It is concluded that the protochlorophyllide molecules *in vivo* are organized into energy-transferring units. Energy transfer from protochlorophyllide to chlorophyll(ide) indicates that at least some of the newly formed chlorophyll(ide) remains in close proximity to the protochlorophyllide. From the measured efficiencies of energy transfer between protochlorophyllide and chlorophyll(ide) *a* at different degrees of photoconversion, Kahn *et al*. (1970) estimated that the energy-transferring units contain at least four chromophores. A similar conclusion was reached by Vaughan and Sauer (1974) from the variation of fluorescence polarization during photoconversion. The fluorescence polarization at 690 nm of chlorophyllide (excitation wavelength 670 nm) decreased from 0.38 at 6% conversion to 0.17% at 50% and 0.11% at 100% conversion. The decrease in polarization is explained by an increase in energy transfer between chlorophyllide *a* molecules. On the assumption that each unit contains, on the average, one molecule of

C-668 during the early stages of photoconversion (Fig. 3), Thorne (1971a) estimated that the units may contain as many as 20 chromophores.

The kinetics of photoconversion of protochlorophyllide in a leaf or in the isolated protochlorophyllide–protein complex from bean seedlings approximate to second order, but first-order kinetics might be expected since protochlorophyllide is the photoreceptor for its own conversion (Boardman, 1966). The second-order kinetics are now explained in terms of a competition between the photoconversion process and transfer of excitation energy from protochlorophyllide to chlorophyllide a (Thorne and Boardman, 1972; Nielsen and Kahn, 1973). The energy transfer rate increases with the degree of photoconversion, and the rate of photoconversion decreases.

Chlorophyll b, which is considered to be formed from chlorophyll a (Shlyk, 1971), is detectable after 10 minutes of illumination of etiolated seedlings (Thorne and Boardman, 1971) or in a comparable dark period following photoconversion of protochlorophyllide to chlorophyllide a by a brief illumination of seedlings (Fradkin et al., 1966). The lack of a fluorescence emission band from newly formed chlorophyll b and the observation, at an early stage of greening, of energy-transfer from chlorophyll b to chlorophyll a (Fradkin et al., 1966; Thorne and Boardman, 1971) and carotenoid to chlorophyll a (Butler, 1961a; Thorne and Boardman, 1971) provides further evidence that chlorophyll is organized into energy-transferring units at an early stage of chloroplast development.

4. Shibata Spectral Shift in Relation to Phytylation and Structural Changes

It has been suggested that the Shibata spectral shift from 682 to 672 nm following photoconversion of protochlorophyllide in an etiolated leaf is due to the phytylation of chlorophyllide a (Klein, 1962; Sironval et al., 1965), but the experimental evidence is contradictory. Phytylation has been reported to take place simultaneously with the Shibata shift (Sironval et al., 1965), after the shift (Akoyunoglou and Michalopoulos, 1971), or over a period longer than the shift (Wolff and Price, 1957; Boardman, 1967). In some recent studies, Henningsen and Thorne (1974) observed a good correlation between the time-courses of esterification and that of the conversion of Chl-682 to Chl-672 in bean and barley seedlings at a number of temperatures. Studies with mutants of barley also show a relationship between the spectral shift and phytylation of chlorophyllide a. Mutants $alb-f^{17}$ and $xan-j^{59}$ did not show the spectral shift from 682 to 672 nm, and esterification of chlorophyllide a could not be detected (Henningsen and Thorne, 1974). Some mutants showed a more rapid spectral shift and a faster esterification than the wild-type seedlings.

However, the Shibata shift in isolated etioplasts does not seem to be connected with the phytylation of chlorophyllide a (Horton and Leech, 1975). After 3.5 hours of illumination the spectral shift of etioplasts was nearly complete, but only about 13% of the chlorophyllide was esterified. Horton and Leech (1975) made the interesting observation that the Shibata shift in etioplasts is inhibited by ATP. The possible significance of this finding in relation to the structure of the protochlorophyllide–protein complex is discussed later.

Treffry (1970) reported that there was no loss in the regularity of the prolamellar body (tube transformation; Section III, B) if etiolated leaves were illuminated at 0°. Photoconversion of protochlorophyllide occurred at 0°, but the subsequent esterification was inhibited. Treffry suggested that tube transformation is associated with an increase in the hydrophobicity of the membrane due to some phytylation of chlorophyllide a.

Dispersal of the prolamellar body tubules into the perforated lamellae (vesicle dispersal; Section III, B) has been investigated in relation to the spectral shift from 682 to 672 nm. Similar time courses for the two phenomena were observed at a number of temperatures, suggesting at least a causal relationship between them (Henningsen, 1970). Freezing and thawing of leaves, or appropriate heat treatment, causes disruption of the prolamellar bodies and results in spectral shifts to shorter wavelengths (Butler and Briggs, 1966; Henningsen, 1970). If the freezing and thawing is carried out prior to illumination, the photochlorophyllide maximum shifts from 650 to 635 nm, and subsequent illumination causes the direct formation of Chl-672 from PChl-635. Freezing and thawing of leaves immediately after illumination causes a shift in the chlorophyllide a maximum from 682 nm to 672 nm. Mutants of barley (e.g., $alb-f^{17}$ and $xan-j^{59}$), which lack the spectral shift from 682 to 672 nm and phytylation, show tube transformation, but prolamellar body dispersal is inhibited (Henningsen, 1970).

A correlation between the spectral shift and transformation of the crystalline prolamellar bodies into loosely knit membrane structures has been reported for isolated etioplasts (Horton and Leech, 1975). ATP inhibited both the transformation and the spectral shift.

B. PROTOCHLOROPHYLLIDE–PROTEIN COMPLEX

1. Soluble Protochlorophyllide Holochrome

The protochlorophyllide of dark-grown bean seedlings can be extracted as a protochlorophyllide–protein complex, termed protochlorophyllide holochrome, and purified (see review in Boardman, 1966, for early work). The protochlorophyllide in the complex retains the ability to be photoconverted to chlorophyllide a, although there is some inactivation of the holochrome during extraction and purification. It can be stabilized against inactivation by a high concentration of glycerol or sucrose (Boardman, 1966; Schultz and Sauer, 1972). Photoconversion of protochlorophyllide to chlorophyllide a involves the addition of two hydrogen atoms to the 7,8 positions of the porphyrin ring, but the nature of the hydrogen donor is unknown. Photoconvertibility is lost by heating the holochrome or treating it with protein-denaturing agents, suggesting that a specific binding of the pigment to the protein is essential for photoconversion. The temperature coefficient of photoconversion is relatively low for an enzymic reaction, and photoconversion can be observed down to $-70°C$. Boardman (1962b) suggested that photoconversion involves a restricted collision process within the holochrome between the hydrogen donor and the protochlorophyllide.

The protochlorophyllide holochrome from bean leaves after substantial purification gave a single symmetrical band in the analytical ultracentrifuge. The molecular weight ratio (M_r) of the holochrome was 600,000 $\pm$ 50,000 and its density was 1.37 g/ml, which is indicative of the density of a protein (Boardman, 1962a). Further purification was achieved by zone electrophoresis to give a preparation that contained one protochlorophyllide per M_r 600,000. Schopfer and Siegelman (1968) purified the protochlorophyllide holochrome by ion-exchange and gel-filtration chromatography. Molecular weight determination by gel filtration showed two particle sizes with apparent M_r 300,000 and 550,000. The larger particle contained two chromophores and appeared to be a dimer of the smaller particle, which contained one chromophore.

The absorption maximum of protochlorophyllide in the soluble holochrome is at 638–639 nm at room temperature. At $77°K$, the peak is at 638 nm, but with a slight shoulder at 650 nm (Kahn et $al.$, 1970). The fluorescence emission spectrum of the holochrome resembles those of the leaf and prolamellar body membranes, except that the peak is at 652 nm, compared with 655 nm for the leaf, and the holochrome spectrum shows a shoulder at 643 nm. The fluorescence excitation spectrum of the holochrome at $77°K$ indicates energy transfer from the 638 nm form of protochlorophyllide to the 650-nm form, even though the proportion of the 650-nm form is low. Immediately after photoconversion, the absorption maximum of the holochrome is at 678 nm, which shifts to a shorter wavelength (672–675 nm). The spectral shift can be rapid (Boardman, 1966) or take 10–15 minutes to complete (Schopfer and Siegelman, 1968). It is completely inhibited by 2 M sucrose (Schultz and Sauer, 1972).

Extraction of a soluble protochlorophyllide holochrome by aqueous buffers appears to be successful only for etiolated bean leaves. Particulate polydisperse preparations are obtained from other etiolated plants. Protochlorophyllide holochrome is associated with the prolamellar body membranes in the etioplast, and the success in solubilizing it from bean leaves may possibly be due to lipase action in bean homogenates. For instance, it is known that bean leaves contain high levels of galactolipase (McCarty and Jagendorf, 1965).

2. Saponin Protochlorophyllide Holochrome

Henningsen and Kahn (1971) obtained a soluble photoactive protochlorophyllide holochrome by extraction of barley leaves with 1.2% saponin. The saponin holochrome was partly purified by ammonium sulfate fractionation and Sephadex gel filtration. Its absorption maximum was at 644 nm, and it showed a high degree of photoconvertibly (70–80%) to the saponin chlorophyllide a holochrome with an absorption maximum at 678 nm. Gel filtration of the saponin protochlorophyllide holochrome on Sephadex G-100 indicated an apparent M_r of 63,000. Spectrofluorimetry gave no evidence for energy transfer between protochlorophyllide and chlorophyllide a in partly photoconverted preparations of barley saponin holochrome. This is in distinct contrast with the fluorescence properties of the soluble bean protochlorophyllide holochrome. Henningsen and Kahn (1971) suggested that the protochlorophyllide holochrome obtained from barley represents an active subunit with a single chromophore per M_r 63,000 of protein. The high fluorescence polarization of saponin protochlorophyllide holochrome (Henningsen et al., 1974), the first-order kinetics of its photoconversion (Nielsen and Kahn, 1973) and the lack of exciton interaction between pigment molecules (Henningsen et al., 1973) are consistent with a single protochlorophyllide per holochrome molecule. Bean protochlorophyllide holochrome extracted with 1.2% or 3.6% saponin has an apparent molecular weight, as determined by gel filtration, of 170,000 (Henningsen et al., 1974). From a comparison with the barley saponin holochrome, it was suggested that the bean preparation may represent a dimer or timer. However, fluorescence spectroscopy of the bean saponin holochrome did not provide any evidence for energy transfer between chromophores.

Photoconversion of barley saponin protochlorophyllide holochrome to saponin chlorophyllide a holochrome absorbing at 678 nm does not alter its apparent molecular weight (Henningsen and Kahn, 1971). If the saponin chlorophyllide a holochrome is allowed to stand at room temperature the absorption maximum shifts from 678 nm to 672 nm, the shift being completed in rather less than 1 hour. The saponin chlorophyllide a holochrome now has an M_r of 29,000 instead of 63,000 (Henningsen *et al.*, 1974). It was suggested that the subunits of the chlorophyll(ide) *a* holochrome dissociate into a colorless "photoenzyme" and a chlorophyll(ide) *a*-carrier protein. In an earlier study, Bogorad *et al.* (1968) had obtained evidence from sucrose density gradient centrifugation for a dissociation of soluble chlorophyllide *a* holochrome from bean seedlings. The term "photoenzyme" (Nadler and Granick, 1970) is used to denote the protein that is essential for the photoconversion of protochlorophyllide to chlorophyllide *a*. The photoenzyme appears to be used repeatedly (Nadler and Granick, 1970; Sundquist, 1970; Süzer and Sauer, 1971; Thorne, 1971b), as the number of sites for the photoconversion of protochlorophyllide remains constant over several hours of greening including the period of rapid chlorophyll synthesis. It has been proposed that the chlorophyll-carrier protein transports chlorophyll from the photoenzyme to sites on the developing thylakoids (Bogorad *et al.*, 1968; Henningsen *et al.*, 1974). Whether the chlorophyll–protein is incorporated into the thylakoid or the chlorophyll is transferred to the thylakoid is not known.

The gel filtration behavior on Sephadex G-100 of "chlorophyll holochrome" extracted by saponin from etiolated barley seedlings that had been greened for at least 15 minutes is consistent with the hypothesis that the chlorophyll has been incorporated into the developing thylakoid membrane. The chlorophyll holochrome does not enter the gel, indicating an apparent molecular weight in excess of 100,000 (Henningsen *et al.*, 1974). Furthermore, plastid cytochromes are now associated with the holochrome, whereas saponin protochlorophyllide holochrome and its conversion products (Chl-678 and Chl-672) are separated from the cytochromes by gel filtration.

C. Molecular Structure of the Protochlorophyllide and Chlorophyllide Holochromes in Relation to Their Spectroscopic Properties

The molecular basis for the different spectroscopic forms of protochlorophyllide and chlorophyllide *a* in the early stages of greening is not established. Krasnovsky and Kosobutskaya (1952) were the first to relate the *in vivo* spectral characteristics of protochlorophyllide and chlorophyllide *a* with different states of aggregation of the pigments. The 650-nm form of protochlorophyllide was attributed to an aggregated form of protochlorophyllide, and the 635-nm form to the monomeric porphyrin (cf. Butler and Briggs, 1966; Fradkin *et al.*, 1969). Studies on protochlorophyllide dissolved in organic solvents lend support to this view. For example, when protochlorophyllide is dissolved in solvents of low dielectric constant, such as benzene or chloroform, its absorption peak slowly shifts from 634 nm to 651 nm (Seliskar and Ke, 1968). The spectral shift is accompanied by a decrease in the quantum yield of fluorescence and an increase in light scattering, which indicate that the red shift in solution is due to pigment aggregation. Alternatively, the pigment forms *in vivo* have been attributed to different modes of binding of pigment to protein (Boardman, 1966).

1. Circular Dichroism

On the basis of circular dichroism (CD) measurements, Schultz and Sauer (1972) and Mathis and Sauer (1972, 1973) proposed that photoactive protochlorophyllide is present in dimers in the etiolated leaf, and complete photoconversion involves a two-step light reaction to produce a dimer of chlorophyllide *a*, which then dissociates into monomeric chlorophyllide *a*. They suggested that the intermediate pigment form absorbing at 676 nm at partial photoconversion (Litvin and Belyaeva, 1971) is a dimer of chlorophyllide *a* and protochlorophyllide, which dissociates in the dark to give C-668. The CD spectrum of the unilluminated protochlorophyllide holochrome showed evidence for pigment–pigment interaction by virtue of the size of the signal, its multiple character, and reversal of sign in the long-wavelength region. It differed markedly from the CD spectrum of monomeric protochlorophyllide in diethyl ether, but resembled the CD spectrum of protochlorophyllide in carbon tetrachloride. However, the resolution of the holochrome CD spectrum was not good enough to permit a conclusion as to whether protochlorophyllide exists as dimers in the holochrome or as a weak exciton interacting array. The CD spectrum of chlorophyllide *a* holochrome also showed evidence for pigment–pigment interactions, but it differed from that of chlorophyll *a* dimers in nonpolar solvents. The geometry of pigment interaction in the holochrome apparently is different from dimer interactions.

The CD spectra did not evolve in a linear manner during the course of photoconversion (Mathis and Sauer, 1972). When the photoconversion was less than 20%, the CD spectrum consisted of a single weak-positive component, which developed into a double CD spectrum on further illumination. This indicated a lack of exciton interaction in the postulated dimer between protochlorophyllide and chlorophyllide *a*, which is understandable because of the difference in energy levels of the long-wavelength transitions of protochlorophyllide and chlorophyllide *a*. There are spectroscopic observations, however, that argue against the dimer model proposed by Schultz and Sauer (1972) and Mathis and Sauer (1972).

2. Absorption and Fluorescence Properties

During photoconversion of protochlorophyllide, the absorbance of homogenates at 678.5 nm was proportional to chlorophyllide a concentration; hypo- or hyperchromism could not be detected at partial photoconversion (Mathis and Sauer, 1972). The fluorescence quantum efficiency of chlorophyllide a remained constant during the photoconversion, provided the fluorescence was excited at 670 nm within the chlorophyllide a band (Vaughan and Sauer, 1974; Thorne, unpublished observations). On the other hand, if fluorescence is excited at a wavelength absorbed by protochlorophyllide, the quantum efficiency of chlorophyllide a changes with the degree of photoconversion, owing to energy transfer from protochlorophyllide to chlorophyllide a. Kahn and Nielsen (1974) observed that the 635-nm form of protochlorophyllide in an etiolated leaf is photoconverted at the same rate as the 650-nm form, irrespective of whether the actinic beam is at 630, 640, or 670 nm. This indicates that light absorbed by the 650-nm or the 635-nm band does not affect the spectral character of the remaining protochlorophyllide. This would not be expected for the simple dimer model, since PChl in a mixed PChl–Chl dimer should have a different absorption maximum than the PChl in a PChl–PChl dimer.

3. Model for Organization of Protochlorophyllide

On the current evidence, we put forward the following model for the organization of protochlorophyllide in the prolamellar body membranes of the etiolated leaf. Protochlorophyllide holochrome is postulated to consist of a number of subunits (each containing one chromophore), which are organized in such a manner that there is weak exciton interaction between the molecules of protochlorophyllide. If the molecular weight of the subunit is about 60,000 (as determined for the saponin protochlorophyllide holochrome) and that of the whole complex 550,000–600,000, there would be 9–10 pigment molecules per molecule of protochlorophyllide holochrome. Admittedly the purest preparation of holochrome achieved so far has on the average 2 protochlorophyllides per protein molecule. It is possible that even the purest preparations were still contaminated with a colorless protein (fraction 1 protein of leaves) of the same molecular weight as the holochrome. There is also the possibility that protochlorophyllide is lost from the holochrome during its purification.

We would attribute the 635- and 650-nm bands of protochlorophyllide *in vivo* to a band splitting due to the weak exciton interactions of the array. During the early stages of photoconversion, i.e., at partial photoconversion, the band at 676 nm could be due to a single chlorophyllide molecule in each array interacting with the remaining protochlorophyllide molecules in the array (Thorne, 1971a). Some evidence for this proposal is provided by the studies of O. F. Nielsen (1975) with a barley mutant (alb-f^{17}) which has considerably less protochlorophyllide than the wild type. In the mutant leaves, the chlorophyllide a absorption maximum is located at 682–684 nm, even at partial photoconversion. If the mutant is supplied with exogenous δ-aminolevulinic acid, it forms excess (inactive) protochlorophyllide. The presence of the protochlorophyllide causes the peak of the newly formed chlorophyllide a to shift to 675 nm. Unlike the protochlorophyllide from which it is derived, the 682 nm form of chlorophyllide in the wild-type leaves does not exhibit an absorption band splitting, although CD suggests weak exciton interaction.

We return to the question as to whether the spectral properties of protochlorophyll(ide) and chlorophyll(ide) can be attributed solely to pigment–pigment interactions or whether coordination to protein also plays a role. In an earlier review on protochlorophyll, Boardman (1966) outlined the coordination properties of metalloporphyrins. It was suggested that the linkage of certain amino acid side chains of protein, either to the carbonyl group of ring V of the protochlorophyllide or to the Mg might be expected to cause a red shift of the absorption maximum of protochlorophyllide. The phenolic side chain of tyrosine, the imino group of histidine, the ε-amino of lysine and the thiol group of cysteine were identified as possible coordination ligands to the Mg. Amide groups of glutamine and asparagine or the phenolic hydroxyl of tryosine would have the potential ability to hydrogen bond to the keto group of ring V.

Even if the spectral properties of the pigment are due to weak exciton interactions and not to direct coordination to protein, it is highly likely that the protein conformation and the interactions between pigment and protein are of the upmost importance in maintaining the relative orientations of the pigments in the array. Both the rapid spectral shift of newly formed chlorophyllide a from 678 to 682 nm and the slower Shibata shift from 682 to 672 nm may be due to conformational changes in the protein. We have already mentioned that the spectral shift from 682 to 672 nm precedes the doubling of the fluorescence yield, which suggests that the dissociation of the pigment array is not coincident with the spectral shift but in a subsequent event. Horton and Leech (1975) suggested that the inhibition, by ATP, of the spectral shift and of prolamellar body transformation in isolated etioplasts is due to the inhibition of a conformational change in the protochlorophyllide holochrome. The observation of a spectral shift (from 678 to 672 nm) in the saponin protochlorophyllide holochrome suggests that a change in pigment interactions is not the cause of the blue shift. It has been proposed from CD studies with the saponin holochrome that the spectral shift is due to a conformational relaxation of the holochrome protein, initiated by the photoconversion (Foster et al., 1971).

Aging of etioplasts causes a loss of the 650-nm band of protochlorophyllide and an increase in the 635-nm band. Horton and Leech (1972) observed that this spectral change is also inhibited by ATP. Perhaps, ATP helps to maintain the holochrome protein in a particular, but intrinsically unstable, conformational state.

In the water-soluble protochlorophyllide holochrome, which absorbs maximally at 638–640 nm, exciton interaction is apparently diminished as compared with protochlorophyllide in vivo, although some band splitting is still observed at 77°K and the main fluorescence band arises from PChl-650 (Kahn et al., 1970).

V. Chlorophyll–Protein Complexes of Mature Chloroplasts

Thylakoid membranes may be fragmented by nonionic detergents, sonication, or the French pressure cell into subchloroplast fragments that have different photochemical and chemical properties (Boardman, 1970). The small fragments have a higher Chl a:Chl b ratio than do intact thylakoids and mainly PS I properties, whereas the larger fragments have a lower Chl a: Chl b ratio and are enriched in PS II. A more complete solubilization of thylakoid membranes is obtained with anionic detergents that destroy the photochemical activities, but allow separation of discrete chlorophyll–protein complexes either by sodium dodecyl sulfate–polyacrylamide gel electrophoresis (SDS-PAGE) or SDS–hydroxyapatite chromatography (Thornber, 1975).

A. SOLUBILIZATION BY SDS AND SEPARATION BY SDS-PAGE

Chloroplast membranes may be solubilized by the anionic detergents, SDS or sodium dodecylbenzene sulfonate (SDBS), and these extracts (without prior extraction of lipids) are separable into three chlorophyll-containing bands on polyacrylamide gels (Ogawa et al., 1966; Thornber et al., 1967a) or agar (Sironval et al., 1967). The zones of lowest and intermediate electrophoretic mobility are chlorophyll–protein complexes, termed complex I and complex II, respectively, while the third zone migrating to the front consists of pigments and lipids complexed to the detergent, termed "free pigment." Thornber et al. (1967a) estimated that some 20% of the membrane protein was in complex I and 49% in complex II, indicating that a major part of the membrane protein is involved in the organization of chlorophyll. The proportions of chlorophyll found in the two complexes and the "free pigment" zone depend on the ratio of SDS:Chl or SDBS:Chl used for the initial solubilization, the time and method employed for electrophoresis, and the temperatures used for solubilization of the membranes and electrophoresis. With most mature higher-plant chloroplasts, complex I contains some 10–15% and complex II some 40–60% of the total chlorophyll. Reelectrophoresis of either complex removes more chlorophyll and carotenoids from the complexes, particularly in the case of complex II (Leggett-Bailey and Kreutz, 1969). Therefore, part of the pigment in the "free pigment" zone is presumed to arise from these complexes (Thornber, 1975). Initial studies indicated that complex I had a high Chl a:Chl b ratio of 7–12, whereas complex II had a low ratio of 1.1–1.8. Complex I was enriched in β-carotene and contained no violaxanthin or neoxanthin; complex II contained all four carotenoids, but was enriched in lutein (Ogawa et al., 1966; Thornber et al., 1967b). Both complexes contained traces of phospholipids and galactolipids as well as carbohydrates. The complexes differed in their amino acid compositions, and each possessed a high proportion of hydrophobic amino acids residues ($\sim$ 63% for complex I and $\sim$ 60% for complex II). Analytical ultracentrifugation gave sedimentation coefficients of 9 S for complex I and 2–3 S for complex II (Thornber et al., 1967b). No photochemical activities were observed for either complex, but the similarities of the Chl a:Chl b ratios with those of the digitonin fractions led Ogawa et al. (1966) to suggest they might be the complexes of the PS I- and PS II-enriched digitonin subchloroplast fractions. This indeed was the case, since Thornber et al. (1967a) and Sironval et al. (1967) demonstrated that the digitonin subchloroplast fragments, D-10 and D-144, were enriched in complex II and complex I, respectively.

In 1969, Leggett-Bailey and Kreutz observed a light-induced absorbance change due to P700 in complex I, showing that this chlorophyll–protein complex contained the reaction center of PS I; it is termed chlorophyll–protein complex I (CP I). More recently, it was found that complex II is absent in a *chlorina* barley mutant which contains no chlorophyll b (Thornber and Highkin, 1974; Genge et al., 1974; Anderson and Levine, 1974b; Henriques and Park, 1975). Since this Chl b-less mutant has both PS I and PS II activities, Thornber and Highkin (1974) made the important conclusion that complex II is not an essential complex of PS II, and they suggested that this complex functions rather in a light-harvesting capacity. They renamed complex II (variously known also as chlorophyll–protein complex II, and PS II chlorophyll–protein complex) by the more appropriate name, light-harvesting chlorophyll a/b–protein complex (LHCP) which is used here.

This simple electrophoretic procedure for chloroplast membranes solubilized by SDS without prior lipid extraction, provides a rapid method for the detection and qualitative estimation of complex I and the light-harvesting chlorophyll–protein complex. However, this procedure detects only chlorophyll–protein complexes that are stable to SDS. It is postulated that chloroplasts have a third chlorophyll–protein complex containing the reaction center of PS II, but it is not seen on SDS-PAGE, presumably because the chlorophyll is more easily removed by SDS.

B. PREPARATIVE METHODS FOR ISOLATION OF COMPLEXES

In order to isolate larger amounts of the chlorophyll–protein complexes, SDS-solubilized extracts of chloroplast membranes have been chromatographed on hydroxyapatite (Thornber and Alberte, 1976). This procedure was successful with the blue-green alga *Phormidium luridum* and yielded a P700–chlorophyll *a*–protein complex, (P700 CP) amounting to some 70% of the total chlorophyll. The complex had a red-wavelength maximum at 677 nm and contained 1 P700 per 40 chlorophyll *a* molecules (Thornber, 1969; Dietrich and Thornber, 1971). *Phormidium* was a favorable starting material, since blue-green algae have no Chl *b*, and consequently no LHCP. Kung and Thornber (1971) applied the hydroxyapatite method to SDS extracts of higher-plant chloroplast membranes for the preparation of LHCP and CP I. The latter had a red-wavelength maximum at 671 nm and resembled complex I on SDS-PAGE rather than the blue-green algal complex. It contained a variable amount of P700. More recently, Thornber (1975) has preferred to use Triton X-100 extracts rather than SDS extracts for the preparation of CP I (Shiozawa *et al.*, 1974) by hydroxyapatite chromatography, since the Triton X-100 complex has a red-wavelength maximum at 677 nm and contains photochemically active P700. Kan and Thornber (1976) have isolated LHCP from SDS extracts of *Chlamydomonas reinhardii* thylakoids. It had absorption maxima at 670 and 652 nm, and contained equimolar proportions of Chl *a* and Chl *b*.

An alternative approach to the isolation of chlorophyll–protein complexes has evolved as an extension of the methods used to separate subchloroplast fragments enriched in either PS I or PS II. The earlier methods, reviewed by Boardman (1970), leading to a partial fractionation of the photosystems used nonionic detergents, either digitonin (Boardman and Anderson, 1964; Anderson and Boardman, 1966) or Triton X-100 (Vernon *et al.*, 1966) or mechanical procedures, such as sonication (Jacobi and Lehmann, 1968) or the French pressure cell (Michel and Michel-Wolwertz, 1969). Subsequently, these subchloroplast fragments were subjected to further extended detergent treatments, thereby permitting the isolation of smaller particles containing discrete chlorophyll–protein complexes (Brown, 1973).

After prolonged treatment of a spinach subchloroplast fraction with digitonin, Wessels *et al.* (1973) isolated by sucrose density gradient centrifugation, three chlorophyll-containing particles: PS I particles (F_I), particles containing the reaction center of PS II (F_{II}), and particles containing the light-harvesting complex (F_{III}). Subsequently, Wessels and Borchert (1975) further purified these particles by DEAE-cellulose chromatography. F_I had a Chl a:Chl b ratio of 8–10, a Chl:P700 ratio of 110–130, and a high PS I activity. F_{II} had a similar Chl a:Chl b ratio, no P700 and high PS II activity, whereas the accessory complex, F_{III}, had a low Chl a:Chl b ratio of 1.3–1.5 and neither PS I nor PS II activities. Since F_{III} does not have activity, it is impossible to assess whether it was still associated with the purest F_I and F_{II} fractions, a question of relevance in deciding whether Chl b is associated exclusively with LHCP.

By extended Triton X-100 fractionation of the original TSF-1 and TSF-2 fractions of spinach chloroplasts (Vernon *et al.*, 1971), a PS II, reaction-center fraction, designated TSF-2a, was obtained (Ke *et al.*, 1974). It had a high PS II activity if supplied with an artificial electron donor, a Chl a:Chl b ratio of 28, and contained P680, C550, and Cyt b-559, all enriched about 10-fold, relative to unfractionated chloroplasts. Recently, TSF-1 with a Chl a:Chl b ratio of 6, a Chl:P700 of 40, and high PS I activity has been further purified to give a P700–chlorophyll a–protein complex which has no PS I cytochromes (Ke *et al.*, 1975). Interestingly, Ke *et al.* have isolated a cytochrome complex containing cytochromes b-563 and f, plastocyanin, and an iron–sulfur protein. They suggest that this complex may be active in cyclic photophosphorylation and may be analogous to complexes III and IV of the mitchondrial respiratory chain.

Shiozawa *et al.* (1974) purified P700 CP by hydroxyapatite chromotography of chloroplast membranes, solubilized by Triton X-100. The complex contained Chl a and β-carotene and had a Chl/P700 ratio of 40, but it still contained cytochromes and possibly other components of PS 1; no photochemical activities were reported. It had a sedimentation coefficient of 13 S, which is higher than the value of 9 S for the SDS chlorophyll–protein complex of a blue-green alga or spinach beet (Thornber, 1975).

Two other isolation procedures have been reported recently. Malkin (1975) used the zwitterionic detergent lauryl dimethylamine oxide and DEAE-cellulose chromatography. The P700 CP had a Chl:P700 ratio of 40, and P700 photooxidation was observed at room and liquid nitrogen temperatures. The complex also showed the electron paramagnetic resonance (EPR) signals of the primary electron acceptor of PS I (Bearden and Malkin, 1976). Nelson and Bengis (1975) isolated a P700 CP from Swiss chard by treating digitonin PS I particles depleted of cytochromes (Nelson and Racker, 1972) with Triton X-100, followed by differential centrifugation, DEAE-cellulose chromotography, and density-gradient centrifugation. The complex was active in NADP reduction if provided with ferredoxin, ferredoxin–NADP reductase, and plastocyanin. It had a Chl a:Chl b ratio greater than 40, a Chl:P700 ratio of 80–100 and contained β-carotene and about 4 nonheme irons per P700 (Bengis and Nelson, 1975). When the active complex was treated with 0.5% SDS and subjected to sucrose density-gradient centrifugation, a P700 chlorophyll a–protein complex comprising a single M_r 70,000 polypeptide was obtained. The treatment with SDS, however, abolished NADP photoreduction.

C. Pigment Composition and Size of Complexes

Despite success in the isolation of the reaction-center complex of PS I and some success with the reaction-center complex of PS II and the light-harvesting complex, there is considerable difficulty in estimating the amount and composition of the complexes in $vivo$. First, some 30–50% of the chlorophyll of an SDS extract runs as free pigment on SDS-PAGE, and it is not known whether this chlorophyll is attached to the complexes in $vivo$. Second, there is little agreement about the composition of isolated complexes, since these vary according to the method of separation, even for complexes from the same species.

1. Chlorophyll–Protein Complex I

Although SDS-PAGE procedures yield CP I with a Chl a/Chl b ratio of 5–12 (Ogawa et al., 1966; Thornber et al., 1967a; Chua et al., 1975; Nakamura et al., 1976), other methods (Genge et al., 1974; Brown et al., 1975) yield CP I that has no Chl b. Since repeated electrophoresis removes Chl b, Thornber (1975) proposed that the reaction-center complex contains Chl a only. Similarly, with the Triton X-100 PS I fraction (TSF-la) the Chl a:Chl b ratio increases during purification as the Chl:P700 ratio decreases, again indicating that Chl b may not be an integral part of the complex (Vernon and Klein, 1975). Finally, the ubiquitous distribution of CP I (Brown et al., 1975) makes it improbable that Chl b is a component. Hence this complex probably contains only Chl a and β-carotene in a molar ratio of Chl:β-carotene of 20–30 (Thornber, 1975).

The Chl a:P700 ratio of CP I in $vivo$ is unknown, since it is uncertain how much chlorophyll has been stripped from the complex by the detergent treatments. Several preparations have a Chl:P700 ratio of 40 (Shiozawa *et al.*, 1974; Ke *et al.*, 1975; Malkin, 1975). The PS I reaction-center preparation active in NADP reduction (Bengis and Nelson, 1975) has a Chl *a*:P700 ratio of 80, and the P700 reaction-center complex has a Chl *a*:P700 ratio of 40 per two M_r 70,000 polypeptides (Nelson and Bengis, 1975).

The molecular weight of CP I is not established. Calibrated SDS-PAGE gave apparent molecular weights of 100,000 for a tobacco CP I (Kung and Thornber, 1971), 110,000 for a spinach beet CP I (Thornber *et al.*, 1967b), and 112,000–140,000 for a pea CP I (Eaglesham and Ellis, 1974). However, estimates of the molecular weights of chlorophyll–protein complexes by this method are inaccurate. The binding of SDS to intrinsic proteins is frequently anomalous (Grefrath and Reynolds, 1974). It is expected that complexes that contain chlorophyll must retain at least part of their tertiary structure and would not necessarily bind the usual 1.4 g of SDS per gram of protein. Indeed Chua *et al.* (1975) demonstrated directly by Ferguson plots that the migration of CP I was anomalous. Removal of the lipids from the complex by heating or lipid extraction gives a single polypeptide with an apparent M_r of 62,000–70,000 (Section V, D). Spinach beet CP I and a blue-green algal CP I contained 14 Chl *a* molecules per M_r of 110,000 of protein (Thornber *et al.*, 1967b; Thornber, 1969). *Chlamydomonas* CP I has some 8–9 chlorophylls (surprisingly with a Chl *a*:Chl *b* ratio of 5) attached to a M_r 64,000 polypeptide, according to Chua *et al.* (1975), while Bar-Nun *et al.* (1977) find some 30 Chl *a* molecules bound to an M_r 64,000 polypeptide, No P700 data are published yet for *Chlamydomonas* CP I.

From the subunit size, chlorophyll content, and Chl:P700 ratio, it is obvious that each subunit cannot have a P700. A recent model (Thornber *et al.*, 1977) for CP I is based on data of *Phormidium luridum* CP I, which shows a marked similarity to the higher-plant complex. Since there is only one P700 per 314,000 gm of protein, it is apparent that there are at least two different entities in the CP I zone on SDS-PAGE. These must be of identical or almost identical size, since serial slices through the SDS-PAGE zone did not reveal any differences in the Chl:P700 ratios (Dietrich and Thornber, 1971). The blue-green algal CP I has a major subunit of M_r 48,000 and a minor subunit of 46,000 by the SDS-PAGE system of Hoober (1970), and N-terminal amino acid analyses (Thornber *et al.*, 1977) are in agreement with two polypeptides. Thornber *et al.* (1977) propose that the P700–Chl *a*–protein complex consists of two trimers, one containing two M_r 48,000 subunits and one 46,000 subunit, and the other composed of three 48,000 subunits. Each subunit in the model contains 7 Chl a molecules, but only the M_r 46,000 subunit contains P700. This model bears analogy with the structure of a bacteriochlorophyll–protein complex (Fenna and Mathews, 1975; see Section VI, E), which consists of a trimer of subunits each containing the same number of bacteriochlorophyll molecules. Bengis and Nelson (1975) estimate 1 P700 per two M_r 70,000 polypeptides, each of which would have some 20 Chl a molecules. They suggested that cooperation of two subunits might be necessary for P700 formation, or, alternatively, that some of P700 has been lost during purification and each M_r 70,000 polypeptide might have a P700.

The amino acid composition of the SDS-isolated blue-green algal CP I is very similar to that of SDS-PAGE CP I of higher plants (Thornber *et al.*, 1977), indicating little difference in composition between the eukaryotic and prokaryotic complexes, except for their histidine and cysteine content.

2. Light-Harvesting Chlorophyll a:b–Protein Complex

LHCP after hydroxyapatite chromatography has equiomolar amounts of Chl a and Chl b (Thornber, 1975). However, some SDS-PAGE methods give Chl a:Chl b ratios of the LHCP band which exceed 1 (Ogawa *et al.*, 1966; Thornber *et al.*, 1967a; Nakamura *et al.*, 1976). Quick electrophoresis in the cold or lower SDS:Chl ratios during extraction also give Chl a:Chl b ratios greater than 1 (1.3–1.8) (Anderson, unpublished observations). As mentioned earlier, LHCP is photochemically inactive and has neither P700 nor P680, the reaction-center chlorophyll of PS II. The chlorophyll: carotenoid ratio is 6 for both the spinach beet and *Chlamydomonas* complexes with all 4 carotenoids present, lutein being in the highest amount. Traces of phospholipids and galactolipids were found in LHCP, but it is not known whether they are an integral part of the complex (Thornber, 1975).

LHCP as isolated by SDS–PAGE or SDS–hydroxyapatite chromatography has M_r 27,000–35,000 (Kung and Thornber, 1971; Eaglesham and Ellis, 1974; Machold, 1974; Kan and Thornber, 1976). This is considerably lower than the molecular weight of CP I. As mentioned earlier, estimates of the molecular weight of these complexes by SDS–PAGE are likely to be unreliable owing to the high lipid content of the complexes, which should affect the binding of SDS and also alter the hydrodynamic behavior of the SDS–chlorophyll–protein complexes. Bar-Nun *et al.* (1977) have shown that the free mobility of LHCP is characteristically higher than that of its completely denatured polypeptide components. Further, it is not established whether one or two polypeptides belong to the complex (see Section V, D). As is the case for CP I, it is unknown how many subunits of this complex would be associated together *in vivo*. A dimer of LHCP has been isolated by some SDS–PAGE procedures (Hiller *et al.*, 1974; Remy *el al.*, 1977).

Chlorophyll appears to be more easily removed from LHCP than from CP I (Leggett-Bailey and Kreutz, 1969). Originally, Thornber *et al.* (1967b) estimated 2 chlorophylls per M_r 35,000. However, a more recent estimate (Kan and Thornber, 1976) for the *Chlamydomonas* complex, which seems to be more stable than those of higher plants, gives 6 chlorophylls per M_r 29,000 polypeptide. An SDS–PAGE *Chlamydomonas* complex had a Chl:protein weight ratio of 0.09 and consisted of two subunits of M_r 24,000 and 22,000. This would give an assumed M_r of 57,000 if the complex were composed of 1 molecule of each subunit and 6 Chl *a* and 6 Chl *b* molecules (Bar-Nun *et al.*, 1977). However, the Chl:protein ratio indicates only about half this amount of chlorophyll.

The amino acid composition of spinach beet and *Chlamydomonas* LHCP as reported by Thornber (1975) is not very different from that of the total solubilized chloroplast membranes, which is not suprising, as LHCP is the main intrinsic protein of mature chloroplast thylakoids.

The molecular composition and weight of the presumed reaction-center complex of PS II is not known. Since it is not seen under the usual SDS–PAGE employed for separation of chlorophyll-containing complexes, it is impossible to screen for its presence or amount by this method (Section V, A). As reported earlier, Wessels and his colleagues (Wessels *et al.*, 1973; Wessels and Borchert, 1975) have used digitonin to extract and purify a PS II reaction-center preparation with a Chl *a*:Chl *b* ratio of 7–9, and Ke *et al.* (1974) obtained a PS II reaction-center preparation with a Chl *a*:Chl *b* ratio of 28 by Triton treatment. Vernon and Klein (1975) favor the view that the PS II reaction-center complex does not contain Chl *b in vivo*. Wessels and Borchert (1975) did not comment on the Chl *b* in their F_{II} fraction. If all the chlorophyll *b* were attached only to the light-harvesting complex, as postulated by Thornber and Highkin (1974) and considered likely by Thornber (1975), 18% of the chlorophyll of F_{II} would belong to LHCP.

D.　Polypeptides of Thylakoids and Complexes

In general, two different methods have been used for the examination by SDS–PAGE of the polypeptides of chloroplast membranes. Either SDS solubilization of lipid-extracted membranes, which is the conventional method used with other biomembranes, or SDS solubilization of membranes without prior extraction of lipids, in order to resolve the chlorophyll–protein complexes. In the latter case, the customary step of boiling the SDS-solubilized proteins for a minute or so to stop proteolysis and to disperse aggregated polypeptides into monomers must be omitted, since heating removes the chlorophyll from the complexes. Although SDS-PAGE of membrane extracts which still contain lipids may not resolve polypeptides with molecular weight lower than 18,000, it nevertheless is useful in order to establish the presence or the absence of the two major chlorophyll–protein complexes. Moreover, lipid extraction removes some of the polypeptides.

Several approaches have been used for the identification of chloroplast thylakoid polypeptides. These include examination of the various subchloroplast fragments produced by detergent and mechanical fragmentation methods (Remy, 1971; Levine *et al.*, 1972; Klein and Vernon, 1974; Anderson and Levine, 1974a; Apel *et al.*, 1975; Nolan and Park; 1975; Nelson and Bengis, 1975; Wessels and Borchert, 1975); comparison of polypeptides from mutants, which are either pigment deficient or have specific lesions in their electron-transport chains (Machold, 1972; Levine *et al.*, 1972; Levine and Duram, 1973; Anderson and Levine, 1974a; Chua and Bennoun, 1975; Henriques and Park, 1975; Chua *et al.*, 1975; Machold *et al.*, 1977); comparison of polypeptides of etiolated and developing chloroplasts of higher plants (Cobb and Wellburn, 1973; Remy, 1973a; Hiller *et al.*, 1973; Guignery *et al.*, 1974; Hofer *et al.*, 1975; Remy and Bebee, 1975; Lutz, 1975; G. Nielsen, 1975); examination of green algae (Hoober, 1970, 1972; Eytan and Ohad, 1970, 1972; Beck and Levine, 1974; Bar-Nun and Ohad, 1975; Bar-Nun *et al.*, 1977) and other algal classes (Vernon and Klein, 1975).

Most of these methods resolve from 15 to 25 chloroplast membrane polypeptides. Improved resolution of some 50 polypeptide bands has been obtained with thylakoid preparations, after prior lipid extraction with chloroform–methanol, using slab gel SDS–PAGE and a discontinuous buffer system (Fig. 5) (Henriques *et al.*, 1975; Henriques and Park, 1975). Similarly, Chua and Bennoun (1975) separated at least 33 polypeptides from purified *Chlamydomonas reinhardii* membranes (from which the coupling factor had been removed) by gradient SDS–PAGE combined with a discontinuous buffer system. These latter two methods show that the earlier results of 20 or so polypeptides were due to inadequate resolution. However, it should be pointed out that some of the bands may be artifacts of SDS-PAGE and due to aggregates or partially unfolded polypeptides (Wallach and Winzler, 1974). Thus far, identification of thylakoid polypeptides has been tenuous except for those of CF_1 and the main polypeptide of LHCP (Thornber, 1975; Anderson, 1975a).

A promising approach, used to advantage with erythrocyte membranes, consists in analyzing the way selective agents, such as nonionic detergents, chaotropic agents, and proteases, preferentially liberate

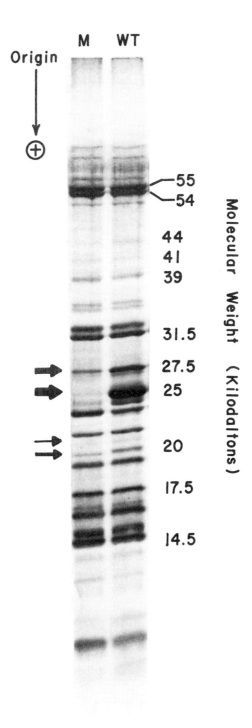

FIG. 5. Slab-gel electrophoretic patterns (SDS-PAGE) of thylakoid membranes from a barley mutant lacking Chl *b* (M) and wild-type barley (WT). Arrows indicate differences between the mutant and wild-type membranes. From Henriques and Park (1975).

certain polypeptides from the membrane (Steck, 1974). Machold (1975) exhaustively treated purified thylakoids of tobacco (protein:Chl ratio of 4) with 6 M guanidine hydrochloride to remove the extrinsic proteins, which comprise some 40–45% of the total thylakoid protein (gel 1, Fig. 6). The guanidine hydrochloride-insoluble membrane residue (protein:Chl ratio of 2) retained its basic membrane structure as shown by electron microscopy. Solubilization in SDS, followed by SDS–PAGE (gel 2, Fig. 6) showed that the membrane residue contained all the polypeptides of CP I and LHCP, and these comprised some 70% of the total intrinsic proteins. This demonstrates unequivocally that these two proteins are the main intrinsic proteins of chloroplast thylakoid membranes. In addition, two major bands at M_r 46,000 and 43,000 were resolved and there were some bands below 20,000. These additional polypeptides may come from the "membrane sector" of the chloroplast ATPase and the reaction–center chlorophyll–protein complex of PS II. Machold (1975) made the interesting observation that if the lipids were removed from the guanidine hydrochloride-insoluble membrane, the protein moiety of LHCP could then be solubilized in guanidine hydrochloride. Apel et al. (1976) showed that *Acetabularia mediterranea* thylakoids on treatment with EDTA and pronase lost 60% of their total membrane protein. The basic membrane structure and the chlorophyll–protein complexes, however, were retained and the polypeptides of the complexes were the dominant intrinsic proteins. Digestion of tobacco thylakoids with trypsin resulted in the removal of all the proteins except those of the pigment complexes and one other polypeptide, and the residue also retained its basic membrane structure (Süss et al., 1976).

Lipids are expected to play an important role in the stabilization of the complexes in the membrane, and lipid extraction of membranes prior to SDS–PAGE may remove all or part of the complexes. For example, Henriques and Park (1976a) showed that extraction of spinach thylakoids with chloroform–methanol (1:2 v/v) led to solubilization of one-third of the membrane protein including some 70% of the polypeptide(s) of LHCP. Chua et al. (1975) found that the polypeptide of CP I is soluble either in 90% acetone or chloroform–methanol (2:1 v/v). In order to determine the polypeptides of the complexes, it is preferable to use SDS-PAGE with thylakoids which have not been lipid extracted, and then reelectrophorese each isolated complex on SDS–PAGE after removal of chlorophyll by heating with SDS or lipid extraction. Even so, the polypeptide compositions of the complexes remains controversial, with different SDS–PAGE procedures giving variable results, even for complexes isolated from the same species.

CP I of SDS-solubilized thylakoids is clearly resolved at a position of low electrophoretic mobility by SDS–PAGE. After chlorophyll removal, the polypeptide moiety has a higher mobility corresponding to M_r 69,500

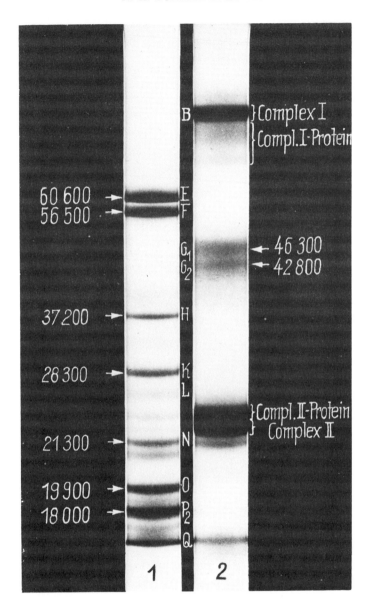

FIG. 6. SDS–PAGE patterns of polypeptides from EDTA-treated tobacco thylakoids: extracted polypeptides (gel 1) and insoluble residue (gel 2) after treatment with 6 M guanidine hydrochloride. From Machold (1975).

for tobacco (Machold, 1975). Similarly, the P700 activity of the PS I reaction-center preparation of Nelson and Bengis (1975) is associated with an M_r 70,000 polypeptide. Sometimes two bands are resolved, as happened after reelectrophoresis of the lipid-treated complex of *Chlamydomonas* (Anderson and Levine, 1974b). Chua *et al.* (1975) showed, however, that the heat-treated *Chlamydomonas* complex has only one polypeptide (M_r 66,000) similar to that of spinach and Chinese cabbage CP I (M_r 64,000). Although there is some agreement that CP I has a single polypeptide with an apparent M_r 60,000–70,000, Bar-Nun *et al.* (1977) resolved two bands at M_r 65,000 and 63,000 in 7.5% polyacrylamide gels, and one band at 64,000 in 10% polyacrylamide gels. SDS–PAGE systems using urea, such as that of Hoober (1970), may give two bands (Anderson and Levine, 1974a; Thornber *et al.*, 1977).

In many studies the presence of the polypeptide of CP I has been inferred, not proved. Since all the lipid-extracted subchloroplast fragments, enriched in PS I, are also enriched in polypeptides in the M_r 60,000–54,000 region, Levine *et al.* (1972) related these polypeptides to PS I, terming them group 1 polypeptides. Anderson and Levine (1974b) showed that some of them belonged to CP I. PS I fractions prepared by Triton X-100 treatment of thylakoids or by passage through French pressure cell show a similar enrichment in this region (Klein and Vernon, 1974; Nolan and Park, 1975). This assignment is complicated by the fact that several other polypeptides are located in this region (Henriques and Park, 1976b); namely the large M_r 54,000 subunit of ribulose-diphosphate carboxylase (Kawashima and Wildman, 1970) and the 59,000 and 56,000 subunits of the chloroplast coupling factor, CF_1 (Nelson *et al.*, 1973). Further, as mentioned, the polypeptide of CP I is soluble in chloroform–methanol (2:1, v/v) and 90% acetone (Chua *et al.*, 1975).

The polypeptide composition of the reaction-center complex of PS II is unknown. The Triton X-100 reaction-center preparation of PS II (TSF-2a) shows intensified bands at M_r 63,000 and 44,000 (Vernon and Klein, 1975), while the digitonin PS II fraction (F_{II}) shows intensified bands at 27,000, 44,000, and 54,000 (Wessels and Brochert, 1975). Some circumstantial evidence for the association of a particular polypeptide with the PS II reaction center comes from the study of Chua and Bennoun (1975), which shows that several *Chlamydomonas* mutants lacking the PS II reaction center lack an M_r 47,000 polypeptide.

The behavior of LHCP on SDS–PAGE is curious; when the pigments are removed from the complex, the protein moiety exhibits a somewhat lower electrophoretic mobility than that of the complex itself (Machold, 1975). Some SDS–PAGE procedures (Remy *et al.*, 1977) show a single band in the region of M_r 25,000–22,000 for the complex, while others resolve two bands. In several higher plants, Anderson and Levine (1974b) found that the chlorophyll zone extended over two polypeptides termed $2b$ and $2c$ ($\sim M_r$ 26,000 and 24,000). When the pigment was on the complex, the amount of $2b$ exceeded that of $2c$, but after pigment removal the reverse was true. Bar-Nun *et al.* (1977) showed that the heat-treated LHCP of *Chlamydomonas* is composed of two subunits of M_r 24,000 and 22,000. However, it is not yet established whether the complex possesses one or two polypeptides. The study of Apel *et al.* (1975) with *Acetabularia mediterrania* is interesting in this regard. These authors isolated a PS II -enriched fraction, which gave two chlorophyll-containing bands at positions of M_r 67,000 and 21,500. Reelectrophoresis of the M_r 67,000 band gave two chlorophyll-containing polypeptide bands at positions of 67,000 and 21,500 and an additional band at 23,000, suggesting that the 67,000 complex was unstable and dissociated into two subunits. The two subunits of *Acetabularia* LHCP were isolated by preparative SDS–PAGE (Apel, 1977). Their amino acid compositions were similar, but the cyanogen bromide fragments exhibited some differences in molecular weight, suggesting differences in the primary structure of the two polypeptides. While it seems likely that the algal LHCP comprises two polypeptides, it is not established with higher plants whether the minor polypeptide associated with LHCP is actually a component of this complex.

A question of interest is whether or not the polypeptide(s) of LHCP are present in the chlorophyll b-less barley mutant. There is agreement that LHCP itself is absent in the *chlorina* barley mutant (Thornber and Highkin, 1974; Genge *et al.*, 1974; Anderson and Levine, 1974b), and several authors suggested that the polypeptide(s) associated with LHCP are also absent (Anderson and Levine, 1974a; Genge *et al.*, 1974; Thornber and Highkin, 1974). Henriques and Park (1975) using high-resolution SDS-PAGE showed that the major M_r 25,000 band was absent in the barley mutant and bands at 27,500 and 20,000 were reduced (Fig. 5). Machold *et al.* (1977) showed that four allelic *chlorina* barley mutants that are capable of photosynthesis, all lack Chl b and LHCP. The polypeptides of well washed, lipid-extracted thylakoids of wild-type and mutant plants were compared by SDS–PAGE. Similar polypeptide patterns were obtained for wild-type and mutant membranes on 10% polyacrylamide gels. Differences were observed, however, on 11% polyacrylamide gels containing 5 M urea. Thylakoids and LHCP from the wild type gave two polypeptides (IIa and IIb) with IIa present in greater amount. Coelectrophoresis of LHCP polypeptides with the thylakoid polypeptides of the *chlorina* barley showed that the mutant contains the minor polypeptide IIb, but not IIa. The studies of Henriques and Park (1975) and Machold *et al.* (1977) suggest that one of the polypeptides associated with LHCP may be present in the *chlorina* barley mutants.

The polypeptide composition of the membranes of the etioplast has been examined by several investigators, but with no consensus as to the number of polypeptides or their identification with the polypeptides of the thylakoid membrane. The chlorophyll–protein complexes are absent from etioplast membranes, but it is not established whether their polypeptides are present. Some workers claim that most (Hoober, 1970, 1972) or some (Cobb and Wellburn, 1973; Eytan and Ohad, 1972; Wellburn and Cobb, 1975) of the membrane polypeptides are synthesized during greening, while others claim that all or most of the polypeptides, including those of the chlorophyll–protein complexes, are already present in etioplasts (Argyroudi-Akoyunoglou and Akoyunoglou, 1973; Remy, 1973a; Lutz, 1975). During greening under continuous illumination, Alberte *et al.* (1972) showed that LHCP was present after 2 hours of illumination in jackbean leaves, whereas it took 6 hours for CP I to be detected by SDS–PAGE. With greening maize, Guignery *et al.* (1974) showed a similar sequence for appearance of the complexes, but with barley and pea, CP I and LHCP appeared together at 2–3 hours of illumination (Hiller, unpublished observations). After 3 hours of greening of barley, SDS–PAGE showed that 12% of the chlorophyll was in CP I and 18% in LHCP: for pea, 9% was in CP I and 15% in LHCP.

Greening of etiolated seedlings in intermittent light produces chloroplasts with unstacked primary thylakoids containing little or no Chl *b* and no detectable LHCP. Thus, flashed bean and pea leaves (2 minutes light–98 minutes dark) (Argyroudi-Akoyunoglou *et al.*, 1972; Hiller *et al.*, 1973; Davis *et al.*, 1976) and microsecond-flashed wheat leaves, (Remy and Bebee, 1975) have CP I only. These chloroplasts showed PS II activity with an artificial electron donor and, after a few minutes of continuous illumination, excellent O_2 evolution (Remy and Bebee, 1975; Remy, 1973b). Yet, it took 10 hours of continuous illumination before the Chl *a*:Chl *b* ratio had decreased to 3.5, and both membrane stacking and LHCP were detected. Remy (1973b) was unable to determine whether the polypeptide(s) of this complex are present in the primary thylakoids of the flashed material.

In studies with greening algae, Eytan and Ohad (1970, 1972) and Hoober (1970, 1972) showed that a dramatic increase of LHCP occurred at about the same time as increased synthesis of chlorophyll. It is clear that continuous light activates the synthesis of Chl *b* and LHCP, but the temporal sequence of polypeptide synthesis is not established (Anderson, 1975a). From studies on the *de novo* formation of the complexes in *Chlamydomonas,* Bar-Nun *et al.* (1977) suggested that chlorophyll is synthesized simultaneously with the polypeptide moiety. The complexes were detected only in membranes where simultaneous synthesis of both chlorophyll and the corresponding polypeptides occurred. Bar-Nun *et al.* (1977) speculated that the association of the polypeptide and the chlorophyll may occur prior to or during the actual insertion of the complexes into the developing membrane, resulting in a thermodynamically more stable molecular conformation of the polypeptide. The polypeptides of CP I are synthesized in the chloroplast, possibly on thylakoid-bound ribosomes. In contrast, the light-dependent synthesis of LHCP polypeptides is under nuclear control and carried out by cytoplasmic ribosomes (Anderson, 1975a; Thornber and Alberte, 1976). Bar-Nun *et al.* (1977) suggested that the association of chlorophyll with the corresponding polypeptide of LHCP may have a role in the transport of the polypeptide across the chloroplast envelope.

In view of the importance of the chlorophyll–protein complexes as the main thylakoid intrinsic proteins and the role of the polypeptides in the localization of chlorophyll molecules, it is important to establish the location of the complexes within the membrane. Specific group labeling reagents, antibody labeling, and hydrolytic enzymes have not been extensively used with thylakoids (Trebst, 1974; Anderson, 1975a), and little information is available for the location of the complexes. Apel *et al.* (1976) showed that treatment of *Acetabularia mediterranea* thylakoids with EDTA and pronase removed some 60% of the membrane protein. Thylakoids were labeled by the iodination technique before and after EDTA treatment in order to determine the accessibility of the chlorophyll–protein complexes at the membrane surface. After EDTA treatment, the M_r 23,000 polypeptide of LHCP was labeled by the iodination technique, but the M_r 21,500 polypeptide of LHCP and the polypeptide of the CP I were unlabeled, and it was postulated that they were buried in the membrane.

E. Occurrence of the Chlorophyll–Protein Complexes

Chlorophyll *a* is found throughout the entire plant kingdom in photosynthetic tissues that evolve oxygen and is essential for photosynthesis. The reaction-center molecules of PS I and PS II, P700 and P680, respectively, are inferred to be Chl *a*. In contrast, Chl *b*, the main accessory pigment of higher plants and green algae is not essential for photosynthesis. Algae, other than the green ones, possess no Chl *b* and use different accessory pigments, and some Chl *b*-less mutants of higher plants are known to be fully competent photosynthetically.

CP I is thought to occur in all plants that contain P700. Thornber (1975) lists its occurrence in angiosperms, gymnosperms, and a wide range of algal species. Brown *et al.* (1975) have proposed that this protein complex is ubiquitous in the plant kingdom. In the organisms examined, 4–30% of the chlorophyll is associated with CP I after SDS–PAGE (Brown *et al.*, 1975). For higher plants, CP I accounts for 10–18% of the chlorophyll and is located in both grana and stroma thylakoids. *Scenedesmus* mutant 8, which lacks P700, has no CP I (Gregory *et al.*, 1971), and an *Antirrhinum* mutant impaired in PS I activity, which may lack P700, also has none of this complex (Herrmann, 1971). Two Mendelian mutants of *Chlamydomonas* that are deficient in PS I activity lack P700, CP I and its specific M_r 66,000 polypeptide (Chua *et al.*, 1975). However, CP I may not be essential for PS I activity, since its appearance in greening seedlings lags behind the onset of PS I activity (Boardman, 1977a).

LHCP (or aggregates thereof) is the major chlorophyll–protein complex of most higher plants and green algae that contain Chl *b*, and it accounts for 30–60% of the chlorophyll of those organisms. Its occurrence in angiosperms, gymnosperms, and green algae has been documented by Thornber (1975). As mentioned earlier, it is not essential for photosynthetic activity of either PS I or PS II.

On the assumption that the reaction center of PS II is a P680 Chl *a*–protein complex, we would expect the complex to be ubiquitous throughout the plant kingdom. As mentioned earlier, the proportion of chlorophyll in the complex cannot be estimated, since it is not observed on SDS–PAGE. It cannot exceed 20–25% of the total chlorophyll, and it may well be considerably lower, since some of the chlorophyll of the free pigment zone may come from the other complexes.

F.   MODELS FOR COMPLEXES IN THE PHOTOSYNTHETIC UNIT

Figure 7 shows some possible models for the organization of the chlorophyll–protein complexes in the photosynthetic unit of higher plants. It is assumed that each photosynthetic unit contains 400 chlorophyll molecules, with one reaction-center chlorophyll (P700) of PS I and one reaction-center chlorophyll of PS II.

Model *a* in Fig. 7, redrawn from the paper by Thornber *et al.* (1977), contains four chlorophyll–protein complexes, designated P700 CP (reaction-center Chl *a* complex of PS I), RC II (reaction-center Chl *a* complex of PS II), LHCP with a Chl *a*:Chl *b* of 1, and CP, a third Chl *a*–protein. CP is included to account for the long-wavelength fluorescence emission of chloroplasts, which comes from PS I (Section VI, B). Thornber *et al.* (1977) proposed that CP is unstable to SDS–PAGE and its Chl *a* molecules are released into the free pigment zone. They also stated: "A

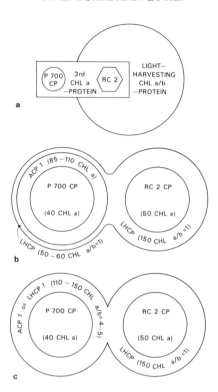

FIG. 7. Models for the organization of the chlorophyll–protein complexes in the photosynthetic unit of higher plants. Model a is redrawn from Thornber *et al.* (1977). See text for explanation.

small portion of the Chl *a* in the free pigment zone must arise from dissociation of the complex containing the chlorophylls of the PS II reaction-center'' (RC II). Thornber *et al.* (1977) ascribed 10-20% of the chlorophyll in the free pigment zone to RC II, and 80–90% to the third Chl *a* protein. Model *a* features a continuous array of light-harvesting chlorophyll, rather than a separate package model in which light-harvesting chlorophyll is added to each photosystem. However, Thornber *et al.* (1977) suggested that CP is primarily associated with PS I and serves as a light-harvesting complex for that photosystem. It is shown in the model as interacting also with RC II to account, according to Thornber *et al.* (1977), for the photochemical properties of the Chl *b*-less mutants of barley, which lack LHCP.

The alternative models that we put forward in *b* and *c* of Fig. 7 are separate package models, but with some interaction between the two photosystems to permit transfer of excitation energy. In model *b*, PS II consists of a reaction-center complex, containing 50 Chl *a* molecules, and LHCP with 150 chlorophyll molecules (Chl *a*:Chl *b* ratio of 1). PS I is composed of P700 CP with 1 P700 and 40 Chl *a* molecules, an antenna Chl *a*–protein complex (ACP I) of some 85–110 Chl *a* molecules, and LHCP with 25–30 Chl *a* and 25–30 Chl *b* molecules. Model *b* is similar to model *a* in that it consists of four chlorophyll–protein complexes. The fractionation of the photosynthetic unit by digitonin or Triton X-100 into PS I and PS II, and the finding that PS I from the digitonin procedure contains approximately one-half as much chlorophyll per molecule of P700 as do chloroplasts (Boardman, 1970), seems more readily explained by a separate package model. Fractionation studies with shade-plant chloroplasts (Anderson *et al.*, 1973) in which all the fractions contain relatively more Chl *b* than the corresponding fractions from spinach would be consistent with the notion that LHCP is added to both photosystems. In model *b*, the third Chl *a*–protein complex is restricted to PS I.

Model *c* also contains four chlorophyll–protein complexes, but it differs from models *a* and *b* in that two of the complexes contain Chl *b*. LHCP is shown with a Chl *a*:Chl *b* ratio of 1, as in the other models, but a specific PS I light-harvesting complex (ACP I or LHCP I) containing Chl *a* and Chl *b* is shown, with a Chl *a*:Chl *b* ratio of 4 to 5. The main reason for proposing model *c* is to account for the low amount of LHCP in SDS–PAGE of PS I subchloroplast fragments prepared by the digitonin method (Section VII, B). In all the models, it is proposed that the antenna chlorophyll–protein complex of PS I is the source of the long-wavelength fluorescence of chloroplasts at liquid-nitrogen temperature.

As mentioned in Section VII, C, the mutant of barley (*chlorina* 2) which lacks LHCP contains about two-thirds of the chlorophyll of the wild-type per molecule of P700. If all the chlorophyll of LHCP were absent, the mutant should have one-half of the chlorophyll of the wild type per molecule of P700. It seems, therefore, that the mutant chloroplasts contain at least 90% of the Chl *a* molecules that are associated with LHCP in the wild type. A similar conclusion is reached from a comparison of the relative photochemical activities per milligram of Chl *a* of the wild type and mutant at saturating light intensities (Highkin and Frenkel, 1962; Boardman and Highkin, 1966). It would appear that the mutant chloroplasts contain a Chl *a*-light-harvesting complex which substitutes for LHCP. This could account for the higher proportion of a form of Chl *a* absorbing at about 670 nm in the mutant (Boardman and Thorne, 1968).

VI. Organization of Pigment Molecules *in Vivo* and in the Chlorophyll–Protein Complexes

A. FORMS OF CHLOROPHYLL

Absorption and fluorescence spectroscopy of chloroplasts and algae indicate that chlorophyll *in vivo* is in a different chemical or physical state than chlorophyll after extraction into organic solvents. As compared with Chl *a* monomers in a dissociating polar solvent, such as acetone, chlorophyll *in vivo* exhibits a broader absorption band and its maximum is red-shifted. The *in vivo* spectrum of chlorophyll in the red region has been computer deconvoluted into a number of Gaussian components. Better resolution of components is obtained at 77°K, and the low-temperature spectra are fitted by four major forms of Chl *a* having peaks at 662, 670, 677, and 684 nm and two minor forms at about 693 an 704 nm (French *et al.*, 1972) (Fig. 8). Chl *b* is fitted by a peak at 650 nm. In the earlier studies it was assumed that a small band at 640 nm was due to Chl *b*, but several pieces of evidence now indicate that the band is not a form of Chl *b* (Brown *et al.*, 1974).

Table I compares the proportions of the chlorphyll forms in chloroplast fragments of spinach. Chloroplasts were passed through the French pressure cell, and the fragments were separated by sucrose density gradient centrifugation. Fraction 1, which is derived from stroma thylakoids and is highly enriched in PS I, contains more of the long-wavelength forms of chlorophyll *a* and less of Ca-677 than does the unfractionated material. Fraction 2, which is mostly representative of grana thylakoids and is somewhat enriched in PS II, contains a lower percentage of the long-wavelength forms and more of Ca-662.

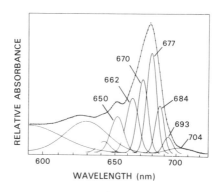

FIG. 8. Resolution of the absorption spectrum of chloroplasts into gaussian components. The observed data are plotted as points, and the line through them is the sum of the component curves. (By courtesy of Dr. C. S. French.)

TABLE I
PROPORTIONS OF CHLOROPHYLL a COMPONENTS IN CHLOROPLAST FRAGMENTS[a]

Fraction	Ca-662	Ca-670	Ca-677	Ca-684	Ca-693	Ca-705
	Chl a components (% of total Chl a)					
Spinach unfractionated preparation	20.5	24.6	37.4	9.8	5.3	2.4
Spinach fraction 1	19.0	21.1	30.3	19.7	6.9	3.0
Spinach fraction 2	25.1	25.5	37.0	7.8	4.6	0
Tobacco CP I	23	24	25	21	6	—
Tobacco LHCP	34	36	23	7	—	—

[a] From French et al. (1972) and Brown et al. (1974).

The chlorophyll–protein complexes extracted by SDS have also been analyzed for chlorophyll forms (Brown et al., 1974). The complexes contain the same major forms of Chl a as do chloroplasts and the French press fractions, but the percentage of Ca-677 is lower in the complexes; LHCP is enriched in Ca-662 and Ca-670, and CP I in Ca-684 and Ca-693 (Table I). The proportions of the spectral forms in CP I were about the same for a diverse range of plants (Brown et al., 1975). In an earlier section, the various spectroscopic forms of protochlorophyllide were discussed in terms of pigment–pigment interactions and pigment–protein interactions. We concluded that pigment–pigment interactions were probably of prime importance in determining the spectral properties of protochlorophyllide, but we also indicated that weak bonding of porphyrin to protein might be expected to influence the spectral properties of the protochlorophyllide. Similar arguments can be advanced to explain the spectral properties of the chlorophyll–protein complexes. Among others, Krasnovsky in the U.S.S.R. and Katz in the U.S.A. have been strong proponents of the view that the spectral properties of chlorophyll in vivo are due to pigment–pigment interactions. Recently, Cotton et al. (1974) compared computer deconvolutions of spectra of Chl a at room termperature in a number of solvents with those of sonicated preparations of Tribonema aequale, a green alga containing Chl a, and a PS I reaction-center preparation of Anabaena variabilis. They found that absorption spectra, either of solutions of Chl a in hexane or of the algal preparations can be fitted with precisely the same Gaussian components. Chl a is thought to form oligomers, $(Chl_2)_n$, in hexane with a chlorophyll dimer as the building block. Over the concentration range from 1 μM to 0.1 M the Chl a spectra were deconvoluted at room temperature into Gaussian components at 628,

659, 662, 678, and 703 nm. Increasing the size of the oligomer by increasing the chlorophyll concentration increased the proportion of the 678-nm component relative to the 662-nm component. Cotton *et al.* (1974) attributed the variation in the relative proportions of the 678- and 662-nm components to a tilting and rotation of the planes of the chlorins with respect to one another, owing to aggregation and coordinate interactions between the molecules of the dimer building block.

However, it is plausible to consider that the specific interactions of the chlorophyll molecules *in vivo* are influenced to a large extent by some coordination of chlorophyll molecules to amino acid side chains of protein. Even if the spectral properties of chlorophyll *in vivo* were due solely to interacting pigment molecules in a hydrophobic environment, it seems very likely that the protein plays an important role in providing the hydrophobic environment and determining the relative orientations of the chlorophylls to one another. This point is well illustrated by the recent X-ray structure of a water-soluble bacteriochlorophyll–protein complex (Fenna and Matthews, 1975) (Section VI, E).

B.   FLUORESCENCE PROPERTIES OF CHLOROPHYLL *in Vivo*

The fluorescence emission spectrum of chloroplasts at room temperature has a main peak at 683 nm and a broad band at 720–740 nm due to vibrational satellites. The fluorescence emanates only from Chl *a* molecules. Chl *b in vivo* transfers its excitation energy to Chl *a* with high efficiency and does not fluoresce itself. The intensity of fluorescence of Chl *a in vivo* is low compared with Chl *a* in a disaggregating organic solvent, and varies with the redox state of Q, the primary electron acceptor of PS II (Duysens and Sweers, 1963). In dark-adapted chloroplasts, Q is oxidized and the fluorescence (F_o) is substantially quenched. Illumination of chloroplasts in the absence of an electron acceptor reduces Q and the fluorescence rises in a biphasic manner to a steady-state level (F_M) which is 4–5-fold greater than the level in dark-adapted chloroplasts (Duysens and Sweers, 1963). More recent studies show that the fluorescence yield of Chl *a* in algae and chloroplasts depends not only on the redox state of Q, but also on the distribution of excitation energy between the two photosystems (see the chapter by Arntzen in this volume). The early work showed that the Chl *a* in PS I is weakly fluorescent, whereas that in PS II was capable of a much stronger fluorescence (Govindjee, 1975). Measurements of the fluorescence yields of subchloroplast fragments enriched in PS I and PS II, respectively, confirmed the weak fluorescence of PS I and that the variable fluorescence arises from PS II (Boardman *et al.*, 1966).

The initial fluorescence (F_0) of dark-adapted chloroplasts was originally thought to arise from PS I, but the fluorescence of PS I particles suggests that a considerable part arises from PS II (Boardman et al., 1966). Butler and Kitajima (1975a) observed that dibromothymoquinone quenches F_0 and F_M in a similar fashion and concluded that the major part of F_0 arises from PS II and is the same type of fluorescence as the fluorescence of variable yield ($F_v = F_M - F_0$).

At 77°K, chloroplasts show a three-banded fluorescence emission spectrum with peaks at 685, 695, and 735 nm (denoted F_{685}, F_{695}, and F_{735}, respectively). The intensity of F_{735} is greatly enhanced at 77°K, compared with room temperature. Measurements of fluorescence emission and excitation spectra of several organisms and of subchloroplast fragments enriched in PS I and PS II suggest that F_{685} and F_{695} originate mainly from PS II and F_{735} from PS I (Boardman, 1970).

Kinetics of fluorescence induction are also observed at liquid-nitrogen temperature (Murata, 1968). The fluorescence induction curves of F_{685} and F_{695} are similar, but F_{695} shows a greater $F_M:F_0$ ratio (about 5) than does F_{685} (about 3). F_{735}, which appears to emanate from a form of Chl a in PS I absorbing at 705 nm (Butler, 1961b), shows a much lower induction ($F_M/F_0 = 1.3$). Murata (1968) suggested that the chlorophyll molecules responsible for F_{695} are closer to the photochemical reaction center of PS II than those responsible for F_{685}. The small variable fluorescence of F_{735} is considered to be mainly due to spillover of energy from PS II to PS I (Murata, 1968; Butler and Kitajima, 1975b), although some will be due to the satellite vibrational levels of F_{685} and F_{695}. The lack of variable fluorescence from PS I subchloroplast particles at 77°K supports this conclusion (Vrendenberg and Slooten, 1967). Butler and Kitajima (1975c) proposed that F_{685} arises from the LHCP and F_{695} from the Chl a molecules of the reaction-center complex of PS II. However, chloroplasts of the Chl b-less mutant of barley also emit fluorescence at 685 nm at 77°K, so that the presence of LHCP is not essential for F_{685} (Boardman and Thorne, 1968). In fact, the shape of the fluorescence emission spectrum of the mutant chloroplasts at 77°K is similar to that of wild-type chloroplasts.

2. Fluorescence Spectra of Chlorophyll–Protein Complexes

The fluorescence emission spectrum of CP I from *Phormidium* at room temperatures showed a peak at 682 nm and shoulders at 692, 720, and 740 nm (Mohanty et al., 1972). At 77°K, fluorescence peaks were observed at 691 and 720 nm with a slight shoulder at 695 nm and a more distinct shoulder at 720 nm. A somewhat similar emission spectrum for CP I from pea leaves *(Pisum sativum)* at 77°K was published by

Herrmann *et al.* (1974), except that a shoulder was not observed at 695 nm and the far-red peak was located at 730 nm with a shoulder at 745 nm. The ratio ($F_{730}:F_{680}$) was approximately 0.7. This is in distinct contrast with the fluorescence emission spectrum of PS I subchloroplast particles, where the 680 peak is very small compared with the far-red peak (Boardman *et al.*, 1966; Ke and Vernon, 1967; Ogawa and Vernon, 1969). The large fluorescence band at 680 nm from CP I at 77°K could reflect an alteration in the organization of the chlorophyll molecules in the PS I antenna, resulting in a decrease in the efficiency of energy transfer from the forms of Chl *a* absorbing around 680 nm to the far-red form of Chl *a*. On the other hand, it may be due to the separation of the reaction-center complex (CP I) from an antenna chlorophyll–protein complex, which contains the far-red form of Chl *a* (cf. Fig. 7).

The fluorescence emission spectrum of LHCP shows bands at 680 nm and about 740 nm, but the band at 740 nm ($F_{740}/F_{680} < 0.3$) is much smaller than in CP I (Butler and Kitajima, 1975b). The spectrum of LHCP resembles that of subchloroplast particles enriched in PS II (Boardman, 1972; Ke and Vernon, 1967).

C. ORIENTATION OF PIGMENT MOLECULES

1. Linear Dichroism and Polarized Fluorescence

The earlier measurements of the linear dichroism of individual chloroplasts showed that a long-wavelength form of Chl *a* ($\lambda > 695$ nm) was oriented with the planes of the chlorin rings parallel to the plane of the thylakoids (Olson *et al.*, 1964a). There was little evidence for the orientation of the bulk of the Chl *a*, which absorbs in the 670–680-nm waveband. Similar conclusions were reached from measurements of the polarization of fluorescence of individual chloroplasts (Olson *et al.*, 1964b). However, recent studies, in which chloroplasts were oriented by a magnetic field (Geacintov *et al.*, 1972; Breton *et al.*, 1973) or by a mechanical method (Breton *et al.*, 1973) indicate substantial orientation of the form of Chl *a* absorbing at 681 nm and less orientation of Chl *a*-670. The magnetic method for alignment of the chloroplasts could be criticized on the basis that it may cause orientation of the chlorophyll molecules (as already pointed out by Geacintov *et al.*, 1972), but this seems unlikely since similar results were obtained when chloroplasts were aligned by spreading on a quartz plate. From measurements of the dichroic ratio of mechanically aligned chloroplasts, Breton *et al.* (1973) concluded that the directions of the Q_y-transition moments (red band) of the Chl *a*-680 lie closer to the lamellar plane with an angle to the normal greater than 60°–65°, and the directions of the Q_x-transition moments (Soret band) are tilted out of the membrane plane at an angle of 48° to the normal. This conclusion of a uniform angle of tilt of 48° for Chl *a* *in vivo* must be regarded as tentative, since it is based on the assumption that the transition moments of Chl *a*, corresponding to a particular absorption band, are distributed isotropically around the normal to the plane of the thylakoid membrane. The assumption may not be valid for chlorophyll *in vivo*, even though it appears to be true for chlorophyll in artificial lipid membranes (Cherry *et al.*, 1972). As already discussed, the arrangement of chlorophyll molecules in the chlorophyll–protein complexes probably depends on interactions between protein and pigment as well as on pigment–pigment interactions.

Junge and Eckhof (1974) studied the linear dichroism of the light-induced absorption change at 705 and 430 nm, due to the photooxidation of P700. The dichroism was maximum for excitation at wavelengths > 690 nm, thus supporting the earlier work showing a higher degree of orientation of the long-wavelength forms of chlorophyll. The dichroism of the absorption change was similar at 430 nm and 705 nm. Junge (1975) discussed this finding in relation to the orientation of the porphyrin rings of the dimer in P700, and concluded that the porphyrin rings are oriented more or less parallel to the plane of the thylakoid membrane.

Mathis *et al.* (1976) arrived at a similar conclusion for the orientation of the reaction-center chlorophyll of PS II. They measured the dichroism of the rapid light-induced absorption increase at 825 nm, due to the oxidized (chlorophyll radical-cation) reaction center.

Whitmarsh and Levine (1974) measured the fluorescence polarization of chlorophyll in intact cells of *Chlamydomonas reinhardii*. The degree of polarization (P) varied with excitation wavelength from 1.3% at 600 nm to 4.6% at 673 nm. Inhibition of electron flow with 3-(3,'4'-dichlorophenyl-1,1-dimethylurea (DCMU) led to an increase in fluorescence lifetime and a large decrease in P for excitation wavelengths between 600 and 660 nm. This result was explained by increased energy transfer by the Förster mechanism among chlorophyll molecules in PS II absorbing below 660 nm. The chlorophyll molecules absorbing at 673 nm showed little change in P on closure of the PS II reaction-center traps by DCMU. Thus increasing the probability of energy transfer does not affect the fluorescence polarization, excited at 673 nm. Whitmarsh and Levine (1974) interpreted the higher fluorescence polarization for excitation at 673 nm as indicative of some order of the Q_y-transition dipoles of chlorophyll absorbing at 673 nm. Because DCMU only affects P at excitation wavelengths shorter than 660 nm, Whitmarsh and Levine (1974) proposed that the chlorophyll molecules in PS II exist in discrete groups characterized by different absorption maxima and different degrees of fluorescence polarization, and excitation energy is not shared between some groups absorbing at different wavelengths.

Becker *et al.* (1976) observed that the degree of fluorescence polarization of spinach chloroplasts, excited at 670–680 nm, decreased from as high as 14% to 5% when the chloroplasts were oriented in a magnetic field. The high polarization value of unoriented chloroplasts was considered to be due to optical anisotropy of the membranes. For unoriented chloroplasts, Becker *et al.* (1976) divided P into two contributions; P_{AN}, due to the optical anisotropy, and P_{IN}, the intrinsic polarization, which reflects the degree of mutual orientation of the chlorophyll molecules within the membrane and the degree of energy transfer between them. From a study of the excitation wavelength dependence of P_{IN} for oriented chloroplasts, they concluded that the carotenoid and chlorophyll Q_y-transition moments are partially oriented with respect to each other within a photosynthetic unit.

2. Circular Dichroism of Chloroplasts

In contrast with their lack of linear dichroism, randomly oriented asymmetric molecules show circular dichroism (CD) in their absorption bands because they absorb left- and right-circularly polarized light to slightly different extents (Sauer, 1975). However, the CD of an asymmetric molecule is enhanced if the molecules are asymmetrically arranged. A chlorophyll molecule has five asymmetric carbon atoms, three in the porphyrin macrocycle and two in the phytyl chain. It shows a weak CD in disaggregating solvents. In dry solutions of solvents, such as CCl_4, Chl *a* and Chl *b* show enhanced CD and a band splitting consistent with exciton interaction between the individual molecules of dimers (Dratz *et al.*, 1966).

Chloroplasts of higher plants exhibit an intense CD spectrum in the red region, characterized by a positive component at 683–686 nm and negative ones at 672–677 nm and 650–652 nm (Brody and Nathanson, 1972; Gregory *et al.*, 1972). The smaller negative band is attributed to Chl *b* since it is absent from chloroplasts of the *chlorina* mutant of barley, which lacks Chl *b* (Dratz *et al.*, 1966; Gregory and Raps, 1975). The chloroplast CD spectrum is characteristic of a split-exciton spectrum typical of chlorophyll dimers, although its magnitude on a chlorophyll basis is greater than that of Chl *a* dimers in solution (Raps and Gregory, 1975). In the far-red region (> 700 nm), however, the spectrum falls too slowly for a simple-exciton curve, suggesting that an additional component with a simple maximum at 700–710 nm may be present (Gregory, 1975).

There is a striking decrease by an order of magnitude in the CD of chloroplasts on fragmentation by sonication or detergent treatment (Gregory *et al.*, 1972; Philipson and Sauer, 1973). Philipson and Sauer (1973) attributed the large CD signal of intact chloroplasts to differential light scattering of circularly polarized light, not to the intrinsic CD of chloroplasts. However, Gregory and Raps (1974) measured the differential scattering and showed that both the intensity and characteristic shape of the CD remains after correction for differential scattering. They proposed that the CD of whole chloroplasts reflects the intrinsic organization of pigment molecules in the grana stacks due to membrane pairing. Barley mutant chloroplasts lacking Chl b, which show only 50% of the membrane stacking of wild-type chloroplasts, exhibited a smaller CD signal, about one-half that of wild-type chloroplasts (Gregory and Raps, 1975). Thus the size of the CD signal in the mutant appear to be adequately explained by its lower degree of membrane stacking, although Gregory and Raps (1975) implied that the thylakoid system is mechanically weaker in the mutant, due to a looser pairing of the thylakoids in the grana. Electron microscopy, however, provides no evidence for a looser pairing (Goodchild et al., 1966). Faludi-Dániel et al. (1973) observed that the CD signal of the agranal bundle sheath chloroplasts of maize was about 8-fold smaller than those of grana-containing mesophyll chloroplasts. However, bundle-sheath chloroplasts were fragmented during their isolation, and the comparison with whole mesophyll chloroplasts is complicated. Raps and Gregory (1975) reported smaller CD signals for algae than for higher plant chloroplasts. They concluded that Chl a "bulk pigment" in chloroplasts is organized, probably in the form of closely associated dimers, and that the organization is dependent on thylakoid stacking. Gregory (1975) examined the orientation of the pigment molecules in relation to the plane of the thylakoid membrane. Chloroplasts were oriented by flow, either at right angles or in the direction of the light path. The amplitude of the CD was maximum when the light path was in the plane of the membranes, suggesting that the preferred direction of the red-transition moment was at right angles to the plane of the thylakoids. When chloroplasts were illuminated in the presence of an electron acceptor, such as ferricyanide, there was a slow decrease in the CD signal, amounting to about 25% of the dark signal. The original CD level was slowly restored on turning off the light.

3. Circular Dichorism of Chlorophyll–Protein Complexes

The CD spectrum of CP I shows a negative component at about 690 nm and a positive component at about 676 nm (Brody and Nathanson, 1972; Scott and Gregory, 1975) (Fig. 9). Thus the signs of the components are reversed compared with the corresponding bands in intact chloroplasts, and the CP I spectrum is less intense. A substantial part of the CD spectrum of CP I can be fitted to a curve obtained by subtracting two Gaussian curves of equal magnitude (Scott and Gregory, 1975). The

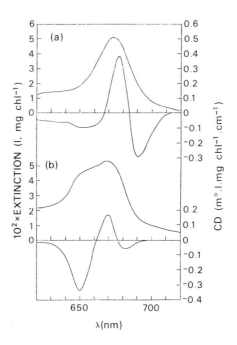

FIG. 9. Absorption and circular dichroism (CD) spectra of CP I (a) and LHCP (b). From Scott and Gregory (1975.)

double CD is typical of a split-exciton spectrum of a dimer. The center of the fitted curve is about 685 nm, whereas the absorption maximum of CP I is at 674–676 nm. Thus, most of chlorophyll does not appear to be organized into dimers. Scott and Gregory (1975) concluded that of the 20 or 50 molecules of Chl a present in CP I, only one or two pairs would be generating the split-exciton signal. PS I particles isolated by the digitonin method gave a similar CD spectrum, but the peaks of the negative and positive components were shifted to shorter wavelength compared with those of CP I (Brody and Nathanson, 1972). Philipson et $al.$ (1972) examined light-induced changes, in both absorption and CD, of preparations of PS I particles enriched about 10-fold in P700, as compared with chloroplasts. The particles (HP 700) were obtained from spinach chloroplasts by a modification of the Triton X-100 method of Yamamoto and Vernon (1969). The reversible photooxidation of P700 gave a two-banded light-minus-dark difference absorption spectrum, with troughs at about 700 nm and 680–685 nm (Döring et $al.$, 1968). The light-minus-dark CD difference spectrum showed a positive peak at 696.6 and a negative one at 688 nm, which is consistent with an exciton interaction between two Chl a molecules in the PS I reaction center. Photooxidation of one of the chlorophyll molecules destroys the exciton interaction and results in the loss in absorption at 700 nm and about 680 nm, and the appearance of a band at 686 nm due to the remaining unoxidized chlorophyll molecule in P700$^+$ (Philipson et al., 1972). Thus the CD spectra of CP I and PS I subchloroplast particles appear to be due to exciton interaction between the chlorophyll molecules of the reaction-center of PS I. Although the bulk of the chlorophylls in these preparations enriched in the reaction center of PS I appear to give a negligible CD signal, they may play an important role in determining the specific geometric arrangement of the two exciton-interacting chlorophylls in the reaction center.

The CD spectrum of LHCP (Fig. 9), although superficially similar to that of CP I, is explained not by a split-exciton interaction, but as a sum of the CD components due to Chl a and Chl b (Scott and Gregory, 1975). In order to account for the size of the CD signal of LHCP compared with that of purified Chl a and Chl b, Scott and Gregory (1975) proposed that the CD contribution of each chlorophyll is enhanced by pigment–protein interactions. However, a satisfactory explanation is needed to account for the positive CD of Chl a in LHCP, whereas Chl a in solution and Chl b in the complex give negative CD signals. Garay et al. (1972) observed a positive CD component at about 670 nm and a negative one at 650 nm, when the extracted leaf pigments were transferred to ethyl ether saturated with water. The concentration of pigments was relatively high, so that some aggregation of Chl a may have occurred.

The chloroplast fragments (so-called quantasomes) prepared by soni-cation of chloroplasts (Dratz et al., 1966) show a three-peaked CD spectrum, which roughly corresponds to the sum of the spectra of CP I and LHCP (Scott and Gregory, 1975). The CD measurements provide evidence for the conclusion that CP I and LHCP exist in vivo and are not artifacts of the SDS extraction. Garay et al. (1972) compared the CD spectra of PS I subchloroplast particles obtained by digitonin fragmenta-tion of normal and carotenoid-deficient mutants of maize. The CD spectra of the mutants resembled that of normal chloroplasts in the red region, but there were differences in the 450–520 nm region, due to differences in carotenoid composition.

D. MODEL FOR REACTION-CENTER CHLOROPHYLL OF PHOTOSYSTEM I

As mentioned above, the light-induced changes in absorbance and CD of the reaction-center chlorophyll of PS I are interpreted by exciton interaction in a chlorophyll dimer. Electron spin resonance spectroscopy (ESR) and electron nuclear double resonance spectroscopy (ENDOR) provide good evidence for an unpaired electron being delocalized over two pigment molecules (Katz and Norris, 1973). It is believed that the chlorophyll species responsible for the *in vivo* radical is the species (P700) that bleaches in the light. The model of the reaction-center chlorophyll of PS I proposed by Katz and Norris (1973) consisted of two Chl *a* molecules, linked by a water molecule that coordinates to the Mg of one Chl *a* molecule and hydrogen bonds to both the ring V keto C=O and the carbomethoxy C=O of the other Chl *a* molecule. Fong (1974) suggested that the two Chl *a* molecules are bonded by two water molecules. Recently, Shipman *et al.* (1976) proposed a further model, which was based on exciton-theoretical considerations and the infrared and visible absorption spectra of a 700-nm absorbing species of Chl *a* in ethanol and toluene. In the model, two Chl *a* molecules are held together by two linking R¹—O—H ligands, each of which is simultaneously coordinated via the oxygen lone-pair electrons to the Mg of one Chl *a* and hydrogen bonded to the keto C=O of the other Chl *a*. Π—Π van der Waals stacking interactions between the two Chl *a* molecules are also involved. The red shift in the absorption maximum of the dimer to near 700 nm is attributed to a combination of the Π—Π stacking, strong hydrogen bonds to ring V keto carbonyls, and transition density interactions (Shipman *et al.*, 1976). An important feature of this model is that it suggests an explicit role for protein in the formation of the special chlorophyll dimer. The following protein side chains are suggested as possible ligands; those of arginine, (N—H), cysteine (S—H), hydroxylysine (O—H or N—H), hydroxyproline (O—H), lysine (N—H), serine (O—H), threonine (O—H), and tyrosine (O—H).

E. THREE-DIMENSIONAL STRUCTURE OF BACTERIOCHLOROPHYLL–PROTEIN COMPLEX

Recently, the three-dimensional structure of a water-soluble bacteriochlorophyll–protein complex from *Prosthecochloris aestuarii* was determined by X-ray crystallography at 2.8 Å resolution (Fenna and Matthews, 1975, 1977). Although this review is devoted to chlorophyll–protein complexes and excludes coverage of the extensive literature on bacteriochlorophyll–protein complexes, we consider it worthwhile to include an account of the arrangement of bacteriochlorophyll (BChl) molecules in the *Prosthecochloris* BChl–protein complex. This is the first chlorophyll-containing protein for which the structure is known, and it may prove to be a good model for the organization of chlorophyll in the chlorophyll–protein complexes discussed in this review.

The function of the BChl–protein in *Prosthecochloris aestuarii* is to mediate the transfer of excitation energy from the light-harvesting chlorobium chlorophyll to the photochemical reaction center. The complex, M_r about 150,000, contains 21 ± 2 molecules of BChl and is a trimer of identical subunits (Olson, 1978; Thornber *et al.*, 1978). The arrangement of the polypeptide chain and the BChl molecules in one subunit is illustrated in Fig. 10. Fenna and Matthews (1977) emphasize that the polypeptide chain topology should be regarded as tentative at this stage.

In overall shape, the subunit resembles a hollow cylinder, one end of which is closed, mainly by two strands of the chain which traverse the "top" of the subunit. The other end of the cylinder is open, but this becomes occluded when the subunits associate to form the trimer. The wall of the cylinder, which in the trimer is exposed to the solvent, is composed mainly of 15 strands of β-sheet, mainly antiparallel. The β-sheet does not form a completely closed cylinder, and part of the polypeptide envelope consists of several helices, which occur in loops emanating from strands toward the ends of the extended β-sheet.

The seven BChl molecules of the subunit are located within the hollow cavity, an ellipsoid of axial dimensions 45 × 35 × 15 Å. The average center–center nearest neighbor distance between the BChl rings is 12 Å.

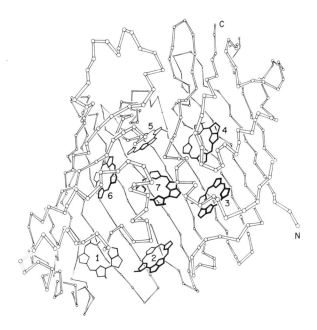

FIG. 10. Arrangement of the peptide chain and bacteriochlorophyll molecules in a subunit of the bacteriochlorophyll–protein complex from *Prosthecochloris aestuarii*, as determined from X-ray crystallography at 2.8 Å resolution. From Fenna and Matthews (1977).

The closest distance between BChl rings in adjacent subunits of a trimer is 24 Å. Five of the BChls in each subunit are buried, while two (1 and 2) are partially exposed, but are surrounded by protein in the trimer. Although the planes of the rings lie roughly parallel to one another, the Q_y-transition moments are not oriented in any way.

The phytyl chains are mostly extended, except for BChl 1, where the chain is bent into a U-shape loop. They are close together, sandwiched between five of the BChl rings on one side and the other two rings and the β-sheet protein on the other. The chains of BChls 4, 5, and 6 are parallel and form a planar structure in close contact with the inwardly directed side chains of the β-sheet. It seems probable that the inwardly directed side chains of the β-sheet will prove to be largely hydrophobic and the outwardly directed ones hydrophilic.

The orientations of the BChl rings appear to be determined by their interactions with the polypeptide chain. The Mg atoms of six of the BChl molecules interact with amino acid residues of the protein, liganding being restricted to one side of a BChl ring. Access to a sixth ligand is prevented by the phytyl chains traversing one face of each of the rings. BChl 2 does not seem to be liganded to protein, but a water molecule could be coordinated to the Mg (Fenna and Matthews, 1977). The electron density evidence suggests that BChls 3 and 7 are liganded to histidine residues, but definitive conclusions on liganding awaits knowledge of the amino acid sequence of the polypeptide. Fenna and Matthews (1977) pointed out that liganding does not always occur on the same side of the BChl ring.

There are further interactions between the BChl molecules and the protein, involving hydrogen bonding between the polypeptide chain and oxygen-containing ring substituents. There is no evidence, however, for chemical bonding between neighboring BChl molecules except for hydrophobic interactions between the phytyl chains. Thus, distribution and the orientations of the BChl molecules are determined by specific interactions with the protein, rather than with one another. The absorption and CD spectral properties of the BChl–protein complex seem consistent with exciton interactions between the BChl molecules (Fenna and Matthews, 1975, 1977; Olson, 1978).

F. PROPOSED ORGANIZATION OF CHLOROPHYLL AS BOUNDARY LIPID

Thornber et al. (1977) have emphasized the similarities between CP 1 of higher plants and CP I of the blue-green alga, *Phormidium luridum*. Their proposed model for CP I, where each subunit of a trimer contains seven Chl a molecules (Section V, C) bears close analogy to the BChl–protein complex of *Prosthecochloris*. Thus the organization of chlorophyll in CP I may be similar to that of bacteriochlorophyll in the BChl complex, with the chlorophyll molecules buried in the hydrophobic interior of the protein. However, the chlorophyll which is readily released into the free pigment zone on SDS–PAGE, or even the chlorophyll of LHCP may be organized in a different manner. Anderson (1975b) suggested that the phytyl chains of chlorophylls might be associated *in vivo* with the hydrophobic exterior of the major intrinsic proteins of the thylakoid membrane. In such a model, chlorophyll forms part of the boundary lipids of these proteins. The hydrophilic edge of the chlorin ring, adjacent to the phytyl chain, is postulated to interact at the membrane surface with the exposed hydrophilic segment of the intrinsic protein, and the more hydrophobic portion of the tetrapyrrole would be buried in the protein. Such an organization of chlorophyll molecules might be more easily perturbed by SDS than the arrangement of pigment molecules in the BChl–protein complex, and may account for the release of chlorophyll into the free pigment zone during SDS-PAGE. The chlorophyll–protein complexes are distinguished from the BChl–protein complex by their nonextractibility in aqueous buffers in the absence of detergents. Their outer protein surface is likely to be less hydrophilic than that of the BChl–protein complex.

VII.   Chlorophyll–Protein Complexes and Membrane Structure

A.   MEMBRANE STRUCTURE AND CHLOROPHYLL CONTENT

Many studies with chlorophyll-deficient mutants and plants grown under different light regimes have shown a correlation between the total amount of chlorophyll per unit area of leaf and the extent of grana formation. A higher chlorophyll content is often associated with a higher proportion of Chl *b* to Chl *a*, and it has been suggested that Chl *b* may play a role in grana formation. For example, shade plants contain more chlorophyll than do sun plants and more Chl *b* relative to Chl *a* (Kirk and Tilney-Bassett, 1967; Boardman, 1977b). The chloroplasts of three extreme shade plants from the floor of a dense rain forest contained 4–5 times more chlorophyll than did spinach chloroplasts, and their Chl *a*:Chl *b* ratio was 2.3, compared with 2.8 for spinach (Anderson *et al.*, 1973). They contained very large grana stacks, and their total length of stacked grana regions relative to unstacked stroma thylakoids was double that of spinach chloroplasts. A similar inverse relationship between the extent of grana formation and the Chl *a*:Chl *b* ratio was observed for a particular sun species (*Atriplex patula*) grown under three different light intensities (Björkman *et al.*, 1972) and for pea and maize plants grown under different light–dark regimes (Smillie *et al.*, 1975).

Woo *et al.* (1971) found an inverse relationship between the Chl *a*:Chl *b* ratio and membrane stacking in mesophyll and bundle sheath chloroplasts of five C_4-plants. The bundle sheath chloroplasts showed less membrane stacking and less Chl *b* compared with mesophyll chloroplasts. During the greening of dark-grown seedlings, grana formation in the etioplasts correlates with the period of rapid chlorophyll synthesis and the decline in the Chl *a*:Chl *b* ratio (Boardman *et al.*, 1971). Mutant chloroplasts often have less chlorophyll, a higher ratio of Chl *a*:Chl *b* and reduced membrane stacking as compared to wild-type chloroplasts (Schmid *et al.*, 1966; Highkin *et al.*, 1969; Keck *et al.*, 1970; Smillie *et al.*, 1975).

But there is no quantitative relationship between the amount of Chl *b* and the degree of membrane stacking. For example, bundle-sheath chloroplasts of maize and sorghum show little or no membrane stacking, but have a Chl *a*:Chl *b* ratio of 5–6 (Woo *et al.*, 1971), whereas a pea mutant with Chl *a*:Chl *b* ratio of 10–18 shows considerable stacking (Highkin *et al.*, 1969). It is concluded from studies of mutants of barley, which are devoid of Chl *b*, that Chl *b* is not essential for membrane stacking. Mutant *chlorina* 2 was reported to have about 50% of the degree of grana formation of the wild-type (Goodchild *et al.*, 1966), but three other Chl *b*-less mutants (which are allelic with *chlorina* 2; von Wettstein *et al.*, unpublished observations) were similar to the wild-type in thylakoid structure and grana formation (Sagromsky and Döbel, 1974). The lower degree of thylakoid stacking in *chlorina* 2 may have been due to the high light intensity under which the plants were grown.

B. LIGHT-HARVESTING CHLOROPHYLL–PROTEIN AND CHLOROPHYLL *b*

Plants grown under a high light intensity have a lower percentage of LHCP on SDS–PAGE than do low-light plants (Brown *et al.*, 1974, 1975). For example, high-light soybeans with a Chl *a*:Chl *b* ratio of 2.6 gave 57% of the chlorophyll in LHCP, while low-light plants had a ratio of 2.1 and 65% LHCP. Brown *et al.* (1974, 1975) concluded that all the Chl *b* can be accounted for by LHCP, and the Chl *a*: Chl *b* ratio is a reflection of the same amount of LHCP. However, in their calculations they assumed a Chl *a*:Chl *b* ratio of 1 for LHCP (as found with the purified complex) and the Chl *a*:Chl *b* ratio of the band on the gel was not reported. As mentioned earlier, the Chl *a*:Chl *b* ratio of the LHCP band can vary markedly, and Chl *b* is often found in the free-pigment zone. A higher percentage of LHCP correlated with a higher Chl:P700 ratio, or less P700 per unit of chlorophyll (Brown *et al.*, 1974, 1975). The extreme shade plants showed a higher Chl:P700 ratio than some species, but there was no significant variation of the Chl:P700 ratio of *Atriplex patula* grown under different light intensities (Boardman *et al.*, 1975).

The mutants of barley lacking Chl *b* are devoid of LHCP (Thornber and Highkin, 1974; Genge *et al.*, 1974; Anderson and Levine, 1974b; Henriques and Park, 1975; Machold *et al.*, 1977), and the chlorophyll-deficient tobacco mutant (SU/su) with a Chl *a*:Chl *b* ratio of 3.5–4 has little LHCP (Remy and Bebee, 1975). A similar finding was reported for *Chlamydomonas* mutants (ac-5 and ac-31) with little Chl *b*. Polypeptides 2b and 2c, which belong to LHCP (Anderson and Levine, 1974b), were missing from SDS extracts of the mutants (Levine and Duram, 1973). *Euglena gracilis* has a high Chl *a*:Chl *b* ratio and little LHCP (Genge *et al.*, 1974; Brown *et al.*, 1975), while blue-green and red algae, which lack Chl *b*, are devoid of LHCP (Thornber and Olson, 1971; Vernon and Klein, 1975). Grana are not present in these algae, although in *Euglena gracilis* the thylakoid membranes may be paired over long areas under high-salt conditions (Schwelitz *et al.*, 1972; Miller and Staehelin, 1973).

LHCP is absent from etioplasts, and its formation during greening of etiolated seedlings correlates with the rapid synthesis of Chl *b* and grana formation (Alberte *et al.*, 1972). Etiolated plants placed in alternating light–dark regimes, either 1-msec light flash–15-minute dark regime (Phung Nhu Hung *et al.*, 1972; Remy, 1973b; Hofer *et al.*, 1975; Remy and Bebee, 1975) or in 2-minute light–2-hour dark regime (Argyroudi-Akoyunoglou *et al.*, 1972; Hiller *et al.*, 1973; Akoyunoglou and Micheli-naki-Maneta, 1975; Armond *et al.*, 1976; Davis *et al.*, 1976) or under continuous far-red light (de Greef *et al.*, 1971; Oelze-Karow and Butler, 1971) show marked chloroplast development, but the membranes, which contain mainly Chl *a*, are usually primary thylakoids only. In some cases, it has been demonstrated that these primary thylakoids, which have PS I and PS II activities, have CP I, but not LHCP. Similarly, greening studies with *Chlamydomonas reinhardii y-1*, showed that membrane development and stacking are associated with a major protein, "L" (Eytan and Ohad, 1970, 1972), which corresponds to the polypeptides of the LHCP.

The foregoing studies indicate a good relationship between Chl *b* and LHCP. However, there are situations where the content of Chl *b* does not correspond with the amount of LHCP, as estimated by SDS–PAGE. Mature sorghum and maize bundle-sheath chloroplasts, which have very low membrane stacking (Laetsch, 1971) and Chl *a*:Chl *b* ratios between 4 and 6, have little or no LHCP (Leggett-Bailey *et al.*, 1971; Genge *et al.*, 1974; Anderson and Levine, 1974a). For maize bundle-sheath chloroplasts, 5.4% of the chlorophyll was in the SDS-PAGE band corresponding to LHCP (Genge *et al.*, 1974). If all the Chl *b* were accounted for by LHCP, some 30–40% of the chlorophyll should have been observed in the LHCP band. Photosystem I subchloroplast fragments obtained by French press or digitonin treatment (Chl *a*:Chl *b* of 5–

6) also gave a low yield of LHCP (Brown et al., 1974; Anderson, unpublished observations).

C. LARGE FREEZE-FRACTURE PARTICLES AND MEMBRANE STACKING

In the fluid lipid–protein mosaic model for the molecular organization of membrane components (Singer, 1971; Singer and Nicolson, 1972; Capaldi, 1974; Bretscher and Raff, 1975) the basic membrane consists of a lipid bilayer in which the intrinsic proteins are intercalated. Their hydrophobic regions are associated with the lipid fatty acid tails by hydrophobic interactions in the lipid bilayer, and their hydrophilic areas are associated with the lipid polar head groups at the membrane surface by electrostatic interactions. Intrinsic proteins may extend into the membrane, or actually extend across it, thereby having two hydrophilic areas, one at each membrane surface.

Since the polypeptides of the two main chlorophyll–protein complexes account for at least 70% of the intrinsic protein of thylakoid membrane (Section V, D), it would seem logical that they must be included in the intrinsic protein vizualized by freeze-fracture microscopy (Anderson, 1975a). In view of the association of large freeze-fracture particles with stacked-membrane areas, Anderson (1975a) proposed that large freeze-fracture particles include LHCP and small freeze-fracture particles include CP I.

The finding that large particles are confined to stacked membranes (Park and Sane, 1971) was the first indication that large particles might play an important role in membrane stacking (Section II, C). Studies with developing chloroplasts also showed a correlation between grana formation and the presence of large particles. The primary thylakoids of dark-grown seedlings exposed to intermittent light–dark cycles had few or no large particles (Remy et al., 1972; Armond et al., 1976). Subsequent exposure of the flashed seedlings to continuous white light led to grana formation, concomitant with an increased rate of chlorophyll synthesis and formation of LHCP. Despite considerable protein synthesis, no change in particle numbers was observed on the freeze-fracture faces, but there was an increase in the size of the particles on the EFs face (Armond et al., 1977). The particle size increased in discrete jumps from 8.0 nm for the flashed seedlings to 10.5, 13.2, and 16.4 nm, the large sizes being predominant after 48 hours of continuous light. Armond et al. (1977) suggested that the 8.0-nm particles seen in the plastids of flashed seedlings represent a "core" complex of PS II, and that during greening aggregates of LHCP are added to this "core" complex to form the "complete" PS II. The particles on the other fracture face (PF_s) increased from 7.0 nm for flashed seedlings to 7.7 nm during continuous illumination.

In view of the relationship between LHCP, grana formation, and large particles, several authors have suggested that the complex has a role in maintaining contact between the stacked membranes (Levine and Duram, 1973; Thornber and Highkin, 1974; Anderson and Levine, 1974b). Anderson (1975a) postulated that this complex would span the membrane and contain a suitable arrangement of negatively charged sites both in amino acid residues and the polar head groups of its boundary lipids at the outer membrane surface. Thus aggregates of this complex would be linked by divalent salt-bridge formation between identical units on adjacent membrane surfaces. Removal of salts would disrupt the salt bridges and cause the membranes to unstack. In the unstacked membranes it is proposed that LHCP would become homogeneously distributed among the membranes (Anderson, 1975a). This has already been demonstrated for the large particles seen by freeze-fracturing (Section II, C). LHCP was shown to bind divalent ions and it was suggested that it may also be involved in the regulation of energy distribution between the PS I and PS II (Arntzen et al., 1977).

While this seems to be an attractive hypothesis, the studies with the barley mutant lacking Chl b show that LHCP is not essential for membrane stacking. The barley mutant (chlorina 2) also shows some cation control of the distribution of energy between the photosystems (Boardman and Thorne, 1976). The observation of Goodchild (Section II, C) that the barley mutant has no large (16 nm) particles on the EF_s fracture face, but an equivalent number of smaller particles suggested a correlation between the lack of large particles in the mutant and its lack of Chl b and LHCP (Boardman et al., 1974, 1975). The barley mutant studies support the view that the large particles consist largely of aggregates of LHCP (Anderson, 1975a).

Although the barley mutant lacks LHCP, it seems likely that some of the Chl a molecules of the mutant are in a light-harvesting complex that substitutes for LHCP. The Chl:P700 ratio of the chlorina 2 mutant is two-thirds that of wild-type plants (Thornber and Highkin, 1974), suggesting that the mutant may contain about one-half of the light-harvesting chlorophyll of the wild type (Boardman and Thorne, 1976). Perhaps, this light-harvesting Chl a of the mutant is complexed in vivo to one of the two polypeptides of LHCP which may be present in the mutant (Section V,D), and the Chl a–protein complex is not stable to extraction of the mutant with SDS.

Maize bundle-sheath chloroplasts, which have a low proportion of LHCP, do not have the large particles on the EF_s fraction face (Goodchild, unpublished observations).

Freeze-fracture studies with sun and shade plants do not show a proportionality between LHCP and large particles (Goodchild et al., 1972; Boardman et al., 1975). The shade-plant chloroplasts with a greater proportion of Chl b had fewer particles per square micrometer on the large-particle fracture face (EF_s). The ratio of the number of the particles on the PF_s face to the EF_s face was 4.0 for the shade species (*Alocasia*) and 1.95 for the sun species (*Atriplex*), grown under a high light intensity. We have already seen that the large particles of stacked thylakoid membranes show a considerable size distribution. If the large particles of shade-plant chloroplasts were biased toward the higher sizes, it might account for the apparent discrepancy between the lower number of large particles per square micrometer of membrane surface and the higher proportion of Chl b. However, particle size measurements on shade-plant chloroplasts showed a similar distribution in size of large particles on the EF_s face to that of spinach chloroplasts (Goodchild, unpublished observations).

D. DIFFERENTIATION OF FUNCTION

From electron microscopic observations, Sane et al. (1970) proposed that the small fragments obtained by passage of chloroplasts through the French pressure cell are fragments of stroma thylakoids and the large fragments are grana that are not disrupted by the French press. Digitonin appears to have a similar initial action in fragmenting stroma thylakoids, but digitonin also release PS I fragments from the grana regions (Arntzen et al., 1972; Boardman, 1972).

Since the small particles obtained by digitonin or French press treatments are highly enriched in PS I (Boardman, 1970; Park and Sane, 1971), Sane et al. (1970) proposed that stroma thylakoids contain only PS I, whereas grana contain both PS I and PS II. Athough this is the simplest explanation of the fractionation experiments and is generally accepted, there remains the possibility that the PS I fragments are released from the stroma lamellae by the French press or digitonin treatments and the PS II of the stroma aggregates to the grana fraction.

It is known that stacked membranes are not essential for PS II activity, even in the higher plant, since the primary thylakoids of seedlings greened in intermittent light/dark cycles possess good PS II activity (Argyroudi-Akoyunoglou et al., 1972; Hiller et al., 1973; Remy, 1973b; Remy and Bebee, 1975; Hofer et al., 1975; Armond et al., 1976). Studies with barley mutants also indicate that stacking of thylakoids is not necessary for high rates of PS II activity (Smillie et al., 1975). The photochemical activities of sun and shade plants suggest that PS II is not confined to grana. The quantum efficiency of sun and shade plant chloroplasts was identical, even though shade plants contain a greater proportion of their thylakoids in grana (Björkman et al., 1972; Boardman, 1977b). On the other hand, shade plants saturate at a lower light intensity than sun plants, and their saturating activities are lower than for sun species. The constant quantum efficiency of sun and shade plants (determined at limiting light intensity) is easily explained if PS I and PS II are present in both grana and stroma thylakoids. The higher saturating intensities and higher activities of sun species could be due to their higher proportion of single stroma thylakoids compared with grana, since, in general, photosynthetic organisms with nonappressed thylakoids saturate at higher light intensities (Smillie *et al.*, 1975; Armond *et al.*, 1976).

Definitive evidence for or against the presence of PS II in stroma thylakoids awaits the development of a method for the localization of PS II activity *in vivo*. For the subsequent discussion, however, we follow the generally accepted view that there is differentiation of photochemical function between the stacked and unstacked membranes of a mature chloroplast, and that the small French press particles represent fragments of stroma lamellae.

The differentiation of structure and function during development of chloroplasts occurs later than the onset of PS I and PS II activities, and correlates with the increased rate of chlorophyll synthesis and the synthesis of LHCP.

It is perhaps rather surprising that the lipid contents (except the pigments) of the stroma fragments and grana are so very similar. Allen *et al.* (1972), using French pressure cell subchloroplast fragments, showed that the relative proportions of the galactolipids and phospholipids are similar for grana and stroma thylakoids. The fatty acid acylation patterns were virtually identical for each lipid. Furthermore, the total lipid:protein ratio of about 1 was the same for stroma and grana thylakoids. However, the Chl:total lipid ratios were different, the grana thylakoids (Chl *a*:Chl *b*, 2.3) containing about 1.4 times more chlorophyll than stroma thylakoids (Chl *a*:Chl *b*, 6). The carotenoid compositions of grana and stroma fragments were different, the grana containing a higher proportion of xanthophylls (Trosper and Allen, 1973).

The striking similarity of glycolipid and phospholipid composition in grana and stroma thylakoids supports the concept of a continuity of the fluid matrix of the membranes in the stacked and unstacked regions of the thylakoid system. The unstacking of grana thylakoids in low-salt media and the restacking on addition of cations (Izawa and Good, 1966) also point to the continuity of the stacked and unstacked membranes of the chloroplast thylakoid system.

Since the two membranes differ significantly in pigment and polypeptide compositions but not in other lipids, it seems a reasonable postulate that the pigment–protein complexes are responsible for the differentiation into grana and stroma. There were conflicting views earlier as to whether PS I of grana and stroma thylakoids were the same or different (see review in Arntzen and Briantais, 1975), but recent work indicates no differences in CP I in stacked and unstacked membranes, either in the Chl:P700 ratio or polypeptides (Brown *et al.*, 1975; Wessels and Borchert, 1975). Further, we have seen that LHCP is largely confined to the grana. It seems plausible, therefore, that, if PS II is confined to grana, LHCP is responsible for such a segregation.

VIII. Concluding Remarks

It seems highly likely that proteins play an important role in determining the environment of chlorophyll molecules in the chloroplast thylakoid membrane and the interactions of the pigments with one another. These interactions are crucial for the efficient transfer of excitation energy from the light-harvesting and antenna pigment molecules to the reaction centers. Seely (1973a) proposed that the different spectral forms of Chl *a* facilitate the transfer of excitation energy to the reaction center and allow the operation of larger light-harvesting units. Computer calculation of trapping probabilities on model arrays containing three spectral forms of light-harvesting chlorophyll together with reaction-center chlorophyll indicated that the trapping rate at the reaction center is increased 4–5 times by having the different spectral forms in the photosynthetic unit, rather than a single form of chlorophyll. The proper orientation of the pigment molecules may increase the trapping probability by a similar factor. For a model array of 344 chlorophyll molecules of seven spectral forms and two reaction-center chlorophylls, representing PS I and PS II, Seely (1973b) calculated a 15-fold increase in the trapping rate of excitation energy. In the Seely model, most of the chlorophyll molecules fell into two groups, which transfer their energy to one or other of the reaction centers. Computer calculations indicated that the rate of transfer of energy between the two reaction centers can be controlled by redirecting the orientation of only six of the chlorophyll molecules which occupy a key position in the array (Seely, 1973b).

The exact molecular interactions responsible for the different spectral forms of Chl *a* are unknown. Liganding between protein side chains and the chlorophyll rings, and exciton interactions between the chlorin rings, which are governed by the center-to-center distance of the rings and their orientations, are expected to influence the spectral properties of Chl *a*, but the relative importance of these two parameters is not determined. The reaction-center chlorophyll of PS I appears to consist of a special dimer of Chl *a*, but the most purified preparations of the reaction-center complex obtained so far contain at least 30 other Chl *a* molecules. These antenna Chl *a* molecules may play an important role in maintaining the specific orientations of the reaction-center chlorophylls.

The relationship between the protochlorophyllide–protein complex of the etioplast tubular membranes and the chlorophyll–protein complexes of the chloroplast thylakoid membranes is a subject for future study. We have seen that the pigment molecules in the etioplast are organized into energy-transferring units, although the size of the units appear to be considerably smaller than those of the chloroplast. It is not known whether chlorophyll molecules are added to preexisting units during the early stages of greening or whether new units are built from the preexisting pigment molecules and the newly synthesized chlorophyll molecules. However, the evidence strongly suggests that the assembly of units of LHCP, which is not required for photochemical activity of PS I or PS II, involves the synthesis of a unique polypeptide that is absent from etioplasts.

There is now compelling evidence that the chlorophyll–protein complexes that are isolated by detergent extraction of thylakoid membranes are not formed by nonspecific association of chlorophyll and thylakoid proteins. But the question remains whether the isolated complexes reflect exactly the organization of chlorophyll *in vivo*. New methods for the isolation and purification of the chlorophyll–protein complexes from the thylakoid membrane are required. Crystallization of the higher plant complexes would enable definitive determination of structure by X-ray crystallography.

Evidence presented in this review indicates that the chlorophyll–proteins are the main intrinsic proteins of the thylakoid membrane, but little is known about their location in the membrane. The relationship between the chlorophyll–protein complexes and the particles seen in freeze-fracture electron microscopy, and the proposed role for LHCP in membrane stacking, need further study.

REFERENCES

Akoyunoglou, G., and Michalopoulos, G. (1971). *Physiol. Plant.* **25**, 324–329.
Akoyunoglou, G., and Michelinaki-Maneta, M. (1975). *Proc. Int. Congr. Photosynth., 3rd Rehovot, 1974* **3**, 1885–1896.
Alberte, R. S., Thornber, J. P., and Naylor, A. W., (1972). *J. Exp. Bot.* **23**, 1060–1069.
Allen, C. F., Good, P., Trosper, T., and Park, R. B. (1972). *Biochem. Biophys, Res. Commun.* **48**, 907–913.
Anderson, J. M. (1975a). *Biochim. Biophys. Acta* **416**, 191–235.
Anderson, J. M. (1975b). *Nature (London)* **253**, 536–537.
Anderson, J. M., and Boardman, N. K. (1966). *Biochim. Biophys. Acta* **112**, 403–421.
Anderson, J. M., and Levine, R. P. (1974a). *Biochim. Biophys. Acta* **333**, 378–387.
Anderson, J. M., and Levine, R. P. (1974b). *Biochim. Biophys. Acta* **357**, 118–126.
Anderson, J. M., Goodchild, D. J., and Boardman, N. K. (1973). *Biochim. Biophys. Acta* **325**, 573–585.
Apel, K. (1977). *Brookhaven Symp. Biol.* **28**, 149–161

Apel, K., Bogorad, L., and Woodcock, C. L. F. (1975). *Biochim. Biophys. Acta* **387**, 568–579.

Apel, K., Miller, K. R., Bogorad, L., and Miller, G. J. (1976). *J. Cell Biol.* **71**, 876–893.

Argyroudi-Akoyunoglou, J. H., and Akoyunoglou, G. (1973). *Photochem. Photobiol.* **18**, 219–228.

Argyroudi-Akoyunoglou, J. H., Feleki, Z., and Akoyunoglou, G. (1972). *Proc. Int. Congr. Photosynth. Res., 2nd, Stresa, 1971* **3**, 2417–2426.

Armond, P. A., Arntzen, C. J., Briantais, J. M., and Vernotte, C. (1976). *Arch. Biochem. Biophys.* **175**, 54–63.

Armond, P. A., Staehelin, L. A., and Arntzen, C. J. (1977). *J. Cell Biol.* **73**, 400–418.

Arntzen, C. J., and Briantais, J. M. (1975). *In* "Bioenergetics of Photosynthesis" (Govindjee, ed.), pp. 51–113. Academic Press, New York.

Arntzen, C. J., Dilley, R. A., and Crane, F. L. (1969). *J. Cell Biol.* **43**, 16–31.

Arntzen, C. J., Dilley, R. A., Peters, G. A., and Shaw, E. R. (1972). *Biochim. Biophys. Acta* **256**, 85–107.

Arntzen, C. J., Armond, P. A., Briantais, J. M., Burke, J. J., and Novitzky, W. P. (1977). *Brookhaven Symp. Biol.* **28**, 316–337.

Bar-Nun, S., and Ohad, I. (1975). *Proc. Int. Congr. Photosynth., 3rd, Rehovot, 1974* **3**, 1627–1638.

Bar-Nun, S., Schantz, R., and Ohad, I. (1977). *Biochim. Biophys. Acta* **459**, 451–467.

Bearden, A. J., and Malkin, R. (1976). *Biochim. Biophys. Acta* **430**, 538–547.

Beck, D. P., and Levine, R. P. (1974). *J. Cell Biol.* **63**, 759–772.

Becker, J. F., Breton, J., Geacintov, N. E., and Trentacosti, F. (1976). *Biochim. Biophys. Acta* **440**, 531–544.

Bengis, C., and Nelson, N. (1975). *J. Biol. Chem.* **250**, 2783–2788.

Berzborn, R. J., Kopp, F., and Mühlethaler, K. (1974). *Z. Naturforsch. C* **29**, 694–699.

Björkman, O., Boardman, N. K., Anderson, J. M., Thorne, S. W., Goodchild, D. J., and Pyliotis, N. A. (1972). *Carnegie Inst. Washington, Year,* **71**, 115–135.

Boardman, N. K. (1962a). *Biochim. Biophys. Acta* **62**, 63–79.

Boardman, N. K. (1962b). *Biochim. Biophys. Acta* **64**, 279–293.

Boardman, N. K. (1966). *In* "The Chlorophylls" (L. P. Vernon and G. R. Seely, eds), pp. 437–479. Academic Press, New York.

Boardman, N. K. (1967). *In* "Harvesting the Sun" (A. San Pietro, F. A. Greer, and T. J. Army, eds), pp. 211–230. Academic Press, New York.

Boardman, N. K. (1970). *Annu. Rev. Plant Physiol.* **21**, 115–140.

Boardman, N. K. (1972). *Biochim. Biophys. Acta* **283**, 469–482.

Boardman, N. K. (1977a). *In* "Encyclopedia of Plant Physiology" (A. Trebst and M. Avron, eds.), Vol. 5., pp. 583–600. Springer-Verlag, Berlin and New York.

Boardman, N. K. (1977b). *Annu. Rev. Plant Physiol.* **28**, 355–377.

Boardman, N. K., and Anderson, J. M. (1964). *Nature (London)* **203**, 166–167.

Boardman, N. K., and Highkin, H. R. (1966). *Biochim. Biophys. Acta* **126**, 189–199.

Boardman, N. K., and Thorne, S. W. (1968). *Biochim. Biophys. Acta* **153**, 448–458.

Boardman, N. K., and Thorne, S. W. (1976). *Plant Sci. Lett.* **7**, 219–224.

Boardman, N. K., Thorne, S. W., and Anderson, J. M. (1966). *Proc. Natl. Acad. Sci. U.S.A.* **56**, 586–593.

Boardman, N. K., Anderson, J. M., Kahn, A., Thorne, S. W., and Treffry, T. E. (1971). *In* "Autonomy and Biogenesis of Mitochondria and Chloroplasts" (N. K. Boardman, A. W. Linnane, and R. M. Smillie, eds.), pp. 70–84. North-Holland Publ., Amsterdam.

Boardman, N. K., Anderson, J. M., Björkman, O., Goodchild, D. J., Grimme, L. H., and Thorne, S. W. (1974). *Port. Acta Biol. Ser. A* **14**, 213–236.

Boardman, N. K., Björkman, O., Anderson, J. M., Goodchild, D. J., and Thorne, S. W. (1975). *Proc. Int. Congr. Photosynth., 3rd, Rehovot, 1974* **3**, 1809–1827.

Bogorad, L., Laber, L., and Gassman, M. (1968). In "Comparative Biochemistry and Biophysics of Photosynthesis" (K. Shibata, A. Takamiya, A. T. Jagendorf, and R. C. Fuller, eds.), pp. 299–312. Univ. Park Press, State College, Pennsylvania.

Bovey, F., Ogawa, T., and Shibata, K. (1974). *Plant Cell Physiol.* **15**, 1133–1137.

Bradbeer, J. W., Ireland, H. M. M., Smith, J. W., Rest, J., and Edge, H. J. W. (1974a). *New Phytol.* **73**, 263–270.

Bradbeer, J. W., Gyldenholm, A. O., Ireland, H. M. M., Smith, J. W., Rest, J., and Edge, H. J. W. (1974b). *New Phytol.* **73**, 271–279.

Branton, D. (1973). In "Freeze-Etching Techniques and Applications" (E. L. Benedetti and P. V. Favard, eds.), pp. 107–112. Soc. Fr. Micros. Electron., Paris.

Branton, D., and Park, R. B. (1967). *J. Ultrastruct. Res.* **19**, 283–303.

Branton, D., Bullivant, S., Gilula, N. B., Karnovsky, M. J., Moor, H., Mühlathaler, K., Northcote, D. H., Packer, L., Satir, B., Satir, P., Speth, V., Staehelin, L. A., Steer, R. L., and Weinstein, R. S. (1975). *Science* **190**, 54–56.

Breton, J., Michel-Villaz, M., and Paillotin, G. (1973). *Biochim. Biophys. Acta* **314**, 42–56.

Bretscher, M. S., and Raff, M. C. (1975). *Nature (London)* **258**, 43–49.

Brody, M., and Nathanson, B. (1972). *Biophys. J.* **12**, 774–790.

Bronchart, R. (1970). *C. R. Acad. Sci., Ser. D* **270**, 1789–1791.

Brown, J. S. (1973). In "Photophysiology" (A. C. Giese, ed.), Vol. 8, pp. 97–112. Academic Press, New York.

Brown, J. S., Alberte, R. S., Thornber, J. P., and French, C. S. (1974). *Carnegie Inst. Washington, Year.* **73**, 694–706.

Brown, J. S., Alberte, R. S., and Thornber, J. P. (1975). *Proc. Int. Congr. Photosynth., 3rd, Rehovot, 1974* **3**, 1951–1962.

Bullivant, S. (1974). *Phil. Trans. R. Soc. London, Ser. B* **268**, 5–14.

Butler, W. L. (1961a). *Arch. Biochem. Biophys.* **92**, 287–295.

Butler, W. L. (1961b). *Arch. Biochem. Biophys.* **93**, 413–422.

Butler, W. L., and Briggs, W. R. (1966). *Biochim. Biophys. Acta* **112**, 45–53.

Butler, W. L., and Kitajima, M. (1975a). *Biochim. Biophys. Acta* **376**, 105–115.

Butler, W. L., and Kitajima, M. (1975b). *Biochim. Biophys. Acta* **396**, 72–85.

Butler, W. L., and Kitajima, M. (1975c). *Proc. Int. Congr. Photosynth., 3rd, Rehovot, 1974* **1**, 13–24.

Capaldi, R. A. (1974). *Sci. Am.* **230**, 26–33.

Cherry, R. J., Kwan Hsu, and Chapman, D. (1972). *Biochim. Biophys. Acta* **267**, 512–522.

Chua, N. H., and Bennoun, P. (1975). *Proc. Natl. Acad. Sci. U.S.A.* **72**, 2175–2179.

Chua, N. H., Matlin, K., and Bennoun, P. (1975). *J. Cell Biol.* **67**, 361–377.

Clark, A. W., and Branton, D. (1968). *Z. Zellforsch. Mikrosk. Anat.* **91**, 586–603.

Cobb, A. H., and Wellburn, A. R. (1973). *Planta* **114**, 131–142.

Colbow, K. (1973). *Biochim. Biophys. Acta* **314**, 320–327.

Cotton, T. M., Trifunae, A. D., Ballschmiter, K., and Katz, J. J. (1974). *Biochim. Biophys. Acta* **368**, 181–198.

Davis, D. J., Armond, P. A., Gross, E. L., and Arntzen, C. J. (1976). *Arch. Biochem. Biophys.* **175**, 64–70.

de Greef, J., Butler, W. L., and Roth, T. F. (1971). *Plant Physiol.* **47**, 457–464.

Dietrich, W. E., and Thornber, J. P. (1971). *Biochim. Biophys. Acta* **245**, 482–493.

Döring, G., Bailey, J. L., Kreutz, W., Weikard, J., and Witt, H. T. (1968). *Naturwissenschaften* **55**, 219–220.

Douce, R., Holtz, R. B., and Benson, A. A. (1973). *J. Biol. Chem.* **248**, 7215–7222.

Dratz, E. A., Schultz, A. J., and Sauer, K. (1966). *Brookhaven Symp. Biol.* **19**, 303–318.

Duysens, L. N. M., and Sweers, H. E. (1963). In "Microalgae and Photosynthetic Bacteria" (Japanese Society Plant Physiologists, eds.), pp. 353–372. Univ. of Tokyo Press, Tokyo.

Eaglesham, A. R. J., and Ellis, R. J. (1974). Biochim. Biophys. Acta 335, 396–407.

Eriksson, G., Kahn, A., Walles, B., and von Wettstein, D. (1961). Ber. Dtsch. Bot. Ges. 74, 221–232.

Eytan, G., and Ohad, I. (1970). J. Biol. Chem. 245, 4297–4307.

Eytan, G., and Ohad, I. (1972). J. Biol. Chem. 247, 112–121.

Faludi-Dániel, A., Demeter, S., and Garay, A. S. (1973). Plant Physiol. 52, 54–56.

Fenna, R. E., and Matthews, B. W. (1975). Nature (London) 258, 573–577.

Fenna, R. E., and Matthews, B. W. (1977). Brookhaven Symp. Biol. 28, 170–182.

Förster, T. (1959). Discuss. Faraday Soc. 27, 7–17.

Fong, F. K. (1974). Proc. Natl. Acad. Sci. U.S.A. 71, 3692–3695.

Foster, R. J., Gibbons, G. C., Gough, S., Henningsen, K. W., Kahn, A., Nielsen, O. F., and von Wettstein, D. (1971). Proc. Eur. Biophys. Congr. 1st 4, 137.

Fradkin, L. I., Shlyk, A. A., and Kolyago, V. M. (1966). Dokl. Akad. Nauk SSSR 171, 222–225.

Fradkin, L. I., Shlyk, A. A., Kalinina, L. M., and Faludi-Dániel, A. (1969). Photosynthetica 3, 326–337.

French, C. S., Brown, J. S., and Lawrence, M. C. (1972). Plant Physiol. 49, 421–429.

Garay, A., Demester, S., Kovács, K., Horvath, G., and Faludi-Dániel, A. (1972). Photochem. Photobiol. 16, 139–144.

Garber, M. P., and Steponkus, P. L. (1974). J. Cell Biol. 63, 24–34.

Gassman, M. (1973). Plant Physiol. 52, 590–594.

Geacintov, N. E., Van Nostrand, F., Beker, J. F., and Tinkel, J. B. (1972). Biochim. Biophys. Acta 267, 65–79.

Genge, S., Pilger, D., and Hiller, R. G. (1974). Biochim. Biophys. Acta 347, 22–30.

Goodchild, D. J., Highkin, H. R., and Boardman, N. K. (1966). Exp. Cell Res. 43, 684–688.

Goodchild, D. J., Björkman, O., and Pyliotis, N. A. (1972). Carnegie Inst. Washington, Year. 71, 102–107.

Goodenough, U. W., and Staehelin, L. A. (1971). J. Cell Biol. 48, 594–619.

Govindjee, ed. (1975). "Bioenergetics of Photosynthesis." Academic Press, New York.

Grefrath, S. P., and Reynolds, J. A. (1974). Proc. Natl. Acad. Sci. U.S.A. 71, 3913–3916.

Gregory, R. P. F. (1975). Biochem. J. 148, 487–497.

Gregory, R. P. F., and Raps, S. (1974). Biochem. J. 142, 193–201.

Gregory, R. P. F., and Raps, S. (1975). Proc. Int. Congr. Photosynth., 3rd, Rehovot, 1974 3, 1977–1982.

Gregory, R. P. F., Raps, S., and Bertsch, W. (1971). Biochim. Biophys. Acta 234, 330–334.

Gregory, R. P. F., Raps, S., Thornber, J. P., and Bertsch, W. F. (1972). Proc. Int. Congr. Photosynth., 2nd, 1971 2, 1503–1508.

Guignery, G., Luzzati, A., and Duranton, J. (1974). Planta 115, 227–243.

Gunning, B. E. S. (1965). Protoplasma 60, 111–130.

Gunning, B. E. S., and Jagoe, M. P. (1967). In "The Biochemistry of Chloroplasts" (T. W. Goodwin, ed.), Vol. 2, pp. 665–676. Academic Press, New York.

Gunning, B. E. S., and Steer, M. (1975). "Ultrastructure and the Biology of Plant Cells." Edward Arnold, London.

Gyldenholm, A. O., and Whatley, F. R. (1968). New Phytol. 67, 461–468.

Henningsen, K. W. (1970). J. Cell Sci. 7, 587–621.

Henningsen, K. W., and Boynton, J. E. (1969). J. Cell Sci. 5, 757–793.

Henningsen, K. W., and Boynton, J. E. (1974). J. Cell Sci. 15, 31–55.

Henningsen, K. W., and Kahn, A. (1971). *Plant Physiol.* **47**, 685–690.
Henningsen, K. W., and Thorne, S. W. (1974). *Physiol. Plant.* **30**, 82–89.
Henningsen, K. W., Kahn, A., and Houssier, C. (1973). *FEBS Lett.* **37**, 103–108.
Henningsen, K. W., Thorne, S. W., and Boardman, N. K. (1974). *Plant Physiol.* **53**, 419–425.
Henriques, F., and Park, R. B. (1975). *Plant Physiol.* **55**, 763–767.
Henriques F., and Park, R. B. (1976a). *Biochim. Biophys. Acta* **430**, 312–320.
Henriques, F., and Park, R. B. (1976b). *Arch. Biochem. Biophys.* **176**, 472–478.
Henriques, F., Vaughan, W., and Park, R. (1975). *Plant Physiol.* **55**, 338–339.
Herrmann, F. H. (1971). *FEBS Lett.* **19**, 267–269.
Herrmann, F. H., Yusupova, G. A., Giller, Y. E., and Börner, T. (1974). *Stud. Biophys.* **46**, 9–12.
Heslop-Harrison, J. (1963). *Planta* **60**, 243–260.
Highkin, H. R., and Frenkel, A. W. (1962). *Plant Physiol.* **37**, 814–820.
Highkin, H. R., Boardman, N. K., and Goodchild, D. J. (1969). *Plant Physiol.* **44**, 1310–1320.
Hiller, R. G., Pilger, D., and Genge, S. (1973). *Plant Sci. Lett.* **1**, 81–88.
Hiller, R. G., Genge, S., and Pilger, D. (1974). *Plant Sci. Lett.* **2**, 239–242.
Hofer, I., Strasser, R. J., and Sironval, C. (1975). *Proc. Int. Congr. Photosynth., 3rd, Rehovot, 1974* **3**, 1685–1690.
Hoober, J. K. (1970). *J. Biol. Chem.* **245**, 4327–4334.
Hoober, J. K. (1972). *J. Cell Biol.* **52**, 84–96.
Horton, P., and Leech, R. M. (1972). *FEBS Lett.* **26**, 277–280.
Horton, P., and Leech, R. M. (1975). *Plant Physiol.* **55**, 393–400.
Ikeda, T. (1968). *Shokubutsugaku Zasshi* **81**, 517–527.
Izawa, S., and Good, N. E. (1966). *Plant Physiol.* **41**, 544–552.
Jacobi, G., and Lehmann, H. (1968). *Z. Pflanzenphysiol.* **59**, 457–476.
Jeffrey, S. W., Douce, R., and Benson, A. A. (1974). *Proc. Natl. Acad. Sci. U.S.A.* **71**, 807–810.
Joy, K. W., and Ellis, J. (1975). *Biochim. Biophys. Acta* **378**, 143–151.
Junge, W. (1975). *Proc. Int. Congr. Photosynth., 3rd, Rehovot, 1974* **1**, 273–286.
Junge, W., and Eckhof, A. (1974). *Biochim. Biophys. Acta* **357**, 103–117.
Kahn, A. (1968). *Plant Physiol.* **43**, 1769–1780.
Kahn A., and Nielsen, O. F. (1974). *Biochim. Biophys. Acta* **333**, 409–414.
Kahn, A., Boardman, N. K., and Thorne, S. W. (1970). *J. Mol. Biol.* **48**, 85–101.
Kan, K. S., and Thornber, J. P. (1976). *Plant Physiol.* **57**, 47–52.
Katz, J. J., and Norris, J. R. (1973). *Curr. Top. Bioenerg.* **5**, 41–75.
Kawashima, N., and Wildman, S. G. (1970). *Annu. Rev. Plant Physiol.* **21**, 325–358.
Ke, B., and Vernon, L. P. (1967). *Biochemistry* **6**, 2221–2226.
Ke, B., Sahu, S., Shaw, E., and Beinert, H. (1974). *Biochim. Biophys. Acta* **347**, 36–48.
Ke, B., Sugahara, K., and Shaw, E. R. (1975). *Biochim. Biophys. Acta* **408**, 12–25.
Keck, R. W., Dilley, R. A., Allen, C. F., and Biggs, S. (1970). *Plant Physiol.* **46**, 692–698.
Kirk, J. T. O. (1970). *Annu. Rev. Plant Physiol.* **21**, 11–42.
Kirk, J. T. O. (1971). *Annu. Rev. Biochem.* **40**, 161–196.
Kirk, J. T. O., and Tilney-Bassett, R. A. E. (1967). "The Plastids." Freeman, San Francisco, California.
Klein, S. (1962). *Nature (London)* **196**, 992–993.
Klein, S., and Schiff, J. A. (1972). *Plant Physiol.* **49**, 619–626.
Klein, S. M., and Vernon, L. P. (1974). *Photochem. Photobiol.* **19**, 43–49.
Klein, S., Bryan, G., and Bogorad, L. (1964). *J. Cell Biol.* **22**, 433–451.
Krasnovsky, A. A., and Kosobutskaya, L. M. (1952). *Dokl. Akad. Nauk SSSR* **85**, 177.

Kung, S. D. and Thornber, J. P. (1971). *Biochim. Biophys. Acta* **253**, 285–289.
Laetsch, W. M. (1971). In "Photosynthesis and Photorespiration" (M. D. Hatch, C. B. Osmond, and R. O. Slatyer, eds.), pp. 323–349. Wiley (Interscience), New York.
Laetsch, W. M. (1974). *Annu. Rev. Plant Physiol.* **25**, 27–52.
Leggett-Bailey, J., and Kreutz, W. (1969). *Proc. Int. Congr. Photosynth. Res., 1st, 1968* **1**, 149–158.
Leggett-Bailey, J., Downton, W. J. S., and Masiar, E. (1971). In "Photosynthesis and Photorespiration" (M. D. Hatch, C. B. Osmond, and R. O. Slatyer, eds.), pp. 382–386. Wiley (Interscience), New York.
Lemoine, Y. (1968). *J. Microsc. (Paris)* **7**, 755–770.
Levine, R. P., and Duram, H. A. (1973). *Biochim. Biophys. Acta* **325**, 565–572.
Levine, R. P., Burton, W. G., and Duram, H. A. (1972). *Nature (London)* **237**, 176–177.
Liljenberg, C. (1974). *Physiol. Plant.* **32**, 208–213.
Litvin, F. F., and Belyaeva, O. B. (1968). *Biokhimiya* **33**, 928–936.
Litvin, F. F., and Belyaeva, O. B. (1971). *Photosynthetica* **5**, 200–209.
Litvin, F. F., and Krasnovsky, A. A. (1957). *Dokl. Akad. Nauk SSSR* **117**, 106.
Lutz, C. (1975). *Z. Pflanzenphysiol.* **75**, 346–359.
McCarty, R. E., and Jagendorf, A. T. (1965). *Plant Physiol.* **40**, 725–735.
Machold, O. (1971). *Biochim. Biophys. Acta* **238**, 324–331.
Machold, O. (1972). *Biochem. Physiol. Pflanzen.* **163**, 30–41.
Machold, O. (1974). *Biochem. Physiol. Pflanzen.* **166**, 149–162.
Machold, O. (1975). *Biochim. Biophys. Acta* **382**, 495–505.
Machold, O., Meister, H., Sagromsky, H., Hoeyer-Hansen, G., and von Wettstein, D. (1977). *Photosynthetica,* **11**, 200–206.
Mackender, R. O., and Leech, R. M. (1974). *Plant Physiol.* **53**, 496–502.
Malkin, R. (1975). *Arch. Biochem. Biophys.* **169**, 77–83.
Mathis, P., and Sauer, K. (1972). *Biochim. Biophys. Acta* **267**, 498–511.
Mathis, P., and Sauer, K. (1973). *Plant Physiol.* **51**, 115–119.
Mathis, P., Breton, J., Vermeglio, A., and Yates, M. (1976). *FEBS Lett.* **63**, 171–174.
Menke, W. (1962). *Annu. Rev. Plant Physiol.* **13**, 27–44.
Michel, J. M., and Michel-Wolwertz, M. R. (1969). *Prog. Photosynth. Res.* **1**, 115–127.
Miller, K. R. (1976). *J. Ultrastruct. Res.* **54**, 159–167.
Miller, K. R., and Staehelin, L. A. (1973). *Protoplasma* **77**, 55–78.
Miller, K. R., and Staehelin, L. A. (1976). *J. Cell Biol.* **68**, 30–47.
Mohanty, P., Braun, B. Z., Govindjee, and Thornber, J. P. (1972). *Plant Cell Physiol.* **13**, 81–91.
Mühlethaler, K. (1972). *Proc. Int. Congr. Photosynth. Res., 2nd, 1971* **2**, 1423–1429.
Murata, N. (1968). *Biochim. Biophys. Acta* **162**, 106–121.
Murray, A. E., and Klein, A. E. (1971). *Plant Physiol.* **48**, 383–388.
Nadler, K., and Granick, S. (1970). *Plant Physiol.* **46**, 240–246.
Nakamura, K., Ogawa, T., and Shibata, K. (1976). *Biochim. Biophys. Acta* **423**, 227–236.
Nelson, N., and Bengis, C. (1975). *Proc. Int. Congr. Photosynth., 3rd, Rehovot, 1974* **1**, 609–620.
Nelson, N., and Racker, E. (1972). *J. Biol. Chem.* **247**, 3848–3853.
Nelson, N., Deters, D. W., Nelson, H., and Racker, E. (1973). *J. Biol. Chem.* **248**, 2049–2055.
Nielsen, G. (1975). *Eur. J. Biochem.* **50**, 611–623.
Nielsen, O. F. (1975). *Biochem. Physiol. Pflanzen.* **167**, 195–206.
Nielsen, O. F., and Kahn, A. (1973). *Biochim. Biophys. Acta* **292**, 117–129.
Nir, I., and Pease, D. C. (1973). *J. Ultrastruct. Res.* **42**, 534–550.
Nolan, W. G., and Park, R. B. (1975). *Biochim. Biophys. Acta* **375**, 406–421.
Oelze-Karow, H., and Butler, W. L. (1971). *Plant Physiol.* **48**, 621–625.
Ogawa, T., and Vernon, L. P. (1969). *Biochim. Biophys. Acta* **180**, 334–336.

Ogawa, T., Obata, F., and Shibata, K. (1966). *Biochim. Biophys. Acta* **112**, 223–234.
Ogawa, T., Bovey, F., Inoue, Y., and Shibata, K. (1975). *Proc. Int. Congr. Photosynth., 3rd, Rehovot, 1974* **3**, 1829–1832.
Ojakian, G. K., and Satir, P. (1974). *Proc. Natl. Acad. Sci. U.S.A.* **71**, 2052–2056.
Oleszko, S., and Moudrianakis, E. N. (1974). *J. Cell Biol.* **63**, 936–948.
Olson, J. M. (1978). *In* "The Photosynthetic Bacteria" (R. K. Clayton and W. R. Sistrom, eds.). Plenum Press, New York, in press.
Olson, R. A., Jennings, W. H., and Butler, W. L. (1964a). *Biochim. Biophys. Acta* **88**, 318–330.
Olson, R. A., Jennings, W. H., and Butler, W. L. (1964b). *Biochim. Biophys. Acta* **88**, 331–337.
Paolillo, D. J. (1970). *J. Cell Sci.* **6**, 243–255.
Park, R. B., and Biggins, J. (1964). *Science* **144**, 1009–1011.
Park, R. B., and Pfeifhofer, A. O. (1968). *Proc. Natl. Acad. Sci. U.S.A.* **60**, 337–343.
Park, R. B., and Pfeifhofer, A. O. (1969). *J. Cell Sci.* **5**, 299–311.
Park, R. B., and Sane, P. V. (1971). *Annu. Rev. Plant Physiol.* **22**, 395–430.
Penland, J. C., and Aldrich, H. C. (1973). *J. Cell Biol.* **57**, 306–314.
Philipson, K. D., and Sauer, K. (1973). *Biochemistry* **12**, 3454–3458.
Philipson, K. D., Sato, V. L., and Sauer, K. (1972). *Biochemistry* **11**, 4591–4595.
Phung Nhu Hung, S., Lacourly, A., and Sarda, C. (1970). *Z. Pflanzenphysiol.* **62**, 1–16.
Phung Nhu Hung, S., Remy, E., and Moyse, A. (1972). *Proc. Int. Congr. Photosynth. Res., 2nd, Stresa, 1971* **3**, 2407–2416.
Poincelot, R. P. (1973). *Arch. Biochem. Biophys.* **159**, 134–142.
Raps, S., and Gregory, R. P. F. (1975). *Proc. Int. Congr. Photosynth., 3rd, Rehovot, 1974* **3**, 1983–1990.
Remy, R. (1969). *C. R. Acad Sci., Ser. D* **268**, 3057–3060.
Remy, R. (1971). *FEBS Lett.* **13**, 313–317.
Remy, R. (1973a) *FEBS Lett.* **31**, 308–312.
Remy, R. (1973b). *Photochem. Photobiol.* **18**, 409–416.
Remy, R., and Bebee, G. (1975). *Proc. Int. Congr. Photosynth., 3rd, Rehovot, 1974* **3**, 1675–1685.
Remy, R., Hoarau, J., and Lelerc, J. C. (1977). *Photochem. Photobiol.* **26**, 151–158.
Remy, R., Phung Nhu Hung, S., and Moyse, A. (1972). *Physiol. Veg.* **10**, 269–290.
Robertson, D., and Laetsch, W. M. (1974). *Plant Physiol.* **54**, 148–159.
Rosinski, J., and Rosen, W. G. (1972). *Q. Rev. Biol.* **47**, 160–191.
Sagromsky, H., and Döbel, P. (1974). *Biochem. Physiol. Pflanzen.* **166**, 371–376.
Sane, P. V., Goodchild, D. J., and Park, R. B. (1970). *Biochim. Biophys. Acta* **216**, 162–178.
Sauer, K. (1975). *In* "Bioenergetics of Photosynthesis" (Govindjee, ed.), pp. 115–181. Academic Press, New York.
Schmid, G. H., Price, J. M., and Jaffron, H. (1966). *J. Microsc. (Paris)* **5**, 205–212.
Schopfer, P., and Siegelman, H. W. (1968). *Plant Physiol.* **43**, 990–996.
Schultz, A., and Sauer, K. (1972). *Biochim. Biophys. Acta* **267**, 320–340.
Schwelitz, F. D., Dilley, R. A., and Crane, F. L. (1972). *Plant Physiol.* **50**, 166–170.
Scott, B., and Gregory, R. P. F. (1975). *Biochem. J.* **149**, 341–347.
Seely, G. R. (1973a). *J. Theor. Biol.* **40**, 173–187.
Seely, G. R. (1973b). *J. Theor. Biol.* **40**, 189–198.
Seliskar, C. J., and Ke, B. (1968). *Biochim. Biophys. Acta* **153**, 685–691.
Shibata, K. (1957). *J. Biochem. (Tokyo)* **44**, 147–173.
Shiozawa, J. A., Alberte, R. S., and Thornber, J. P. (1974). *Arch. Biochem. Biophys.* **165**, 385–397.
Shipman, L. L., Cotton, T. M., Norris, J. R., and Katz, J. J. (1976). *Proc. Natl. Acad. Sci. U.S.A.* **73**, 1791–1794.

Shlyk, A. A. (1971). *Annu. Rev. Plant Physiol.* **22**, 169–184.
Singer, S. J. (1971). *In* "Structure and Function of Biological Membranes" (L. I. Rothfield, ed.), pp. 145–222. Academic Press, New York.
Singer, S. J., and Nicolson, G. L. (1972). *Science* **175**, 720–731.
Sironval, C., Michel-Wolwertz, M. R., and Madsen, A. (1965). *Biochim. Biophys. Acta* **94**, 344–354.
Sironval, C., Clijsters, H., Michel, J.-M., Bronchart, R., and Michel-Wolwertz, M.-R. (1967). *In* "Le Chloroplaste" (C. Sironval, ed,), pp. 99–123. Masson, Paris.
Smillie, R. M., Nielsen, N. C., Henningsen, K. W., and von Wettstein, D. (1975). *Proc. Int. Congr. Photosynth., 3rd, Rehovot, 1974* **3**, 1841–1860.
Sprey, B., and Laetsch, W. M. (1976). *Z. Pflanzenphysiol.* **78**, 360–371.
Steck, T. L. (1974). *J. Cell Biol.* **62**, 1–19.
Strotmann, H. H., Hesse, K., and Edlemann, K. (1973). *Biochim. Biophys. Acta* **314**, 202–210.
Stumpf, P. K. (1975). *In* "Recent Advances in the Chemistry and Biochemistry of Plant Lipids" (T. Galliard and E. I. Mercer eds.), pp. 95–113. Academic Press, New York.
Süss, K. H., Schmidt, O., and Machold, O. (1976). *Biochim. Biophys. Acta* **448**, 103–113.
Süzer, S., and Sauer, K. (1971). *Plant Physiol.* **48**, 60–63.
Sundquist, C. (1970). *Physiol. Plant.* **23**, 412–424.
Sundquist, C. (1973). *Physiol. Plant.* **28**, 464–470.
Thornber, J. P. (1969). *Biochim. Biophys. Acta* **172**, 230–241.
Thornber, J. P. (1975). *Annu. Rev. Plant Physiol.* **26**, 127–158.
Thornber, J. P., and Alberte, R. S. (1976). *In* "The Enzymes of Biological Membranes" (A. Martonosi, ed.), Vol. 3, pp. 163–190. Plenum, New York.
Thornber, J. P., and Highkin, H. R. (1974). *Eur. J. Biochem.* **41**, 109–116.
Thornber, J. P., and Olson, J. M. (1971). *Photochem. Photobiol.* **14**, 329–341.
Thornber, J. P., Gregory, R. P. F., Smith, C. A., and Leggett-Bailey, J. (1967a). *Biochemistry* **6**, 391–396.
Thornber, J. P., Stewart, J. C., Hatton, M. W. C., and Leggett-Bailey, J. (1967b). *Biochemistry* **6**, 2006–2014.
Thornber, J. P., Alberte, R. S., Hunter, F. A., Shiozawa, J. A., and Kan, K. S. (1977). *Brookhaven Symp. Biol.* **28**, 132–148.
Thornber, J. P., Trosper, T. L., and Strause, C. E. (1978). *In* "The Photosynthetic Bacteria" (R. K. Clayton and W. R. Sistron, eds.), in press. Plenum, New York.
Thorne, S. W. (1971a). *Biochim. Biophys. Acta* **226**, 113–127.
Thorne, S. W. (1971b). *Biochim. Biophys. Acta* **226**, 128–134.
Thorne, S. W., and Boardman, N. K. (1971). *Plant Physiol.* **47**, 252–261.
Thorne, S. W., and Boardman, N. K. (1972). *Biochim. Biophys. Acta* **267**, 104–110.
Trebst, A. (1974). *Annu. Rev. Plant Physiol.* **25**, 423–458.
Treffry, T. (1970). *Planta* **91**, 279–284.
Trosper, T., and Allen, C. F. (1973). *Plant Physiol.* **51**, 584–585.
Tweet, A. G., Gaines, G. L., and Bellamy, W. D. (1964). *Nature (London)* **202**, 696–697.
Vaughan, G. D., and Sauer, K. (1974). *Biochim. Biophys. Acta* **347**, 383–394.
Vernon, L. P., and Klein, S. M. (1975). *Ann. N. Y. Acad. Sci.* **244**, 281–296.
Vernon, L. P., Shaw, E. R., and Ke, B. (1966). *J. Biol. Chem.* **241**, 4101–4109.
Vernon, L. P., Shaw, E. R., Ogawa, T., and Raveed, D. (1971). *Photochem. Photobiol.* **14**, 343–357.
Virgin, H. I. (1960). *Physiol. Plant.* **13**, 155–164.
Virgin, H. I., Kahn, A., and von Wettstein, D. (1963). *Photochem. Photobiol.* **2**, 83–91.
von Wettstein, D. (1958). *Brookhaven Symp. Biol.* **11**, 138–159.
Vrendenberg, W. J., and Slooten, L. (1967). *Biochim. Biophys. Acta* **143**, 583–594.

Wang, A. Y. I., and Packer, L. (1973). *Biochim. Biophys. Acta* **305**, 488–492.
Wallach, D. F. H., and Winzler, R. J. (1974). "Evolving Strategies and Tactics in Membrane Research." Springer-Verlag, Berlin and New York.
Wehrli, E., Mühlethaler, K., and Moor, H. (1970). *Exp. Cell Res.* **59**, 336–339.
Wehrmeyer, W. (1965a). *Z. Naturforsch. B* **20**, 1270–1278.
Wehrmeyer, W. (1965b). *Z. Naturforsch. B* **20**, 1278–1288.
Weier, T. E., and Brown, D. J. (1970). *Am. J. Bot.* **57**, 267–275.
Weier, T. E. Stoland, R. D., and Brown, D. L. (1974). *Am. J. Bot.* **57**, 276–284.
Wellburn, A. R., and Cobb, A. H. (1975). *Proc. Int. Congr. Photosynth., 3rd, Rehovot, 1974* **3**, 1647–1673.
Wellburn, F. A. M., and Wellburn, A. R. (1971). *J. Cell Sci.* **9**, 271–287.
Wessels, J. S. C., and Borchert, M. T. (1975). *Proc. Int. Congr. Photosynth., 3rd, Rehovot, 1974* **1**, 473–484.
Wessels, J. S. C., Van Alphen-van Waveren, O., and Voorn, G. (1973). *Biochim. Biophys. Acta* **292**, 741–752.
Whitmarsh, J., and Levine, R. P. (1974). *Biochim. Biophys. Acta* **368**, 199–213.
Wolff, J. B., and Price, L. (1957). *Arch. Biochem. Biophys.* **72**, 293–301.
Woo, K. C., Pyliotis, N. A., and Downton, W. J. S. (1971). *Z. Pflanzenphysiol.* **64**, 400–413.
Yamamoto, H. Y., and Vernon, L. P. (1969). *Biochemistry* **8**, 4131–4137.

Dynamic Structural Features of Chloroplast Lamellae

CHARLES J. ARNTZEN
USDA/ARS; Department of Botany
University of Illinois
Urbana, Illinois

I. Introduction

Comprehensive analysis of the photochemical processes catalyzed by chlorophyll during photosynthesis requires a detailed understanding of the structural organization of the chloroplast membranes. Since the functional components catalyzing the "light reactions" are membrane bound, many investigators have attempted to understand the biosynthesis, turnover and molecular architecture of the lamellar system. Over the last two and one-half decades these studies have led to numerous membrane models. Many early models, and some more recent ones, have suggested somewhat fixed or rigid interactions between the lipid and protein components of the chloroplast membrane (see discussion of the globular subunit theory and the three-layer model by Kirk, 1972). More recently, both biochemical and biophysical techniques, coupled with the freeze-etch procedure for electron microscopy, have supported the "lipid matrix theory" (Kirk, 1972) or "fluid mosaic membrane model" (Singer and Nicolson, 1972). These concepts suggest that the membrane is a lipid continuum in which protein molecules are partially or wholly embedded. These ideas have led to the concept of a dynamic pattern of structural organization. It is now accepted that both lipids and proteins can move laterally along biological membranes (Edidin, 1974; Singer, 1974), thus allowing changing interactions among membrane constituents. This review will attempt to summarize data relating to (1) the classification of various types of protein complexes of chloroplast lamellae, (2) the structural organization of protein complexes within the chloroplast membrane, (3) the evidence for dynamic changes in organization of these complexes, and (4) the relation of these dynamic changes to regulation of functional activity within the plastid.

II. Chloroplast Membrane Substructure: Visualization of Membrane-Protein Complexes

The overall morphological organization of chloroplast lamellae has been studied for many years using thin-sectioning techniques (Arntzen and Briantais, 1975). In recent years, detailed structural features of the membrane surface have been visualized using negative staining and/or

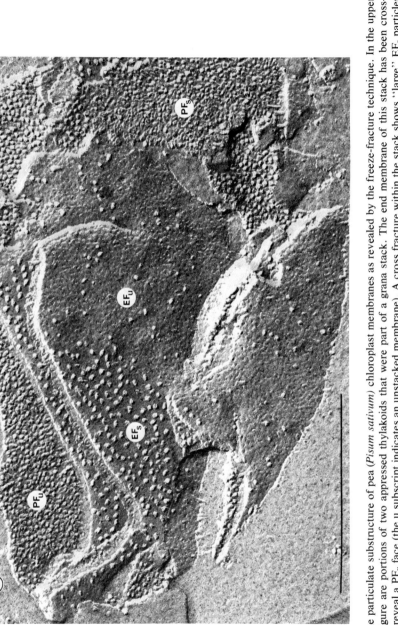

Fig. 1. The particulate substructure of pea (*Pisum sativum*) chloroplast membranes as revealed by the freeze-fracture technique. In the upper left of the figure are portions of two appressed thylakoids that were part of a grana stack. The end membrane of this stack has been cross-fractured to reveal a PF_u face (the u subscript indicates an unstacked membrane). A cross fracture within the stack shows "large" EF_s particles (the subscript s indicates that this is a stacked, or appressed, membrane). It may be noted that the EF_s particles are often oriented in "chains" of three or more particles; this is discussed in Section VIII,B. The semicircular edge of a partition (the region where two grana membranes fuse) can be identified in the left-central portion of the figure as an area of transition from relatively densely packed EF_s particles of the grana stack to the stroma lamellae with widely spaced EF_u particles. EF_u particles are also observed on the end membrane of the circular thylakoid that is partially visible in the lower central portion of the figure. PF_s particles can be seen within a cross-fractured grana stack on the right-hand side of the figure. Micrograph provided through the courtesy of P. Armond. Bar = 0.5 μm.

deep-etching techniques for electron microscopy (Arntzen and Briantais, 1975; Staehelin et al., 1976). For examination of the internal chloroplast membrane structure, the freeze-fracture technique has been extensively utilized in many laboratories. An example of the latter type of preparation is shown in Fig. 1. The types of data obtained by these various techniques and general interpretations of their meanings have been reviewed (Arntzen and Briantais, 1975; Staehelin et al., 1976). An artist's three-dimensional rendition of the combined sets of data describing the morphology, surface detail, and internal structure of three thylakoids is shown in Fig. 2. The nomenclature which has recently been adopted (Branton et al., 1975; Staehelin, 1976; Staehelin et al., 1976) to describe the various membrane regions is diagrammatically indicated in Figs. 2 and 3. This nomenclature is based on the fact that all biological membranes consist of two leaflets, a protoplasmic (P) and an exoplasmic (E) leaflet, and that each leaflet possesses a fracture face (F) and a true surface (S). The designations PF and EF therefore refer to fracture faces, and PS and ES refer to surfaces. The subscripts s and u indicate whether the membrane is in a stacked (grana) or unstacked (stroma lamellae) region, respectively. Further details of this terminology are described by Staehelin and co-workers (1976; Staehelin, 1976).

The structural image of chloroplast lamellae obtained in freeze-fracture electron microscopic studies is that of a relatively smooth membrane continuum interrupted periodically by subunits of various sizes (Fig. 1). It is now generally believed that the subunits observed are protein aggregates and the smooth continuum represents one half of the lipid bilayer forming the matrix of the membrane (Branton, 1971; DaSilva and Branton, 1970; Kirk, 1972). The protein complexes are thought to be asymmetrically localized within the lipid matrix; the freeze-cleaving through the center of the lipid bilayer thus results in a selective segregation of particles to either the protoplasmic or exoplasmic half-membrane leaflet (DaSilva and Branton, 1970; Kirk, 1972). In the case of chloroplast membranes, the freeze-fracture results in segregation of numerous, relatively small subunits to the PF face and fewer, but relatively larger, particles to the EF face (Figs. 1 and 2). In the following sections we will attempt to relate these features of the membrane revealed by electron microscopy to information on protein complexes gained by chemical or kinetic analyses of chloroplast lamellae or submembrane fragments.

III. Chloroplast Membrane Fractionation by Detergents: Isolation of Functionally Active Membrane Protein Complexes

Biochemical characterization of chloroplast membrane proteins has been highly dependent upon the use of surface-active agents to disrupt

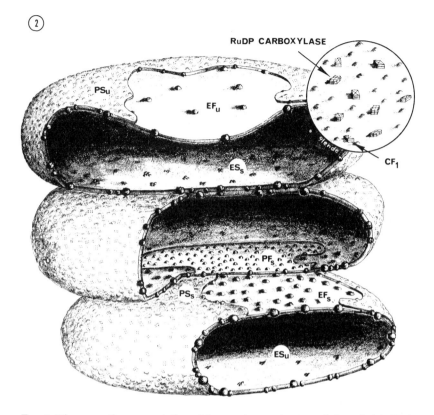

FIG. 2. Diagrammatic representation of the membrane structure of three thylakoids in a grana stack. Within the partition regions (where two thylakoids are appressed), the "large" freeze-fracture particles (EF_s) are more densely distributed than in the unstacked (EF_u) end membranes. The "small" (PF_s or PF_u) particles are rather uniformly distributed throughout the membrane. The CF_1 portion of the coupling factor and RuDP carboxylase are localized on the protoplasmic surface (PS) as shown in the inset; this is discussed in Section IV,A. Substructure of the large freeze-fracture particle is shown on the exposed EF_s and EF_u surfaces in agreement with deep-etching studies (see Fig. 6). The significance of this substructure is discussed in Section IV,C.

the hydrophobic bonds between lipids and proteins and to overcome the protein insolubility in aqueous media. Anionic detergents such as sodium dodecyl sulfate (SDS) or sodium dodecylbenzene sulfonate (SDBS) extensively solubilize chloroplast membranes and release of small particles with sedimentation coefficients from 1.25 to 2.65 (Shibata, 1971). These preparations are of major biochemical value since the solubilized components can be separated into individual polypeptide species by techniques such as polyacrylamide gel electrophoresis (see the chapter by Boardman *et al.* in this volume). Unfortunately, however,

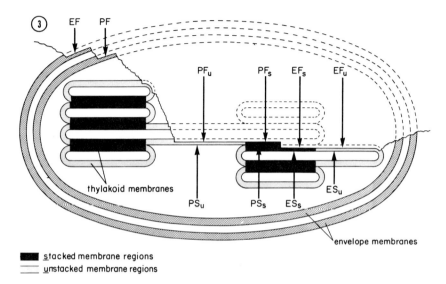

stacked membrane regions
unstacked membrane regions

Fig. 3. Freeze-etching nomenclature of chloroplast membranes. Fracturing of plastid membranes results in either cross-fractures or longitudinal splitting of lamellae exposing the interior region of the membrane (EF or PF faces). Deep-etching (removal of surrounding aqueous medium by sublimation) results in exposure of true membrane surfaces (ES or PS). Subscripts s or u refer to stacked or unstacked membrane regions, respectively. Figure by courtesy of L. A. Staehelin. After Staehelin (1976).

anionic detergent treatments abolish most photochemical activities of the submembranane preparations, presumably because anionic detergents disrupt the tertiary structure of membrane proteins. Anionic detergents therefore cannot be used in attempts at isolating "physiologically native" membrane protein complexes.

Nonionic detergents, such as digitonin or Triton X-100, have been used to fractionate chloroplast membranes into particulate subunits that contain several polypeptides (Boardman, 1968; Vernon et al., 1968). The complexes derived in this manner are protein aggregates that at least partially simulate the physiologically active protein configurations existing in the intact membrane, since photochemical activity of individual partial reactions is retained by the separated subunits and, under certain conditions, the protein complexes can be reconstituted to give membranous fractions that will catalyze whole-chain electron transport (Arntzen et al., 1972; Huzisige et al., 1969; Ke and Shaw, 1972). The following sections will be devoted to describing the different types of functionally active particle preparations that can be derived from chloroplast lamellae.

A. "Complete" Photosystem II Complexes

Chloroplast submembrane fragments enriched in photosystem II (PS II) activity can be prepared using either digitonin or Triton X-100 for membrane disruption (Boardman, 1968; Briantais, 1969; Brown, 1972; Vernon *et al.*, 1968; Wessels and van Leeuwen, 1971). The most highly purified membrane fragments containing PS II and associated light-harvesting pigments have been isolated from chloroplast grana membrane preparations using digitonin fractionation (Arntzen *et al.*, 1972), from isolated chloroplasts using combined digitonin and Triton X-100 treatments (Boardman, 1972), or a digitonin–sonication–Triton X-100 treatment (Huzisige *et al.*, 1969), or through the use of Triton X-100 at relatively high detergent:chlorophyll ratios (Vernon and Shaw, 1971). These preparations have a low (1.8–2.0) chlorophyll (Chl) *a:b* ratio and catalyze a light-induced, DCMU-sensitive reduction of dichlorophenolindophenol (DCPIP). When examined by electron microscopy, these preparations were found to be small membranous sheets (Arntzen *et al.*, 1969; Vernon *et al.*, 1968; Vernon and Shaw, 1971). Polypeptide analysis of PS II preparations by polyacrylamide gel electrophoresis has shown that the submembrane preparations contained many polypeptide species, but are quantitatively enriched in proteins of the 21–30 kdalton range (group II polypeptides) (Klein and Vernon, 1974; Levine *et al.*, 1972; Nolan and Park, 1975; Vernon and Klein, 1975).

B. The Photosystem II Reaction-Center Complex

Vernon and co-workers (1971) subjected their usual Triton X-100-derived, PS II-enriched submembrane preparation to further sucrose gradient fractionation and obtained a small amount of highly active PS II particles. Subsequently, Wessels and co-workers (1973) found a similar particle in digitonin-solubilized fractions purified on sucrose gradients. Both particles had similar characteristics: a Chl *a:b* ratio of 25–28, highly active photoreduction of DCPIP (in the presence of a PS II electron donor such as diphenylcarbazide), an enrichment of cytochrome (cyt) b_{559}, and the presence of low amounts of β-carotene and lutein (Ke *et al.*, 1972, 1974; Vernon *et al.*, 1971; Wessels and Borchert, 1975; Wessels *et al.*, 1973). The particles showed a light-induced absorbance change in C550 and photooxidation of Cyt b_{559} (Ke *et al.*, 1972, 1974; Wessels *et al.*, 1973). The presence of P_{680} (reaction-center chlorophyll of PS II) was detected by spectroscopic absorption changes and by electron paramagnetic resonance (EPR) spectroscopy (Ke *et al.*, 1974). Neither the digitonin- nor Triton-derived particles contained detectable P700 or showed any PS I activity (Ke *et al.*, 1974; Vernon *et al.*, 197; Wessels *et al.*, 1973). When examined by negative staining for electron microscopy, the Triton-derived particles were found to exist as discrete, small (roughly 110 Å in diameter) subunits that tended to aggregate (possibly owing to hydrophobic interactions) (Vernon *et al.*, 1971).

Vernon *et al.* (1971), Ke *et al.* (1974), and Wessels *et al.* (1973; Wessels and Borchert, 1975) have concluded that this small detergent-derived particle isolated from sucrose gradients is the reaction-center complex of PS II. The observations that the primary photochemical activity in this particle closely resembles that of PS II in intact chloroplasts suggest that the particles are equivalent to a structurally distinct protein complex existing in the *in vivo* membrane. Both laboratory groups agree that the PS II reaction-center complex must contain the trapping center plus some secondary electron carriers (such as cytochrome b_{559}) and a few light harvesting chl *a* molecules.

Analysis of the polypeptide composition of the isolated PS II reaction center complexes has shown that they consist of several proteins, mainly in size classes from 27 to 60 kdaltons. The most significant difference between these preparations and the PS II fragments that were Chl *b*-enriched was the absence of two polypeptides in the 22–24 kdalton size range in the reaction-center material (Klein and Vernon, 1974; Wessels and Borchert, 1975).

C. THE PHOTOSYSTEM II LIGHT-HARVESTING COMPLEX (LHC)

Most chlorophylls in chloroplast lamellae are now believed to be associated with protein in pigment–protein complexes (Thornber, 1975). From chloroplasts solubilized by SDS and fractionated either by polyacrylamide gel electrophoresis or on hydroxylapatite, a major pigment–protein enriched in Chl *b* (Chl *a:b* ≅ 1) has been isolated. This complex has been commonly referred to as complex II, or more recently as the light-harvesting Chl *a:b* protein (Thornber, 1975). Several laboratories have provided evidence that the complex contains at least two polypeptides of molecular weights in the 23–27 kdalton size range (Anderson and Levine, 1974a; Apel *et al.*, 1975; Henriques and Park, 1975; see more detailed discussion by Boardman *et al.* in this volume). A barley mutant that does not contain this pigment–protein complex was found to have full photochemical activity but had a reduced-size photosynthetic unit (Thornber 1975). It has also been shown that the plant seedlings greened under conditions of intermittent illumination did not contain this pigment–protein complex and had reduced photosynthetic unit sizes, but did have full photochemical activity (Argyroudi-Akoyunoglou *et al.*, 1971; Armond *et al.*, 1976; Davis *et al.*, 1976; Hiller *et al.*, 1973). These data have led to the view that this pigment–protein complex serves as a light-harvesting complex (LHC) (Arntzen *et al.*, 1976; Thornber, 1975). The LHC is believed to contain most, perhaps all, of the Chl *b* in chloroplast lamellae (Thornber, 1975). Preparations of PS II-enriched submembrane fragments obtained by detergent fractionation are enriched in Chl *b* and protein components of the LHC, whereas PS I preparations have high Chl *a:b* ratios and low LHC content (Anderson and Levine, 1974b). It has also been demonstrated that most PS II activity of chloroplast thylakoids is localized in the appressed membrane (grana) regions of the chloroplast lamellae whereas unpaired membranes (stroma lamellae) have low PS II activity (Arntzen and Briantais, 1975; Park and Sane, 1971). In parallel analyses of grana and stroma lamellae, the distribution of Chl *b* and the LHC follow the disribution of PS II centers (Henriques *et al.*, 1975; Park and Sane, 1971). The physical association of the LHC with the PS II centers and the fact that Chl *b* is primarily a sensitizer of PS II have led to the generally accepted view that the LHC acts as such primarily for PS II (Arntzen *et al.*, 1976).

Since PS II reaction centers can be isolated free of the LHC (Section III,B above), and since the PS II centers are not cosynthesized with the LHC during plastid biogenesis (Arntzen *et al.*, 1976), we suggest that these two components are not obligatorily linked as part of a single membrane structural complex *in vivo*. This suggests that it should be possible to release the LHC by mild detergent procedures. This idea is supported by the recent observations that a photochemically inactive, chlorophyll *b*-containing pigment–protein complex can be purified on a sucrose gradient from samples of digitonin-solubilized chloroplast lamellae. Wessels and Borchert (1975) and Arntzen and Ditto (1976) have demonstrated that this complex has a Chl *a:b* ratio of 1.3–1.5. The latter authors have further demonstrated through the use of SDS–polyacrylamide gels that the pigment-protein isolated by sucrose gradient techniques contains the complex previously referred to as complex II or the light-harvesting Chl *a:b* protein. The LHC isolated through the use of digitonin contains a limited complement of polypeptides; the fraction primarily contains proteins of the 23–27 kdalton size class (Arntzen and Ditto, 1976; Wessels and Borchert, 1975). [These are the same proteins that are removed during purification of the PS II reaction centers from PS II-enriched submembrane preparations (Vernon and Klein, 1975).] Through ultrafiltration techniques, Arntzen and Ditto (1976) demonstrated that the LHC solubilized by digitonin is a discrete structural complex (approximately 50–100 Å in diameter) presumably formed by an aggregation of individual pigment-proteins. Arntzen *et al.* (1976) have found that LHC can be reconstituted with PS II particles to form large membranous complexes.

As a point of clarification it should be emphasized that most studies of the PS II light-harvesting components have used SDS to solubilize the membrane proteins. Thornber (1975) has stressed that there is a single, unique protein associated with chlorophylls (the light-harvesting Chl $a:b$ protein) which can be isolated by these techniques. The concept of a light-harvesting complex is not in conflict with such an idea. It is certainly likely that the pigment-protein described by Thornber is a major component of the LHC. This review intends to emphasize the structural organization of membrane components; the existing data demonstrate that in the intact membrane more than one protein is involved in binding PS II-sensitizing Chl a and b, and that all of these components (protein plus Chl) form a structural aggregate which we term the LHC. It is this native structural complex whose characteristics we wish to define herein, since these characteristics will determine the functional behavior of the light-harvesting chlorophylls *in vivo*.

D. PHOTOSYSTEM I PREPARATIONS: A REACTION-CENTER–ACCESSORY-PIGMENT COMPLEX AND A CYTOCHROME COMPLEX

After digitonin or Triton X-100 treatment of chloroplast lamellae, Boardman and Anderson (1964) and Vernon and co-workers (1968) found that heavy fractions containing undigested membrane fragments and PS II-enriched particles could be pelleted from solution leaving a light PS I fraction. These PS I particles exhibited the ability to photoreduce NADP (in the presence of suitable electron donors), had high Chl $a:b$ ratios (in the range of 4.5–6), were enriched in β-carotene content, and fluoresced in the region of 730 nm at liquid-nitrogen temperatures (Boardman, 1968; Vernon et al., 1968; Vernon and Shaw, 1971). Boardman and Anderson (1967) found that Cyt f and Cyt b_6 were enriched in these preparations. As would be expected, P700 (the reaction-center chlorophyll of PS I) was found to be enriched in PS I preparations (Boardman, 1968; Vernon et al., 1968). The P700:chlorophyll ratios for various preparations usually ranged from 1:100 to 1:250 (Vernon et al., 1971).

In recent years, it has been shown that the usual PS I preparations can be further fractionated on sucrose gradients or by additional differential centrifugations to obtain P700-containing preparations that are free of cytochromes (Ke et al., 1975; Nelson and Racker, 1972; Wessels and Borchert, 1975). The fact that these preparations catalyzed light-induced electron transport in the presence of appropriate electron donors and acceptors and showed a light-induced P700 signal indicates that they contain functional reaction-center chlorophylls and the associated primary electron acceptor. In addition, there are about 110–150 chlorophylls per P700 in the purified complex (Ke *et al.*, 1975; Wessels and Borchert, 1975). These accessory chlorophylls are almost entirely Chl a; the Chl $a:b$ ratio of the preparations was high (>7.5) (Ke *et al.*, 1975; Wessels and Borchert, 1975). Structural characterization of these complexes by electron microscopy has not been reported. The digitonin-derived complex contained several polypeptides that were primarily in the size classes of 10, 14, 47, 21, and 74 kdaltons (Wessels and Borchert, 1975). The complex appears to exist as a "complete" PS I reaction-center structural unit with associated antenna chlorophylls.

In some cases, very highly enriched P700 particles with P700:chlorophyll ratios of 1:25 to 1:35 have been obtained by Triton X-100 treatment of carotenoid-depleted chloroplast lamellae (Vernon *et al.*, 1971). This is apparently achieved by the removal of various light-harvesting pigments, and some associated protein, from the "reaction-center–accessory-chlorophyll" complex described above. It seems likely that the "high P700" fractions described by Vernon and co-workers (1971; Vernon and Shaw, 1971) contain pigment–protein complex I (the P700 Chl "a" protein), which comprises the "core" component of the PS I complex (Thornber, 1975).

As described above, the original PS I preparations, which used differential centrifugation to isolate submembrane fragments, were found to contain Cyt f and Cyt b_6 (Boardman and Anderson, 1967). Wessels and van Leeuwen (1971) and Wessels and Voorn (1972), who used sucrose gradient procedures for purification of digitonin-derived submembrane fractions, found that these cytochromes coseparated as a distinct pink band on their gradients. Nelson and Neumann (1972) found that a Cyt f–b_6 complex could be precipitated with protamine sulfate from digitonin-derived PS I preparations. Besides cytochromes, the particle contained nonheme iron, phospholipids, and carotenoids, but no chlorophyll. The complex had a calculated minimum molecular weight of 103,000. Ke *et al.* (1975) have described an apparently similar complex that was obtained by subjecting Triton-derived PS I subchloroplast particles to sucrose gradient purification. Their complex contained Cyt f and Cyt b_6 plus a bound iron–sulfur protein and a bound plastocyanin (characterized by EPR spectroscopy).

It should be noted that this cytochrome complex can be further dissociated with detergents. Stuart and Wasserman (1973) have used Triton X-100 plus 4 M urea to isolate Cyt b_6. Nelson and Racker (1972) extracted Cyt f–b_6 particles with ethanol and acetone and then treated the preparation with sodium deoxycholate and sodium cholate to solubilize Cyt f. Nelson and Racker (1972) pointed out that Cyt f in the purified f–b_6 particle is in monomeric form whereas further solubilization of the particle results in destruction of Cyt b_6 and polymerization of Cyt f.

Attempts to obtain purified monomeric Cyt f resulted in alteration of its spectral characteristics. We can conclude from these observations that the Cyt f–b_6 particle isolated by mild digitonin or Triton X-100 techniques probably represents something close to the physiologically "native" complex of these components, whereas further purification procedures alter the components. The ease of isolation of the Cyt f–b_6 particle suggests that detergents simply solubilize it from the lipid milieu of the membrane.

E. Peripheral and Integral Membrane Components of the Coupling Factor

The membranes of chloroplasts, mitochondria, and bacteria contain a Mg^{2+}-dependent ATPase that is involved in ATP synthesis (Racker, 1970; see the chapter by McCarty in Volume 7 of this series). From isolated mitochondria, a proton-translocating ATPase complex has been isolated (Serrano et al., 1976) that is reconstitutively active in oxidative phosphorylation and can be incorporated into phospholipid vesicles that will catalyze ATP-driven translocation of protons. The ATPase complex is composed of a soluble "F_1" complex and a "hydrophobic protein fraction" (Kagawa and Racker, 1971; Serrano et al., 1976).

A chloroplast enzyme that has been shown to be involved in phosphorylation is called "coupling factor number 1" or CF_1 (Jagendorf, 1975b). It is analogous to F_1 of mitochondria—a large protein of nearly 350 kdaltons. The enzyme contains five polypeptide subunit chains. It is normally bound on the chloroplast membrane protoplasmic surface (a peripheral enzyme); when released and isolated it can be activated to show ATPase activity (Arntzen and Briantais, 1975; Jagendorf, 1975a).

Recently a "hydrophobic protein" component of the chloroplast ATPase (called F_0) has been isolated by Younis and Winget (1977). The F_0 is an intrinsic membrane component that was solubilized by cholate extraction and purified by ammonium sulfate fractionation. The F_0 can be reconstituted into phospholipid vesicles, where it binds the CF_1 enzyme to form a functional ATPase complex. Vesicles reconstituted with F_0, CF_1, phospholipids, and bacteriohodopsin were capable of light-induced ATP synthesis. Polypeptide analysis of the F_0 preparation by SDS–polylacrylamide gel electrophoresis indicated that it contained six major polypeptide species with apparent weights ranging from 42 to 11 kdaltons (Younis and Winget, 1977).

F. Summary

Two nonionic detergents, digitonin and Triton X-100, have been extensively used to solubilize chloroplast lamellae. Six distinct types of enzyme complexes have been found to be released from the membrane by treatments with these detergents. One of these, CF_1 or the chloroplast ATPase, is bound to the surface of the membrane and can be considered a peripheral protein (Arntzen and Briantais, 1975). [Ribulose-diphosphate (RuDP) carboxylase is also known to be a peripheral protein complex associated with the chloroplast membrane. It is thought to be loosely bound, however, since it can be easily removed by gentle washing (Arntzen and Briantais, 1975). Since the presence or the absence of this complex is not known to influence the energy-coupling reactions of the chloroplast membranes, it will not be considered further in this review.]

Five different integral membrane protein complexes have been characterized from detergent-solubilized chloroplast lamellae; these are described in Table I. We suggest that each of the isolated complexes represents "native" protein aggregates that exist in the membrane as a structural unit and are released from the lipid matrix of intact membranes by the detergent action. This is supported by the fact that the isolated complexes retain enzymic activity in various partial reactions of photosynthesis (the PS I, PS II, and cytochrome complex) or retain the capacity to reconstitute with other membrane components (LHC and the F_0 fraction). With the exception of the recently discovered F_0, all protein complexes described in Table I have been isolated independently by at least two separate laboratories using different preparative procedures. In spite of the fact that different detergents and techniques were used, the preparations show striking similarities. This supports the idea that the complexes are simply solubilized from the membrane and are not artifacts "generated" by the detergents.

If it is accepted that the protein complexes described in Table I represent structural units derived from the intact membranes, we can suggest that they may be identified as defined structural complexes by electron microscopy techniques. The following section will deal with this possibility.

IV. Development of a Membrane Structural Model

A. PERIPHERAL MEMBRANE-PROTEIN COMPLEXES

The surface of thylakoid membranes has been examined using negative staining and deep-etching procedures. In some of the earliest studies, Howell and Moudrianakis (1976a,b) showed that the chloroplast CF_1 (ATPase) and RuDP carboxylase are surface bound. These peripheral complexes can be removed by various washing treatments (Howell and Moudrianakis, 1967b; Strotmann et al., 1973). The surface localization of these enzymes was subsequently verified in several other laboratories

TABLE I

A Summary of the Characteristics of Different Types of Integral Membrane Complexes Derived from Chloroplast Lamellae by Mild Detergent Treatments

Complex	Chlorophyll content	Characteristic Activity	Polypeptide content	Structure	References
Photosystem II reaction-center complex	Chlorophyll $a:b$ ratio > 25 (contains P680 and light-harvesting Chl a)	Light-induced DCPIP reduction in the presence of an electron donor; light-induced C550 spectral change; photooxidation of Cyt b_{559}; P680 bleaching	At least six polypeptides, mainly in the 27–54 kdalton size classes	Discrete, small particles; approximately 100 Å in diameter as shown by electron microscopy	Ke et al. (1974), Ke et al. (1972), Vernon et al. (1971), Wessels and Borchert (1975), Wessels et al. (1973)
Light-harvesting complex (LHC)	Chlorophyll $a:b$ ratio = 1.3–1.5	No photochemical activity; Reversibly binds to PS II complex; binding is elicited by cations	Primarily contains polypeptides in the 22–25 kdalton size class	Particulate structure based on ultrafiltration; particle diameter = 50–100 Å	Arntzen et al. (1976), Arntzen and Ditto (1976), Wessels and Borchert (1975)

	Chlorophyll content	Function	Polypeptides		References
Photosystem I reaction-center/accessory pigment complex	Chlorophyll $a:b > 7.5$ (contains P700; 1 P700/110–150 Chl)	Light-induced P700 oxidation; will oxidize added plastocyanin and Cyt f	At least six polypeptides with weights ranging from 10 to 74 kdaltons	Data not available	Ke et al. (1975), Nelson and Racker (1972), Wessels and Borchert (1975)
Cytochrome f–b_6 complex	Contains only trace amounts of chlorophyll	Contains Cyt f and Cyt b_6, nonheme iron, and bound plastocyanin; Cyt f oxidation can be recoupled to P700 reduction if the PS I complexes are reconstituted	Data not available	Data not available	Ke et al. (1975), Nelson and Neumann (1972), Wessels and van Leeuwen (1971), Wessels and Voorn (1972)
Hydrophobic protein complex of the coupling factor (F_0)	Contains no chlorophyll	Can be reconstituted into phospholipid vesicles where it will bind CF_1 to form an active, proton-translocating ATPase complex	Contains six major polypeptides ranging from 11 to 42 kdaltons	Data not available	Younis and Winget (1977)

(Garber and Steponkus, 1974; Miller and Staehelin, 1976; Oleszko and Moudrianakis, 1974; see review in Arntzen and Briantais, 1975). The localization of these surface-bound components is diagrammatically indicated in Fig. 2.

B. INTERNAL MEMBRANE-PROTEIN COMPLEXES

Arntzen et al. (1969) analyzed the structural organization of detergent-derived PS II and PS I submembrane fragments by freeze-fracture techniques for electron microscopy. The PS I preparations contained only particles that were morphologically identical to those of the PF fracture face of intact thylakoid membranes (the "C" surface in older terminology). Preparations of PS II were enriched in particles that were morphologically similar to the larger freeze-etch particles of the EF fracture face (the "B" surface in older terminology). Arntzen and co-workers related the two different types of membrane particles to the two different functional complexes within the membrane. In light of the fact that the chloroplast membrane can now be shown to contain at least five different protein complexes (Table I), it must be recognized that the conclusions of Arntzen et al. (1969) were overly simplistic. The following discussion will attempt to update the earlier ideas in order to relate recent membrane structural studies to current knowledge of membrane protein complexes.

1. Identification of the "Large" (EF) Freeze-Fracture Particle

In early freeze-fracture studies, the EF face was described as containing relatively large, widely spaced particles (Arntzen and Briantais, 1975). In some preparations, these particles were organized in crystalline arrays (Branton and Park, 1967; Garber and Steponkus, 1976; Staehelin, 1976; Staehelin et al., 1976). Goodenough and Staehelin (1971) obtained high-resolution micrographs of Chlamydomonas reinhardtii chloroplast lamellae from which particle size histograms of the exposed fracture faces were constructed. The histograms revealed that the particles of the EF face ranged in size from less than 60 Å to 200 Å. In higher plant chloroplasts, the EF particles had a similar range of sizes; in addition, there were distinct maxima in the particle-size histograms at approximately 115 Å and 155 Å (Armond et al., 1977; Arntzen et al., 1976; Staehelin, 1976; Staehelin et al., 1976). The finding that the EF particles were so nonuniform in size initially made the idea of relating a single, specific functional role to these particles quite untenable. Recent studies of developing chloroplasts have allowed an interpretation of structure–function interrelationships that overcome these apparent contradictions.

In a study of chloroplast greening in barley, Phung-Nhu-Hung *et al.* (1970) found that the large-size class of EF particles could be detected only after extensive membrane development; the appearance of large particles corresponded to the appearance of grana stacks. During greening of *Euglena* chloroplasts, Ophir and Ben-Shaul (1974) also observed a change in size of the EF particles as the membranes developed. The factor(s) controlling particle size increases was not defined.

Armond and co-workers reinvestigated chloroplast development utilizing seedlings that had greened under intermittent illumination. Plastids that form under these conditions do not contain Chl *b* and have no true grana stacks, but have complete photochemical activities (Armond *et al.*, 1976, 1977). These plastids have reduced photosynthetic unit sizes (Armond *et al.*, 1977; Arntzen *et al.*, 1976); this was explained by the demonstration that the membranes do not contain the LHC of PS II (Davis *et al.*, 1976). The lack of LHC was correlated with the absence of specific polypeptides in polyacrylamide gel separations of the proteins from the incompletely differentiated chloroplast membranes (Armond *et al.*, 1977). During subsequent greening of these plastids after transfer of seedlings to continuous illumination, the polypeptides associated with LHC (two polypeptides of 25–27 kdaltons) were inserted into the membrane. In fully differentiated plastids these polypeptides constituted a major portion of the total protein in the membrane.

Armond and co-workers (1977; Arntzen *et al.*, 1976) analyzed the substructure of developing chloroplast membranes after seedlings grown under intermittent light were transferred to continuous light. During this period, photochemical activities of the membrane remained approximately constant when expressed on a photosynthetic unit basis; all developmental changes in the membrane were related to appearance of the LHC. During the greening process only small changes in particle density were observed on the EF and PF faces examined in freeze-fracture preparations. However, major changes in particle size were noted, particularly in the EF particles of stacked regions (EF_s) of the chloroplast membrane. The size of the EF particles increased in steps, from an initial particle population of uniform size (average diameter of 83 Å) to a final population of membrane subunits with maxima in particle-size histograms at 105 and 164 Å. These data were interpreted as showing that the light-harvesting pigment proteins are added into the differentiating membrane in discrete complexes; the complexes bind in various amounts to a basic "core" complex to form EF particles of a range of particle sizes. It is therefore possible to explain why membrane subunits of the EF face are of different sizes: the basic "core" complex of this face can have one or more LHC associated with it such that each additive unit increases the size of the EF particle.

The idea that the size of the EF particle is a function of the amount of LHC bound to a "core" complex is supported by previous studies of chloroplast mutants. A Chl b-deficient *Chlamydomonas reinhardtii* mutant did not contain the large size class of EF particles (Goodenough and Staehelin, 1971). A partially Chl b-deficient soybean mutant (shown to have reduced LHC content; Arntzen *et al.*, 1976) contained EF particles that were somewhat reduced in size (Keck *et al.*, 1970). Chloroplasts of a barley mutant that contain no pigmented LHC (Thornber, 1975), but do contain one of the two major polypeptide components of the LHC (Henriques and Park, 1975; Machold *et al.*, 1977; see extensive discussion by Boardman *et al.* in this volume), were examined by Miller *et al.* (1976). They showed that the EF particles were reduced from an average diameter of near 160 Å to an average diameter of about 125 Å; this is consistent with a partial loss of the LHC polypeptides of these membranes.

Analysis of membranes of various algae also supports the idea that LHC forms part of the EF particle. Dark-grown *Euglena*, which contain no LHC, have small ($\sim$ 75 Å average diameter) EF particles (Staehelin *et al.*, 1976). Light-grown *Euglena*, which have low Chl b and LHC content (Brown *et al.*, 1975), have EF particles averaging about 120 Å in diameter (Miller and Staehelin, 1973). The EF face of *Spermothamnion* (a red alga that contains no Chl b) shows particles averaging about 100 Å in diameter (Staehelin *et al.*, 1976).

2. The "Core" Subunit of the EF Particle

As described above, chloroplast lamellae that contain no LHC have relatively small ($\sim$ 80 Å) particles on the EF face. While we have no direct biochemical evidence for the identity of the "core" unit, the following lines of evidence suggest that it could be the PS II reaction-center complex. It is thought that the light-harvesting chlorophyll protein complex binds all the Chl b in the membrane (Thornber, 1975). It is generally agreed that Chl b primarily sensitizes PS II (Govindjee and Govindjee, 1975). Detergent fractionation studies have indicated that the light-harvesting pigment protein complex is physically associated with the PS II complex (Arntzen *et al.*, 1974); both digitonin or Triton X-100 solubilization of chloroplast membranes release "heavy" submembrane fragments that are enriched in PS II activity and in Chl b (see Section III,A above). These heavy fragments were shown to be enriched in the EF particles (Arntzen *et al.*, 1969).

The idea that EF particles are sites of localization of PS II centers has been tested with respect to membrane compositional differences between grana membranes and stroma lamellae. This is possible because the freeze-fracture particles of the membranes are nonuniformly distributed and are heterogeneous in size (Arntzen and Briantais, 1975; Park and Sane, 1971; Staehelin, 1976; Staehelin et al., 1976; see Table II, Fig. 4); the distribution of functional units along the membrane can be correlated to membrane substructure.

It is now well recognized that PS II is preferentially localized in stacked membrane regions (Arntzen and Briantais, 1975; Park and Sane, 1971). In an attempt at quantifying the amount of PS II centers in various membrane regions, Armond and Arntzen (1977) used lactoperoxidase-catalyzed iodination of chloroplast membranes to inactivate photosystem centers; the large enzyme reacts only with surface-exposed

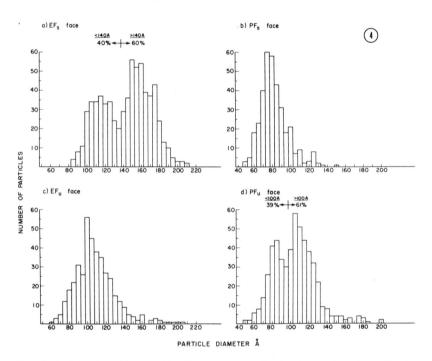

FIG. 4. Particle-size histograms for the membrane subunits observed on freeze-fracture faces of mature spinach chloroplast thylakoids. The number of particles measured (n) for the construction of the histogram and the average size of the particles ($\bar{x}$) are indicated. In histograms showing more than one size-class maximum, the histogram was arbitrarily divided into two subsets for more detailed analysis. (a) EF_s face: all particles, $n = 674$, $\bar{x} = 143.0$ Å; particles < 140 Å, $n = 270$, $\bar{x} = 114.1$ Å; > 140 Å, $n = 404$, $\bar{x} = 162.2$ Å. (b) PF_s face: $n = 354$, $\bar{x} = 82.2$ Å. (c) EF_u face: $n = 412$, $\bar{x} = 109.0$ Å. (d) PF_u face: all particles, $n = 534$, $\bar{x} = 104.3$ Å; particles < 100 Å, $n = 207$, $\bar{x} = 81.8$ Å; > 100 Å, $n = 327$, $\bar{x} = 188.0$ Å. Figure and calculations were provided by L. A. Staehelin; data are from Staehelin (1976).

TABLE II

Comparison of the Size and Distribution of Intramembranous Particles of Chloroplast Membranes to the Distribution and Characteristics of the PS II Units of Higher-Plant Chloroplast Lamellae[a]

Parameter studied	Unstacked membranes (stroma lamellae)	Stacked membrane (grana partitions)	References
Number of PF particles per μm^2	3409 ± 265	3620 ± 286	Staehelin (1976)
Number of EF particles per μm^2	574 ± 47	1495 ± 103	Staehelin (1976)
Proportion of total length of plastid membranes in unstacked or stacked regions	40%	60%	Staehelin (1976)
Proportion of EF particles calculated to be in unstacked or stacked membrane regions	20%	80%	Staehelin (1976)
Proportion of PS II centers found to be localized in unstacked or stacked membranes	20–25% (iodination) 20% (immunochemistry)	75–80% (iodination) 80% (immunochemistry)	Armond and Arntzen (1977) Radunz et al. (1971); Schmid (1972)
Average size of EF particle	109 Å	143 Å	Staehelin (1976)
Chl $a:b$ ratio of isolated membrane fractions	5.4	2.2	Arntzen et al. (1972)
Estimation of PS II unit size based on light intensity (saturation) measurements vs PS II electron-transport rates	Small light-harvesting antennae complement	Normal PS II unit size	Armond and Arntzen (1977)

[a] While PF particles are uniformly distributed along the membrane, EF particles are concentrated in appressed membranes; this concentration is quantitatively correlated with the distribution of PS II centers. The smaller EF particles of the stroma lamellae are also in direct agreement with the fact that there is little Chl b (LHC) in these membranes and that the PS II centers of stroma lamellae have small PS II units (few associated light-harvesting chlorophylls).

membrane components, so only PS II units of stroma lamellae or grana end membranes were inactivated by iodination of "Class II" chloroplasts. Iodination was found to inactivate only 20–25% of the PS II centers in stacked membrane preparations. Schmidt (1972; Radunz *et al.*, 1971) found that antibodies prepared against chlorophyll would inactivate about 20% of the PS II activity of isolated chloroplast lamellae, since the relatively large antibodies probably cannot penetrate into stacked lamellar regions.

The finding that about 20% of the chloroplast PS II activity is localized in stroma lamellae agrees well with the known distribution of the EF particles of chloroplast lamellae (Table II). In recent studies of higher plant chloroplasts it is estimated that 15–20% of the total EF particles are in stroma lamellae (Armond and Arntzen, 1977; Staehelin *et al.*, 1976). It should be noted that the EF particles of stroma lamellae are considerably smaller than those of grana membranes. Armond and Arntzen (1977) have found that PS II activity of isolated stroma lamellae requires much higher light intensities for saturation than that of grana membrane preparations. This evidence for a smaller light harvesting assembly for the PS II units, and the known deficiency of Chl *b* and LHC in stroma lamellae (Park and Sane, 1971), is in agreement with the idea that small EF particles are PS II complexes with little or no attached LHC.

3. The "Small" (PF) Freeze-Fracture Particle

The PF fracture face of chloroplast lamellae is characterized by the presence of numerous, relatively tightly packed particulate subunits. The PF_s particles (in stacked membranes) are uniformly sized with an average diameter of about 80 Å. PF_u faces contain two size classes of particles—about 40% with an average size of 80Å and about 60% with an average size of 118 Å (Fig. 4) (Staehelin, 1976; Staehelin *et al.*, 1976).

It has been demonstrated that digitonin-derived PS I preparations contain particles identical to those of the PF face when examined by freeze-fracture techniques (Anderson *et al.*, 1973; Arntzen *et al.*, 1969, 1972). Unfortunately, the preparations used in these early studies contained a mixture of protein complexes: the PS I reaction-center–antenna-pigment complex, the Cyt f–b_6 complex, and components of the coupling-factor complex (Arntzen *et al.*, 1972; Boardman and Anderson, 1967). No studies to date have attempted to isolate these components separately for study by freeze-fracture techniques; we must assume that all complexes are of approximately the same size based on the relatively uniform particle sizes observed in the preparations of mixed components.

Evidence for the existence of a separate cytochrome-containing structural complex in chloroplast lamellae has been reported in studies of Boardman *et al.* (1972) on plants adapted to growth at different light intensities. "Shade" plants, such as *Alocasia* and *Atriplex*, grown under low light, have a decreased content of Cyt f and Cyt b, on a chlorophyll basis, as compared to plants grown in high light, even though photosynthetic unit sizes on a P700 basis remain constant in both situations. In samples of similar membranes examined by freeze-fracturing, there was a decrease in the number of particles on the PF fracture face in the "low-light" plants which have reduced cytochrome content.

The idea of a portion of the coupling factor complex being identifiable as a subunit of the PF face is, at the present time, largely conjectural. We know that a portion of the enzyme system is a hydrophobic protein complex (F_o) that can be removed from the membrane only by use of detergents (Younis and Winget, 1977). It has been argued that all coupling factor of chloroplast membranes is localized along unpaired membranes (stroma lamellae or grana end membranes) (Miller and Staehelin, 1976). This interpretation is at variance with some data in the literature (Oleszko and Moudrianakis, 1974; discussed in more detail in a review by Anderson, 1975). We can reasonably conclude, however, on the basis of structural studies and of observations that stroma lamellae fractions have high photophosphorylation activities (Arntzen *et al.*, 1971), and that a large portion of the CF_1 is localized in unpaired membranes. It is therefore unlikely that the F_o could be correlated in any way with the EF particles, since EF membrane subunits segregate into stacked membrane regions (Arntzen and Briantais, 1975; Staehelin *et al.*, 1976). It is interesting to note that one size class of PF particles (those with an average diameter of 118 Å) is preferentially localized in the unstacked membranes (Fig. 4). We can speculate that these might be the F_o, if Miller and Staehelin (1976) are correct in asserting that all coupling factor is localized in the unstacked membranes.

C. SUMMARY

Structural studies have unambiguously established that the chloroplast coupling factor component CF_1 is bound to the surface of the membrane (Anderson, 1975; Arntzen and Briantais, 1975; Staehelin *et al.*, 1976). This was shown in the membrane model of Fig. 2.

Interpretations of structural studies of integral membrane complexes are much less precise. In previous reviews it has been argued that the EF particles are sites of localization of PS II centers (Arntzen and Briantais, 1975) or of the light-harvesting complex (CP II) that serves photosystem II (Anderson, 1975). Recent studies conducted by Armond and co-workers (1976, 1977; Arntzen *et al.*, 1976) have provided evidence that both lines of argument are correct; the EF particle is formed from a "core" protein complex (the reaction-center complex of PS II) plus attached aggregates of the pigment-proteins that form a LHC. This is diagrammatically described in Fig. 5. Further discussion of the dynamic interactions between the PS II complex and LHC are included in Sections VI,B and VIII,A.

The particles of the PF face remain the most highly problematic portion of the membrane in terms of identifying their biochemical identity. Since PS I preparations contain subunits of identical size as those of the PF face, it seems highly likely that at least some of these membrane subunits correspond to the PS I reaction-center–accessory-pigment complex and the Cyt f-b_6 complex. We have also suggested that

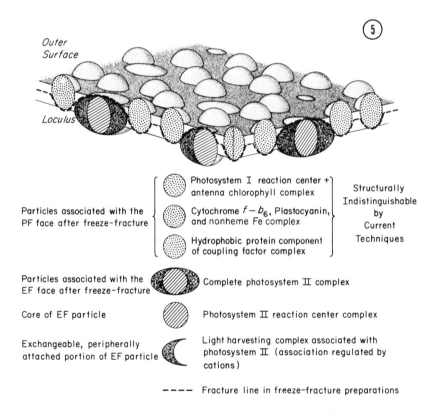

FIG. 5. A diagrammatic interpretation of the organization of functional components in the chloroplast membrane. Integral protein complexes (circular or elliptical structures) are embedded in a lipid bilayer continuum. Freeze-fracture of this bilayer results in segregation of the asymmetrically distributed complexes to either half of the membrane. Five different classes of protein complexes are drawn to be consistent with the data of Table I.

a portion of the particles are the F_0 complex of the chloroplast coupling factor. These conclusions are indicated diagrammatically in Fig. 5.

It should be noted that the peripheral CF_1 protein complex is not shown in Fig. 5. This is partially for artistic clarity in demonstrating positioning of the integral protein complexes; in addition, however, it must be recognized that the mechanism of association of the F_0 and CF_1 is not known (see further discussion by McCarty in Volume 7).

Several generalizations of the membrane model should also be emphasized. Relative sizes of the various parts of the membrane must be considered as approximations. For purposes of constructing the membrane model we have assumed a lipid bilayer thickness of 40 Å based on estimates of Kirk (1971, 1972). The particulate subunits embedded in the lipid phase of the membrane are drawn to scale assuming that the PF particles and the EF "core" complex are 70–80 Å spheres. While these diameters are consistent with measurements of freeze-fractured membranes, they could very easily be overestimates of the true protein complex size. Ruben and Telford (1975) have found that the shadowing procedures used in freeze-etching generate shadow widths that are 1.5–1.8 times larger than the true particle diameter. We have not attempted to correct this error factor due to the uncertainty of the needed correction factor. The correct ratio of particles and true *average* center-to-center spacing of particles has been represented.

We have oversimplified the structure of the membrane in Fig. 5 by drawing all protein complexes in smooth geometric shapes. In fact, high-resolution micrographs reveal that the particles observed are irregular in outline; unfortunately, however, we do not know to what extent these shapes may be generated by the sample preparation procedures. One exception is the subunit structure of the intramembranous particles on the inner (loculus) surface of the membrane. In early shadowed preparations of "quantasomes" (Park, 1965), it was reproducibly shown that the particles of the inner surface have four subunits (tetramers) when the subunits aggregate into crystalline arrays. More recent studies have shown that the ES particles (previously called quantasomes) may have 3–6 subunits (Staehelin, 1976). It is the view of this author that the subunit structure (sometimes appearing as a tetramer) results from aggregation of one or more LHC complexes with a "core" PS II complex; each individual complex appears as a separate subunit in the total supramolecular aggregate forming an EF (or ES) particle. Support for this idea comes from the observations of Miller *et al.* (1976) that barley mutant chloroplasts which contain a partial LHC polypeptide complement (see Section IC,B,1) show less distinct subunit structure.

V. Relating a Structural Membrane Model to Biophysical and Biochemical Properties of the Membrane Constituents

A. THE PHOTOSYSTEM II–LHC COMPLEX

In the membrane model of Fig. 5, the PS II complex is pictured as spanning the lipid bilayer. This positioning is consistent with numerous demonstrations that the primary charge separation in PS II generates an electrical potential across the membrane (Junge, 1975; Witt, 1975). It is generally accepted that components of the oxidizing side of PS II are localized near the inner surface of the membrane (Fowler and Kok, 1974a,b; Izawa and Ort, 1974; Trebst, 1974) whereas the reaction-center chlorophyll and electron acceptors on the reducing side of PS II are more surface localized (Arntzen *et al.*, 1974; Radunz *et al.*, 1971; Renger, 1976; Trebst, 1974). While no attempt has been made to indicate positioning of individual polypeptide components in this model, these observations on asymmetry of membrane constituents can easily be fitted within the existing structural interpretation.

Fluorescence spectroscopy has been frequently used to characterize system II photochemistry. Joliot and co-workers (1973; Dubertret and Joliot, 1974) have carefully analyzed the kinetics of fluorescence inductions to conclude that PS II units represent organized functional entities and that there is only one photoactive chlorophyll per complex. These data are totally compatible with the inclusion of a PS II reaction center complex in our model. Joliot *et al.* (1973) and Melis and Homann (1976) have concluded that PS II units are heterogeneous in size (with respect to the number of light-harvesting chlorophylls per reaction center). This is directly in agreement with our model; we believe that the structural heterogeneity of the EF particles which is observed in chloroplast lamellae can be explained on the basis of various amounts of LHC bound to the PS II centers. In recent years Kitajima and Butler (1975a,b) have used fluorescence and electron-transport analysis to characterize excitation energy transfer among chloroplast pigments. Butler (1976) has presented a functional picture of the pigment bed in which the light-harvesting pigment protein complex physically surrounds the PS II center. This is in direct agreement with our model of the PS II complex (based on structural analysis).

B. INTERACTIONS BETWEEN ELECTRON TRANSPORT CHAINS

It is now well accepted that electron flow from PS II to PS I passes through a series of carriers including plastoquinone, Cyt *f*, and plastocyanin (Trebst, 1974). Flash spectrophotometric studies have been used to demonstrate that the linear electron-transport chains are *not* independent of one another. It is estimated that as many as 10 chains could be interconnected via a common electron carrier—presumably the plastoquinone pool (Siggel *et al.*, 1972; Witt, 1975). Boardman *et al.* (1972) have analyzed electron transport in plants grown at various light intensities and have found that there is no universal stoichiometry between the amount of Cyt f (or other electron carriers) and P700. They proposed that individual electron carriers (such as Cyt f) do not belong to a specific electron-transport chain but are part of a limited pool that is common to several chains. Bouges-Bocquet (1975) has analyzed the recovery of PS I activity after flash-induced charge separation. Her data indicate that there is electron exchange between PS I centers. The electron-transport intermediate connecting the centers is after the plastoquinone pool; Bouges-Bocquet concluded that the common intermediate was plastocyanin.

Schmid and co-workers (1975) have studied the effect of a monospecific antibody against plastocyanin on electron transport. A sigmoidal curve was observed when the amount of antiserum added was plotted versus the amount of inhibition observed in PS I-mediated electron flow. This implies a cooperative effect of plastocyanin in serving various PS I centers. Further analysis of these data (Schmid, personal communication) revealed a Hill interaction coefficient of four, suggesting that four plastocyanin molecules act as common intermediates for any given system I center.

The observations cited above concerning interactions among intermediates in photosynthetic electron transport suggest the following pathway for photosynthetic electron transport:

$$\left\{ \begin{matrix} \text{PS II} \\ \text{reaction center} \end{matrix} \right\} \rightarrow \text{Plastoquinone pool} \rightarrow \left\{ \text{Cyt} f \right\} \rightarrow \text{Plastocyanin pool} \rightarrow \left\{ \begin{matrix} \text{PS I} \\ \text{reaction cente} \end{matrix} \right.$$

It is immediately obvious that this scheme for electron flow is compatible with the membrane structural model of Fig. 5. The electron-transport components indicated in brackets in the electron-transport scheme can be related directly to the PS II reaction-center complex, the Cyt f–b_6 complex, and the PS I reaction-center–accessory-pigment complex of the structural model. We can suggest that a pool of plastoquinone molecules acting as lipid-soluble electron carriers connects various PS II and cytochrome complexes. Plastocyanin may act as an electron shuttling agent between the cytochrome complex and P700, much as Cyt c acts as an electron carrier between complexes III and IV in mitochondria. It has been suggested that plastocyanin is loosely bound at the inner surface of the membrane where it mediates electron flow from Cyt f to P700 (Seidow *et al.*, 1973). It is also known that plastocyanin is very easily released from thylakoids by sonication (Seidow *et al.*, 1973). In both respects, plastocyanin shares biochemical features with Cyt *c* of mitochondria.

C. Lipid-Spanning Protein Complexes of Photosystem I and the Coupling Factor

All of the protein complexes corresponding to the small (PF) freeze-fracture particles of Fig. 5 are diagrammed as spanning the lipid phase of the membrane. This is consistent with the known functioning of the PS I reaction-center complex. Witt (1975) and Junge (1975) have demonstrated that P700-mediated electron transport results in a charge separation across the membrane. It is generally accepted that the electron donor on the oxidizing side of P700 is at or near the inner membrane surface whereas electron acceptors for PS I are near the outer membrane surface (Trebst, 1974).

It is also very likely that the hydrophobic protein (F_0) portion of the chloroplast coupling factor spans the membrane. When the CF_1 (peripheral protein) portion of the coupling factor is removed from the membrane, the F_0 acts as a proton ionophore releasing protons from the interior of the membrane (Racker, 1970). The fact that the F_0 binds CF_1 indicates its exposure at the membrane surface (Younis and Winget, 1977).

The membrane-spanning nature of the Cyt f-b_6 complex remains rather vague at the present time. Horton and Cramer (1974) believe that much of the Cyt f of the membrane is buried in the membrane based on accessibility to ferricyanide, as might be expected if this cytochrome were to act as an electron donor to plastocyanin near the inner surface of the thylakoid. Other details about the structural features of this complex are not available.

D. Stoichiometry of Membrane Components

In chloroplast lamellae there are more particles on the PF face than the EF face; for consistency in relating structure and function, these differences should relate to defined distributions of functional complexes.

As shown in Table II, the number of freeze-fracture particles on the PF face (using an average of PF_u and PF_s faces) is about 3400 particles/μm^2 and an average for the EF face (measured in unstacked lamellae to obtain uniform particle distributions) is about 1100 particles/μm^2. Haehnell (1976) has monitored absorption kinetics of light-induced chloroplast reactions to conclude that there is a ratio of one PS II reaction center to

1.13 PS I reaction centers. In measurements of the amount of coupling factor bound to chloroplast membranes, Strotmann *et al.* (1973) found one CF_1 per 800 chlorophylls (presumably also indicating one F_0/800 chlorophylls). There are thought to be 600 chlorophylls per photosynthetic electron transport chain (based on the finding that 2400 chlorophylls participate in a 4-electron transfer leading to O_2 evolution in flashing light; Govindjee and Govindjee, 1975). These values lead to the conclusion that there is 0.75 coupling factor per PS II unit. In contrast, estimates of Cyt *f* content in mature chloroplasts often fall near values of 1 Cyt *f* per 400 chlorophylls in normal chloroplasts grown at high light (Bennoun and Jupin, 1975; Boardman and Thorne, 1976), that is 1.5 Cyt *f* complexes per PS II unit.

In summary, we can calculate a ratio of [1 PS II complex] : [1.13 PS I complex + 0.75 HF_0 + 1.5 Cyt *f* complex] or 1:3.38. This value is not unreasonably different from the ratio of [1100 EF particles] : [3400 PF particles] (1:3.09). Assuming that the above estimates of content of the various membrane components are reliable approximations, the ratio of numbers of known different functional components of the membrane is very near that expected for ratios of various freeze-fracture particles based on the model of Fig. 5.

VI. Dynamic Structural Changes of Chloroplast Lamellae: Evidence from Microscopic Studies

A. MACROSCOPIC ORGANIZATIONAL CHANGES

Effects of light on chloroplast shape *in vivo* have been observed in many plant species (see review in Murakami *et al.*, 1975). Usually, chloroplasts flatten in the light and become more spherical in the dark (Miller and Nobel, 1972). The light-induced changes appear to be mediated by ion fluxes across the thylakoid membranes (Murakami *et al.*, 1975). The photoinduced plastid flattening was correlated with increased rates of photosynthesis *in vivo* and increased rates of photophosphorylation and CO_2 fixation in chloroplasts isolated from the illuminated leaves (Lin and Nobel, 1971; Nobel, 1968, 1970). Miller and Nobel (1972) demonstrated that overall changes in plastid volume *in vivo* are directly correlated to changes in the volume of individual thylakoids. The width of a thylakoid decreased from 195 Å in the dark to 152 Å in the light. The half-time for volume changes was less than 1 minute; the changes were blocked by uncouplers that affect ion transport across the membranes.

Studies of light-induced structural changes of chloroplasts *in vitro* have shown that there is a flattening of the membrane system in the light and a decrease in the internal thylakoid volume (Dilley and Giaquinta,

1975; Dilley *et al.*, 1967; Murakami and Packer, 1969; Murakami *et al.*, 1975). It is agreed that these changes are mediated by ion fluxes which are known to be coupled to electron transport [uptake of protons and parallel efflux of cations; (Dilley and Giaquinta, 1975; Hind *et al.*, 1974)]. Murakami and Packer (1969, 1970) have found that a decrease in the membrane thickness of isolated chloroplast lamellae can be induced either by illumination or by dark addition of protons to solution. These authors (Murakami and Packer, 1970) concluded that the temporal sequence of events after illumination is as follows: protonation (of fixed negative charges within the membrane as a result of H^+ uptake), changes in the environment within the membrane, change in membrane thickness, change in internal osmolarity accompanying ion movements with consequent flattening of the thylakoid, change in gross morphology of the inner chloroplast membrane system, and change in the gross morphology of the whole chloroplast. In the remainder of this review we will specifically concentrate on the second of these processes, that is, dynamic changes in membrane "environment."

B. MICROSTRUCTURAL CHANGES IN SUBUNIT ORGANIZATION

Wang and Packer (1973) used freeze-fracture analysis of isolated chloroplast lamellae to demonstrate that the particulate subunits of the membrane change in both size and density during sample illumination. They concluded that the protein (or lipoprotein) subunits have fluid mobility within the membrane and change their orientation in response to changing functional states of the membrane.

Wang and Packer (1973) also presented evidence that bivalent cations alter the density of freeze-fracture particles within the membrane. This phenomenon was more extensively examined by Ojakian and Satir (1974) in *Chlamydomonas* chloroplasts. They showed that the large-size class of freeze-fracture particles (EF face) are concentrated within appressed membrane regions in chloroplasts suspended in solutions containing $MgCl_2$. In low-salt solutions, free of $MgCl_2$, the particles were randomly distributed along the membranes. Ojakian and Satir confirmed earlier studies of Izawa and Good (1966) which had shown that chloroplasts suspended in solutions of low-cation content lose all grana stacking. The freeze-fracture studies demonstrated that conversion of stacked to unstacked lamellae is coupled with a dispersion of EF particles from concentrated clusters in appressed lamellae to uniform distribution along the unpaired membranes. It should be emphasized that there is conservation of the EF particles during unstacking; all changes in particle density are due to dispersal of the subunits (Ojakian and Satir, 1974). An example of particle distributions in unstacked membranes is shown in Fig. 6.

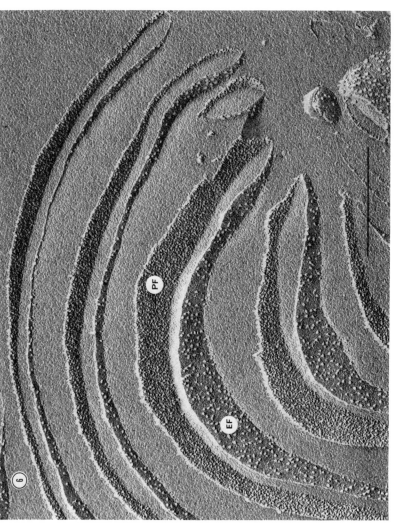

FIG. 6. The membrane substructure (revealed by freeze-fracture) of unstacked pea chloroplast membranes. This sample was from a chloroplast preparation suspended in 50 mM sodium Tricine to cause the loss of grana stacks. The relatively uniform distribution of EF particles in this figure should be compared with the segregation of subunits (EF_s particles) into grana stacks in Fig. 1. By courtesy of P. Armond. Bar = 0.5 μm.

More recently, Staehelin (1976) has characterized the distribution of freeze-fracture particles of stacked and unstacked membranes. He demonstrated that EF particle migration is fully reversible in higher-plant chloroplasts. In plastids suspended in low ionic strength, grana stacking is lost and the EF particles become uniformly distributed throughout the unstacked membranes. Upon readdition of salts to the unstacked membranes, grana stacking is reestablished and the EF particles become concentrated within the appressed membrane regions. During the unstacking–restacking process the density of particles along the PF face remains constant.

Staehelin (1976) observed that there are significant changes in the sizes of the EF particles that occur during the reversible unstacking–stacking process. In experimentally unstacked and then restacked membranes, the average EF_u particle appeared slightly smaller, whereas the average EF_s particle was slightly larger than in control specimens. These observations can be related to the findings of Arntzen et al. (1976; Arntzen and Ditto, 1976) that LHCs show cation-mediated reversible interactions with the PS II complex that forms the "core" of the EF particle. We may suppose that the low cation concentrations used in the unstacking experiments facilitated the dissociation of the LHC from PS II centers and subsequent redistribution of the LHC among different PS II "core" complexes to generate particles of new size classes. Light-induced changes in cation distribution within chloroplast membranes may also have affected LHC distribution on EF_s particles in the studies of Wang and Packer (1973), thus explaining their observations on changes in particle size in the light.

Garber and Steponkus (1976) have reported that cold acclimation of spinach seedlings causes a change in the average size of the EF chloroplast particles. In plastids isolated from freshly harvested green-house-grown leaves, the EF particles were heterogeneous in size, with particle-size maxima on dimension histograms at 100 and 165 Å. If plastids were isolated from leaves of seedlings acclimated to growth at 4°C, however, one size class of EF particles with an average diameter of 140 Å was observed. These EF particles were frequently found to be aggregated into paracrystalline arrays. On the basis of our model in Fig. 5, we can suggest that the cooling treatment (and associated effects on the lipid phase of the membrane) promoted uniform distribution of LHC among PS II centers within the membrane. These observations suggest several experiments. It would be very useful to know if uniformly sized PS II complexes generated in these cold-acclimation experiments result in appearance of homogeneous PS II units as assayed by fluorescence kinetics experiments (Joliot et al., 1973; Melis and Homann, 1976). It would also be of value to know how these changes affect quantum efficiency of the system.

C. SUMMARY

A wide variety of chloroplast structural changes has been observed *in vivo* and *in vitro*. Dynamic interactions among all components within the membranes are indicated by alterations in the thickness of individual membranes and by migration of intramembranous particles. These changes appear to be due to proton and cation fluxes across the membrane and changing local ionic concentrations within the membrane. The ion fluxes can be light induced; similar effects are mimicked by altering ionic conditions with isolated plastids. Large-scale (volume) changes of thylakoids or individual chloroplasts are probably also controlled by ion fluxes and changing osmotic conditions within the various compartments of the plastid.

VII.   Dynamic Structural Changes in the Functional Complexes of the Membrane: Evidence from Biochemical Analysis

A. CONFORMATIONAL CHANGES OF THE PHOTOSYSTEM II COMPLEX

Photoinduced transport of an electron from water to plastoquinone is mediated by an enzyme complex that includes the reaction-center chlorophyll of PS II, its associated primary electron acceptor, plus unidentified electron carriers on the oxidizing side of PS II. Cytochrome 559 is also part of this structural complex. Although this cytochrome undergoes light-induced redox changes, its exact role in the electron-transport process remains unknown (Cramer and Horton, 1975). The water-oxidizing enzyme on the oxidizing side of PS II acts as a charge accumulator. Each electron removed from the enzyme by the PS II center reults in a change in enzyme state (S_0 = fully reduced enzyme; S_1 = one accumulated oxidizing equivalent; $S_2 = 2+$; $S_3 = 3+$; $S_4 = 4+$). It is thought that Mn^{2+} may be involved in the functioning of water-oxidizing enzyme (Kok, 1975).

The functioning of the PS II complex is highly dependent on structural organization of its components. Oxygen evolution is inhibited when the structural integrity of the membrane is perturbed by treatment with chaotropic agents, heat, aging, proteases, or enzymic iodination (Arntzen *et al.*, 1974; Zilinskas and Govindjee, 1976). Membrane stabilization by glutaraldehyde fixation will partially block some of these effects. It has been suggested that a high degree of structural integrity of the PS II components is required to support a transmembrane electron flow (Arntzen *et al.*, 1974; Giaquinta *et al.*, 1974b).

In spite of the strict organizational requirements of the system, there is evidence that the PS II complex is not a static structural entity. Harth *et al.* (1974) found that illumination of chloroplasts at an external pH of 9, in the presence of an uncoupler such as gramicidin or nigericin, causes an irreversible inhibition of the oxygen-evolving enzyme. The inhibition did not occur in dark-treated membranes. Trebst and co-workers (1975; Harth *et al.*, 1974; Reimer and Trebst, 1975) and Cohn *et al.* (1975) have used this experimental system to conclude that oxygen-evolving enzyme is exposed at the inner surface of the thylakoid. Trebst and co-workers believe that the water-splitting enzyme undergoes a light-activated conformational change that disposes it to high pH inactivation. This is supported by recent studies of Briantais *et al.* (1977), who used flashing light to bring about the inhibition. They conclude that it is the S_2 state that is affected by the high pH. These data indicate that state changes of the water-oxidizing enzyme involve conformational changes of the enzyme and perhaps of the entire PS II complex.

Recently, Schmid *et al.* (1976 a,b) have isolated a polypeptide with an apparent molecular weight of 11,000. An antibody prepared against this polypeptide will react with chloroplast membranes causing an inhibition of activity of electron-transport components on the oxidizing side of the PS II center. This inhibition was only observed in the light. Schmid and co-workers have offered the suggestion that photoinduced conformational changes of the membrane may expose the 11,000 molecular weight polypeptide to the membrane surface, where it is susceptible to antibody binding. Even though the water-splitting enzyme may be internally localized in the membrane, antibody binding could disrupt the organization of the PS II complex and thus inhibit oxygen evolution. The idea of a light-induced change in the PS II complex is compatible with observations of Diner and Joliot (1976), who found that the light-induced charge separation across the membrane may activate system II reaction centers.

A series of studies using the relatively small, membrane impermeant diazonium reagent *p*-diazonium benzenesulfonate (DABS) have suggested that light-induced conformational changes occur in components on the reducing side of the PS II reaction center. Giaquinta *et al.* (1973, 1974a, 1975; Giaquinta and Dilley, 1975) found that there was increased binding of DABS to chloroplast membranes in the light; this increased binding was correlated to redox reactions occurring between "Q" (the primary electron acceptor of photosystem II) and plastoquinone. These authors argued that electron transport through this portion of the chain, and perhaps specifically through cytochrome 559, caused a conformational change in the membrane exposing new polypeptides which were highly reactive with DABS. Since a pH shift from high to low pH in the dark elicited similar increases in DABS binding, it was argued that the conformational change involved protonation of fixed negative charge groups on the membrane (Giaquinta and Dilley, 1975) which might be related to membrane structural changes involved in photophosphorylation (Giaquinta et al., 1975). More recently, Lockau and Selman (1976) have reinvestigated the light-induced DABS binding and have shown that photoreduction of the probe rather than a conformational change may be partially responsible for the increased reagent binding in the light.

Independent evidence for a conformational change in Cyt 559 has come from measurements of the redox potential of this component. Cramer and co-workers (1975; Cramer and Horton, 1975) have demonstrated that the cytochrome may function in the light at a midpoint potential considerably more negative than its dark midpoint potential. Horton and Cramer (1975) demonstrated that acid–base transitions can also influence the Cyt 559 potential, thus indicating that protonation of membrane proteins alters the microenvironment of the PS II complex near the cytochrome.

B. Conformational Changes of the Coupling Factor, the Photosystem I Complex, and the Cytochrome f–b_6 Complex

Electron transport in photosynthetic membranes results in the vectorial movement of protons and other ions across the membrane, giving rise to an electrochemical gradient, or protonmotive force (pmf). This pmf is thought to drive ATP synthesis via the activity of the coupling factor complex (Mitchell, 1965). Several lines of evidence suggest that conformational changes of the CF_1 complex are induced by the establishment of the pmf, and that these conformational changes are intimately involved in the chemical processes leading to ATP synthesis. Specific features of these conformational changes have been reviewed by Jagendorf (1975a,b) and by McCarty (in Volume 7) and will not be discussed further here.

There are relatively few studies that have detected or characterized structural changes in the functional components of the membrane that mediate PS I electron transport. Shneyour et al. (1973) have monitored NADP reduction in chloroplasts isolated from chilling sensitive plants. Both the photoreduction of NADP (with reduced DCPIP as electron donor) and ferredoxin–$NADP^+$ reductase activity showed increases in activation energies at 12°C. These increases were determined to be the result of a lipid-phase transition. These data suggest that fluid mobility of the PS I complex is necessary for optimal activity of components on the reducing side of PS I.

Anderson and Avron (1976) have recently demonstrated that photo-synthetic electron transport is required to activate several enzymes participating in the "dark" CO_2 fixation pathway. Modulation of the enzyme activities apparently involves a membrane-bound component of the PS I electron-transport chain preceding ferredoxin and a component at the level of ferredoxin. The activation of the CO_2 pathway enzymes seemed to involve light-induced structural changes of the PS I components that generate membrane-bound vicinal-dithiol groups.

In studies of ferricyanide-induced oxidation of Cyt f in chloroplast membranes, Horton and Cramer (1974) found that the cytochrome is more accessible to ferricyanide in the light than in the dark. These data suggest that there may be a light-induced conformational change in the Cyt f–b_6 complex that alters the exposure of some components to the membrane surface.

C. SUMMARY

There is evidence that at least some components of each of the protein complexes in the chloroplast membrane undergo light-induced conformational changes. The analysis of structural changes in CF_1 and the PS II reaction-center complex are, at present, most carefully documented. In some cases the structural changes probably result from changing ionic concentrations in the membrane, and others may occur directly as a result of redox reactions.

It seems likely that the structural changes detected by biochemical techniques are at least partially related to some of the membrane conformational changes described in the preceding section. All evidence for dynamic organizational features of the membrane are compatible with the membrane model of Fig. 5, in which protein aggregates can have fluid mobility within a lipid matrix. The fact that individual enzymic components can change in position with respect to their degree of surface exposure and interaction with other constituents points out the fact that we cannot confidently construct a membrane model that details exact localization of enzymic components; the highly stylized representation of regular geometric structures in Fig. 5 is necessitated by our current lack of more precise information on the dynamics of membrane structure.

VIII. Regulation of Excitation Energy Distribution: Changing Interactions among Membrane Pigment–Protein Complexes

The two-light reaction formulation for photosynthetic electron transport requires a balanced input of quanta into both photosystems for optimal efficiency. Although the light-harvesting pigment assemblies of

PS I and II are thought to be markedly different, the overall process of photosynthesis is quite wavelength independent. Myers (1971) has suggested that a regulatory mechanism exists which balances the distribution of absorbed excitation energy between the two photosystems. Two possibilities for the action of this mechanism have been hypothesized. Myers and co-workers (Bonaventura and Myers, 1969; Wang and Myers, 1974) suggest that the absorption cross section of each of the photosystems varies to compensate for unequal distribution of excitation energy between the two systems. Murata (1969) proposed that energy transfer from PS II to PS I can be regulated to adjust the distribution of excitation energy. Butler and co-workers (see Butler, 1976) have argued that both of these mechanisms are active and are not mutually exclusive. It has been argued by all these investigators (Bonaventura and Myers, 1969; Butler, 1976; Murata, 1969) that the mechanisms of regulation must involve structural changes of the pigment assemblies of the two photosystems.

To explain the kinetic properties of fluorescence changes associated with excitation energy migration, Butler and Kitajima (1975a) have proposed a tripartite model of the photochemical apparatus which incorporates three types of chlorophyll assemblages: (1) a PS I complex containing P700 and light-harvesting chlorophyll, (2) a PS II complex containing a reaction center and light-harvesting chlorophyll, and (3) a light-harvesting chlorophyll complex. The model has been used to mathematically define the pattern of distribution of energy in the entire apparatus. Based on kinetic analysis of several photosynthetic systems, Butler (1976) has recently presented a revision of the tripartite model in which the light-harvesting chlorophyll is arranged around the PS II center but is in contact with PS I. As will be discussed below, Butler's model is in direct agreement with the membrane formulation of Fig. 5 (which was originally based on entirely different sets of data).

A. CATION REGULATION OF "SPILLOVER": INVOLVEMENT OF THE
     LIGHT-HARVESTING COMPLEX

Murata (1969, 1971) demonstrated that cations regulate the distribution of excitation energy in isolated chloroplasts. He interpreted his data as indicating that "spillover" of energy from PS II to I occurs at low cation levels, whereas spillover is blocked by > 3 mM divalent cations or > 100 mM monovalent cations. Until recently there has been no biochemical evidence that would explain the mechanism of the cation effects.

Gross and Hess (1974) attempted to identify the site of cation by characterizing the binding sites for divalent cations on chloroplast membranes. One site has a dissociation constant of 51 ± 8 μM; binding to this site was correlated with divalent cation induced changes in chlorophyll fluorescence (used as a means of monitoring "spillover") and changes in membrane structure. Davis and Gross (1975) isolated a light-harvesting pigment protein that was solubilized from chloroplast lamellae through the use of sodium dodecyl sulfate (SDS). The isolated protein bound divalent cations with a dissociation constant of 32 μM. Gross and Davis suggested that the light-harvesting pigment protein could be involved in regulation of spillover in the intact membrane.

More extensive evidence for the role of the LHC in regulation of excitation energy distribution has come from studies of developing chloroplast membranes. Armond and co-workers (1976; Arntzen et al., 1976; Davis et al., 1976) have analyzed changing photochemical properties and pigment protein content of chloroplasts greened under intermittent light plus various periods of continuous illumination. Their data show a direct correlation between appearance of the light-harvesting pigment protein (complex II on polyacrylamide gels) and the onset of Mg^{2+} regulation of excitation energy distribution in the greening membranes. Grana stacking, salt-induced changes in membrane substructure, and an increase in the numbers of low-affinity cation-binding sites also appeared in concert with the onset of divalent cation effects (Armond et al., 1976, 1977; Arntzen et al., 1976; Davis et al., 1976).

The correlation between membrane LHC content and divalent cation regulation of "spillover" has also been shown using various higher-plant mutant chloroplasts. Vernotte et al. (1976) found that the Su/su mutant of tobacco contains very small amounts of LHC and shows insignificant cation-induced change in spillover. Arntzen et al. (1976) found that a soybean mutant having a reduced LHC content has reduced Mg^{2+} effects on excitation energy distribution. Boardman and Thorne (1976) have studied a Chl b-less barley mutant that does not contain CP II. They showed that there were small cation-induced changes in excitation energy distribution. Recently, polypeptide analysis of this mutant has revealed that some of the LHC protein is present even though there is no pigmented CP II (see discussion in the chapter by Boardman et al. in this volume). It therefore appears possible that the unpigmented portion of the LHC remaining in this mutant is responsible for the residual cation effects.

It has been recognized for several years that the presence of cations during detergent fractionation of chloroplasts is important for the recovery of a "heavy" PS II fraction (Anderson and Vernon, 1967; Argyroudi-Akoyunoglou, 1976). The mechanism by which cations controlled a "clean" separation of PS I and II was recently elucidated in studies of Arntzen and Ditto (1976). In the absence of cations, it was discovered that detergents solubilize the membrane into separate PS I, PS II, and LHC complexes; the LHC could be purified free of other membrane components by sucrose gradient fractionation. In the presence of cations, the LHC was bound tightly to the PS II complex to form a heavy fraction that could be separated from PS I. The amount of cations (either monovalent or divalent) needed to elicit detergent-resistant PS II–LHC interactions corresponded directly to the cation concentrations that regulate excitation energy distribution between PS II and PS I in the intact membrane. Arntzen and Ditto suggested that the cation-mediated associations between the PS II complex and LHC are primary factors regulating "spillover." These concepts are directly compatible with the membrane model of Fig. 5. The LHC complexes are shown to be separate, discrete membrane subunits which may be isolated free from other membrane components and may, in the intact membrane, undergo structural alterations that modify its association with the PS II complex and thus regulate energy migration through the pigment bed of the membrane. Isaakidou and Papageorgiou (1975a,b) had independently suggested that changing contact between membrane protein subunits could underlie cation induced "spillover" changes based on their analysis of ANS binding in lipid-depleted chloroplasts.

Murata *et al.* (1975) have analyzed changes in excitation energy distribution between PS I and PS II as a function of temperature in *Anacystis nidulans*. Their data demonstrated that the membrane conformational change needed for regulation of energy distribution was sensitive to the physical state of the membrane lipids. This is compatible with the idea that there may be changing interactions by large structural units within the membrane.

It should be pointed out that the cation-elicited structural changes of LHC may change the tertiary structure of the complex only slightly. Seely (1973) has used a computer simulation of the photosynthetic unit to show that slight changes in orientation of a few chlorophylls in the pigment bed could very significantly alter exciton migration.

B. The Physiological Role of Grana Stacking

Several recent reviews have considered the composition and detailed membrane structure of grana stacks (Anderson, 1975; Arntzen and Briantais, 1975; Park and Sane, 1971). It has been shown that most PS II centers and LHCs are concentrated within the appressed lamellar regions. It is recognized, however, that PS II activity does not depend upon stacked lamellae, nor does the absolute activity of any other electron transport or photophosphorylation reaction. It has been suggested that grana stacking may influence the efficiency of the photoreactions.

Anderson et al. (1973) have characterized the activity and structure of chloroplasts from plants grown in extreme shade. The plastids had low Ch1 a:b ratios and very large grana but did not differ from normally grown plastids in PS I or PS II activities. It was suggested that grana formation in these plastids may allow a highly efficient collection of quanta by allowing a higher density of light-harvesting assemblies.

In greening chloroplasts, Kirk and Goodchild (1972) established that the appearance of light-gathering systems is the rate limiting factor in the onset of photosynthetic activity and increased photosynthetic efficiency. Senger et al. (1975) extended these observations in a study of greening in a Scenedesmus mutant. They demonstrated a correlation between formation of stacked membranes and the appearance of maximal quantum efficiency of the photosystems. Armond et al. (1976) also found that the pattern of onset of grana stacking in greening pea membranes corresponded directly to increasing quantum efficiency of noncyclic electron transport. These authors argued that it was not the stacking which was the controlling factor in efficiency, however; rather, the parallel appearance of LHC and Mg^{2+} regulation of excitation energy distribution was the major factor controlling quantum yield. Armond et al. suggested that the greening plastids had poor quantum efficiency because there is unequal photon trapping between unequally sized, partially developed trapping units of PS I and PS II. Appearance of LHC, which allows regulation of energy distribution between the photosystems, was hypothesized to control light-harvesting efficiency.

Two questions remain important. First, does the LHC play a role in the stacking process? Second, is stacking per se involved in cation regulation of energy distribution between the two photosystems?

The evidence for a direct role of the LHC in eliciting membrane appression is now extensive (see review of Anderson, 1975). Anderson and Levine (1974a,b) have shown that polypeptides IIb and IIc (components of the LHC) are deficient in mutant chloroplasts having reduced stacking. There is also a parallel appearance of LHC and stacking during chloroplast greening (Armond et al., 1975; Davis et al., 1976). It should be noted that Miller et al. (1976) have shown that stacking does occur in a barley mutant lacking Chl b. Since this mutant does contain one of the polypeptides of the LHC (see above), it appears that a pigmented LHC is not required for membrane appression.

One very strong supporting argument for the involvement of the LHC in stacking relates to the effects of cations on structural organization of the lamellae. Izawa and Good (1966) demonstrated that chloroplasts suspended in a low-salt medium lose grana stacking. Membrane appression is reestablished upon addition of salts. Davis and Gross (1975) have shown that cations regulate association of LHC prepared from SDS chloroplast extracts. This has been extensively reexamined in a recent study of Burke *et al.* (1978). Isolated LHC (purified on sucrose gradients) from Triton X-100 solubilized, "low-salt" chloroplasts after the procedure of Arntzen and Ditto (1976) exists as small (50–80 Å) subunits in thin-sectioned preparations (see Fig. 7). In the presence of cations, these particles aggregate into a three-dimensional structure which is strikingly similar to an isolated "class II" chloroplast (Fig. 8). The same concentration range of cations that causes aggregation of the purified LHC causes the onset of stacking *in situ*. Burke *et al.* diagrammatically describe these correlations in Figs. 9–12. In "low-salt"

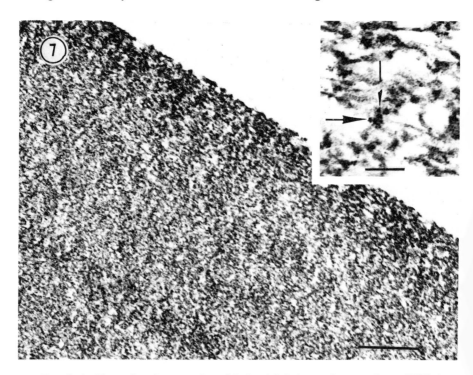

FIG. 7. A thin-sectioned preparation of isolated light-harvesting complexes (LHC) in low-salt solution. LHC was solubilized from "low-salt" chloroplast lamellae by procedures previously described (Burke *et al.*, 1978). To concentrate the material for microscopic preparation, solubilized complexes were prefixed in glutaraldehyde and pelleted by high speed centrifugation. Arrows in the inset figure indicate individual subunits. Association of subunits in this preparation is entirely due to the packing of materials during sedimentation; ultrafiltration of the material prior to centrifugation revealed only small complexes. By courtesy of J. Burke. Bar = 0.5 μm; inset bar = 0.1 μm.

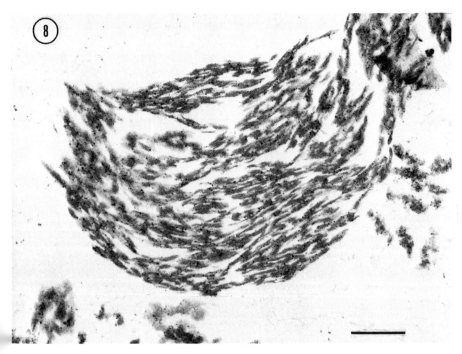

FIG. 8. Isolated light-harvesting complexes (LHC) prepared identically to those of Fig. 7 except that 10 mM MgCl$_2$ was added to the solubilized preparation causing LHC aggregation. The material shown pelleted out of solution at very low centrifugation speeds; the preparation had no photochemical activity and contained only three polypeptides of 23, 25, and 30 Kdaltons. By courtesy of J. Burke. Bar = 0.5 μm.

chloroplasts (and in "low-salt" preparations of LHC) there is no association of the LHC complexes. Upon addition of cations, isolated LHC complexes undergo a conformational change leading to aggregation (Fig. 8); we propose that cations induce LHC association with PS II complexes and between LHC complexes in the intact membrane. The latter process is achieved by cross-linking between two membranes. We have diagrammed the complete PS II complexes (PS II centers and associated LHC complexes) as not linking up directly "head-to-head" across two membranes, but rather as a staggered, connected series. This is consistent with direct structural studies of the EF particle orientation in stacked membranes by Staehelin (1975). This model can explain the fact that EF particles in freeze-fractured grana membrane often appear as linear chains of subunits (see Fig. 1). It may also relate to the cation induced segregation of EF particles into stacked membrane regions (see Section VI,B).

FIGS. 9–12. The effect of cations on the formation of grana stacks.
FIG. 9. Isolated chloroplast lamellae suspended in 10 mM NaCl. No regions of membrane appression are observed. Bar = 0.5 μm.

The question of whether grana stacks are essential for the regulation of "spillover" by cations remains unanswered. At present we can only say that cation inhibition of "spillover" (i.e., increased distribution of excitation energy to PS II centers) is *observed only* when there are stacked membranes (Arntzen *et al.*, 1976; Bennoun and Jupin, 1975; Gross and Prasher, 1974; Telfer *et al.*, 1976). It is possible to increase "spillover" (i.e., increased distribution of excitation energy to PS I centers) without losing stacking (Telfer *et al.*, 1976; Vernotte *et al.*, 1975). It is possible that cation effects on conformation of the LHC are 2-fold: one major effect elicits LHC aggregation (stacking) whereas a second related, minor change influences pigment orientation within the

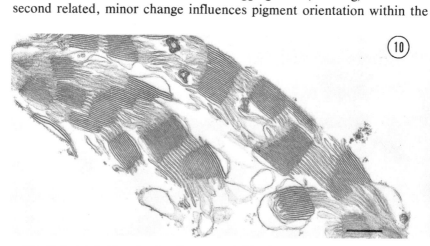

FIG. 10. Isolated chloroplast lamellae suspended in 10 mM NaCl + 3 mM MgCl$_2$. Well defined grana are present. Bar = 0.5 μm.

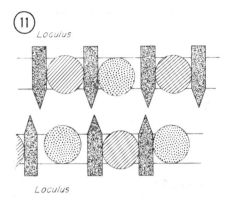

Fig. 11. A highly schematic representation of the proposed organization of LHC (shaded material) and PS II (slanted lines) in "low salt" media. The LHC is suggested to be loosely associated with the PS II center and not interacting with other LHC complexes.

LHC. Circular dichroism studies (Gregory, 1975; Gregory and Raps, 1974) have suggested that chlorophyll movement or chlorophyll–chlorophyll interactions may be related to ion fluxes and thylakoid stacking; definitive interpretations of these data are complicated by light-scattering effects (Philipson and Sauer, 1973; Schooley and Govindjee, 1976). It may be that further studies of isolated pigment–protein complexes will help clarify these problems.

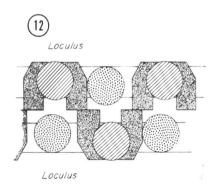

Fig. 12. A diagrammatic indication of the effect of cations (> 3 mM divalent or > 100 mM monovalent) on organization of LHC in the membrane. In the presence of cations, the LHC is suggested to undergo conformational changes that induce close association with the PS II complex, and a concomitant association of LHC units across two membranes. The LHC cross-linking is thought to be the factor inducing grana stacking (in a fashion analogous to the aggregation of subunits observed in purified LHC samples; Fig. 8). By courtesy of J. Burke.

C. ENERGY TRANSFER BETWEEN PHOTOSYSTEM II UNITS

As discussed in Section V,A, PS II units have been identified as discrete structural complexes based on kinetic studies. It is thought that energy transfer occurs between three, and probably more, connected PS II units (Joliot et al., 1973). This conclusion is based on the nonexponential character of the fluorescence induction observed in DCMU-poisoned chloroplasts (Joliot et al., 1973), on studies of the lifetime and yield of chlorophyll fluorescence in vivo (Moya, 1974), and upon the fact that the rate of O_2 evolution is nonlinear with respect to the concentration of open PS II reaction centers (Joliot et al., 1968). Dubertret and Joliot (1974) analyzed the energy transfer between PS II units in greening Chlorella membranes. Their data showed that the first units synthesized are isolated from one another and do not show energy transfer. Further membrane development, including the appearance of Chl b, resulted in membrane changes that allowed electronic excitation transfers. Armond et al. (1976) also found evidence for a gradual onset of energy transfer between PS II units in greening pea chloroplast membranes. They correlated the appearance of energy transfer with appearance of LHC and grana stacks.

Photon transfer between PS II units has been analyzed as a function of cation concentrations in solutions of isolated chloroplasts. Marsho and Kok (1974) found no evidence for energy transfer in plastids in low salts (10 mM KCl), whereas 3 mM MgCl$_2$ (+ 10 mM KCl) induced transfer. Similar data were obtained by Briantais et al. (1973) using fluorescence techniques. It is noteworthy that these same salt concentrations produce marked changes in membrane structure. Chloroplasts in low concentrations of KCl have unstacked membranes, whereas 3 mM MgCl$_2$ will elicit stacking (Figs. 9 and 10). Staehelin (1976) has shown that low Mg^{2+} levels also cause the EF freeze-fracture particles to reaggregate into appressed membrane regions.

As a working hypothesis to explain cation regulation of exciton transfer between PS II units, we suggest that LHC–LHC cross-linking between chloroplast membranes (see Fig. 12) is the mechanism which establishes interaction between units. This interaction could occur either due to shared pigment bed between the two membranes (i.e., exciton transfer between adjacent membranes via linked LHC subunits) or within the plane of the membrane due to close physical proximity of PSII–LHC subunits in the appressed lamellae region. The "obstacles" preventing exciton transfer between PSII units observed in greening studies (Armond et al., 1976; Dubertret and Joliot, 1974) may therefore be due to the spacial separation of the PSII complexes which occurs in the undifferentiated, unstacked membranes lacking LHC.

IX. Conclusions

The goal of this review has been to relate membrane structural data with recently accumulated knowledge of the functional activities of the lamellae, particularly with respect to our growing awareness of the dynamic membrane changes that occur during the photosynthetic process.

A membrane model (Fig. 5) has been developed that portrays the protein components of the membranes as polypeptide complexes embedded in a bimolecular lipid matrix. The two types of subunits visualized in the membrane by freeze-fracture techniques have been related to five different functional complexes (Table I), which account for all known light-harvesting, electron-transfer, and energy-coupling properties of the chloroplast membrane. The model is compatible with a range of known functional features of the membrane. Light-induced charge separations and ion transport can occur via the lipid-spanning protein complexes. Interactions between electron-transport chains can be mediated by soluble pools of electron carriers serving two or more protein complexes. Fluid mobility of discrete structural complexes of pigment-proteins allows for dynamic regulation of energy transfer between photoreactive centers.

It is recognized that the model as it now stands is highly diagrammatic and nonspecific in many details. It must be regarded as a working model, which will certainly see further modification as we learn more about the dynamic processes of photosynthetic energy coupling.

ACKNOWLEDGMENT

Thanks are due to Cathy Ditto, John Burke, and Gerry Miller for their assistance in preparing materials for this review. I am indebted to several colleagues with whom I have interacted in recent years for their discussions on chloroplast membrane structure-function; in particular I thank Drs. P. Armond, J.-M. Briantais, and L. A. Staehelin. This work was supported in part by NSF Grant No. PCM 77-18953.

REFERENCES

Anderson, J. M. (1975). *Biochim. Biophys. Acta* **416**, 191–235.
Anderson, J. M., and Levine, R. P. (1974a). *Biochim. Biophys. Acta* **357**, 118–126.
Anderson, J. M., and Levine, R. P. (1974b). *Biochim. Biophys. Acta* **333**, 378–387.
Anderson, J. M., and Vernon, L. P. (1967). *Biochim. Biophys. Acta* **143**, 363–376.
Anderson, J. M., Goodchild, D. J., and Boardman, N. K. (1973). *Biochim. Biophys. Acta* **325**, 573–585.
Anderson, L. E., and Avron, M. (1976). *Plant Physiol.* **57**, 209–213.
Apel, K., Bogorad, L., and Woodcock, C. L. F. (1975). *Biochim. Biophys. Acta* **387**, 568–579.
Argyroudi-Akoyunoglou, J. H. (1976). *Arch. Biochem. Biophys.* **176**, 267–274.

Argyroudi-Akoyunoglou, J. H., Feleki, Z., and Akoyunoglou, G. (1971). *Biochem. Biophys. Res. Commun.* **45**, 606–614.

Armond, P. A., and Arntzen, C. J. (1977). *Plant Physiol* **59**, 398–404.

Armond, P. A., Arntzen, C. J., Briantais, J.-M., and Vernotte, C. (1976). *Arch. Biochem. Biophys.* **175**, 54–63.

Armond, P. A., Staehelin, L. A., and Arntzen, C. J. (1977). *J. Cell Biol.* **73**, 400–418.

Arntzen, C. J., and Briantais, J.-M. (1975). *In* "Bioenergetics of Photosynthesis" (R. Govindjee, ed.), pp. 51–113. Academic Press, New York.

Arntzen, C. J., and Ditto, C. L. (1976). *Biochim. Biophys. Acta* **449**, 259–274.

Arntzen, C. J., Dilley, R. A., and Crane, F. L. (1969). *J. Cell Biol.* **43**, 16–31.

Arntzen, C. J., Dilley, R. A., and Neumann, J. (1971). *Biochim. Biophys. Acta* **245**, 409–424.

Arntzen, C. J., Dilley, R. A., Peters, G. A., and Shaw, E. R. (1972). *Biochim. Biophys. Acta* **256**, 85–107.

Arntzen, C. J., Vernotte, C., Briantais, J.-M., and Armond, P. (1974). *Biochim. Biophys. Acta* **368**, 39–53.

Arntzen, C. J., Armond, P. A., Briantais, J.-M., Burke, J. J. and Novitzky, W. P. (1976). *Brookhaven Symp. Biol.* **28**, 316–337.

Bennoun, P., and Jupin, H. (1975). *Proc. Int. Congr. Photosynth., 3rd, Rehovot, 1974* **1**, 163–169.

Boardman, N. K. (1968). *Adv. Enzymol. Relat. Areas Mol. Biol.* **30**, 1–79.

Boardman, N. K. (1972). *Biochim. Biophys. Acta* **283**, 469–482.

Boardman, N. K., and Anderson, J. M. (1964). *Nature (London)* **203**, 166–167.

Boardman, N. K., and Anderson, J. M. (1967). *Biochim. Biophys. Acta* **143**, 187–203.

Boardman, N. K., and Thorne, S. W. (1976). *Plant Sci. Lett.* **7**, 219–224.

Boardman, N. K., Björkman, O., Anderson, J. M., Goodchild, D. J., and Thorne, S. W. (1972). *Proc. Int. Congr. Photosynth. Res., 2nd, Stresa, 1971* **2**, 1603–1612.

Bonaventura, C., and Myers, J. (1969). *Biochim. Biophys. Acta* **189**, 366–383.

Bouges-Bocquet, B. (1975). *Biochim. Biophys. Acta* **396**, 382–391.

Branton, D. (1971). *Philos. Trans. R. Soc. London, Ser. B* **261**, 133–138.

Branton, D., and Park, R. B. (1967). *J. Ultrastruct. Res.* **19**, 283–303.

Branton, D., Bullivant, S., Gilula, N. B., Karnovsky, M. J., Moor, H., Mühlethaler, K., Northcote, D. H., Packer, L., Satir, B., Satir, P., Speth, V., Staehelin, L. A., Steere, R. L., and Weinstein, R. S., (1975). *Science* **190**, 54–56.

Briantais, J.-M. (1969). *Physiol. Veg.* **7**, 135–180.

Briantais, J.-M., Vernotte, C., and Moya, I. (1973). *Biochim. Biophys. Acta* **325**, 530–538.

Briantais, J.-M., Vernotte, C., Lavergne, J., and Arntzen, C. J. (1977). *Biochim. Biophys. Acta* **461**, 61–74.

Brown, J. S. (1972). *Photophysiology* **8**, 97–112.

Brown, J. S., Alberte, R. S., and Thornber, J. P. (1975). *Proc. Int. Congr. Photosynth. 3rd, Rehovot, 1974* **3**, 1951–1962.

Burke, J. J., Ditto, C. L., and Arntzen, C. J. (1978). *Arch. Biochem. Biophys.* **187** (in press).

Butler, W. L. (1976). *Brookhaven Symp. Biol.* **28**, 338–346.

Butler, W. L., and Kitajima, M. (1975b). *Proc. Int. Congr. Photosynth. 3rd, Rehovot, 1974* **1**, 13–24.

Butler, W. L., and Kitajima, J. (1975a). *Biochim. Biophys. Acta* **396**, 72–85.

Cohn, D. E., Cohen, W. S., and Bertsch, W. (1975). *Biochim. Biophys. Acta* **376**, 97–104.

Cramer, W. A., and Horton, P. (1975). *Photochem. Photobiol.* **22**, 304–308.

Cramer, W. A., Horton, P., and Wever, R. (1975). *In* "Electron Transfer Chains and Oxidative Phosphorylation" (E. Quagliariello, S. Papa, F. Palmieri, E. C. Slater, and N. Siliprandi, eds.), pp. 31–36. North-Holland Publ., Amsterdam.

DaSilva, P. P., and Branton, D. (1970). *J. Cell Biol.* **45**, 598–605.

Davis, D. J., and Gross, E. L. (1975). *Biochim. Biophys. Acta* **387**, 557–567.

Davis, D. J., Armond, P. A., Gross, E. L., and Arntzen, C. J. (1976). *Arch. Biochem. Biophys.* **175**, 64–70.

Dilley, R. A., and Giaquinta, R. T. (1975). *Curr. Top. Membr. Transp.* **7**, 49–107.

Dilley, R. A., Park, R. B., and Branton, D. (1967). *Photochem. Photobiol.* **6**, 407–412.

Diner, B. and Joliot, P. (1976). *Biochim. Biophys. Acta* **423**, 479–498.

Dubertret, G., and Joliot, P. (1974). *Biochim. Biophys. Acta* **357**, 399–411.

Edidin, M. (1974). *Annu. Rev. Biophys. Bioeng.* **3**, 179–201.

Fowler, C. F., and Kok, B. (1974a). *Biochim. Biophys. Acta* **357**, 299–307.

Fowler, C. F., and Kok, B. (1974b). *Biochim. Biophys. Acta* **357**, 308–318.

Garber, M. P., and Steponkus, P. L. (1974). *J. Cell Biol.* **63**, 24–34.

Garber, M. P., and Steponkus, P. L. (1976). *Plant Physiol.* **57**, 681–686.

Giaquinta, R. T., and Dilley, R. A. (1975). *Proc. Int. Congr. Photosynth. 3rd, Rehovot, 1974* **2**, 883–895.

Giaquinta, R. T., Dilley, R. A., and Anderson, B. J. (1973). *Biochem. Biophys. Res. Commun.* **52**, 1410–1417.

Giaquinta, R. T., Dilley, R. A., Anderson, B. J., and Horton, P. (1974a). *Bioenergetics* **6**, 167–177.

Giaquinta, R. T., Dilley, R. A., Selman, B. R., and Anderson, B. J. (1974b). *Arch. Biochem. Biophys.* **162**, 200–209.

Giaquinta, R. T., Ort, D. R., and Dilley, R. A. (1975). *Biochemistry* **14**, 4392–4396.

Goodenough, U. W., and Staehelin, L. A. (1971). *J. Cell Biol.* **48**, 594–619.

Govindjee and Govindjee, R. (1975). *In* "Bioenergetics of Photosynthesis" (Govindjee, ed.), pp. 1–50. Academic Press, New York.

Gregory, R. P. F. (1975). *Biochem. J.* **148**, 487–497.

Gregory, R. P. F., and Raps, S. (1974). *Biochem. J.* **142**, 193–201.

Gross, E. L., and Hess, S. C. (1974). *Biochim. Biophys. Acta* **339**, 334–346.

Gross, E. L, and Prasher, A. H. (1974). *Arch. Biochem. Biophys.* **164**, 460–468.

Haehnell, W. (1976). *Biochim. Biophys. Acta* **423**, 499–509.

Harth, E., Reimer, S., and Trebst, A. (1974). *FEBS Lett.* **42**, 165–168.

Henriques, F., and Park, R. B. (1975). *Plant Physiol.* **55**, 763–767.

Henriques, F., Vaughn, W., and Park, R. (1975). *Plant Physiol.* **55**, 338–339.

Hiller, R. G., Pilger, D., and Genge, S. (1973). *Plant Sci. Lett.* **1**, 81–88.

Hind, G., Nakatani, H. Y., and Izawa, S. (1974). *Proc. Natl. Acad. Sci. U.S.A.* **71**, 1484–1488.

Horton, P., and Cramer, W. A. (1974). *Biochim. Biophys. Acta* **368**, 348–360.

Horton, P., and Cramer, W. A. (1975). *FEBS Lett.* **56**, 244–247.

Howell, S. H., and Moudrianakis, E. N. (1967a). *J. Mol. Biol.* **27**, 323–333.

Howell, S. H., and Moudrianakis, E. N. (1967b). *Proc. Natl. Acad. Sci. U.S.A.* **58**, 1261–1268.

Huzisige, H., Usiyama, H., Kikuti, T., and Azi, T. (1969). *Plant Cell Physiol.* **10**, 441–455.

Isaakidou, J., and Papageorgiou, G. (1975a). *Arch. Biochem. Biophys.* **168**, 266–272.

Isaakidou, J., and Papageorgiou, G. (1975b). *Arch. Biochem. Biophys.* **175**, 541–548.

Izawa, S., and Good, N. E. (1966). *Plant Physiol.* **41**, 544–553.

Izawa, S., and Ort, D. R. (1974). *Biochim. Biophys. Acta* **357**, 127–143.

Jagendorf, A. (1975a). *In* "Bioenergetics of Photosynthesis" (Govindjee, ed.), pp. 413–492. Academic Press, New York.

Jagendorf, A. T. (1975b). *Fed. Proc. Fed. Am. Soc. Exp. Biol.* **34**, 1718–1722.

Joliot, P., Joliot, A., and Kok, B. (1968). *Biochim. Biophys. Acta* **153**, 635–652.

Joliot, P., Bennoun, P., and Joliot, A. (1973). *Biochim. Biophys. Acta* **305**, 317–328.

Junge, W. (1975). *Ber. Dtsch. Bot. Ges.* **88**, 283–301.

Kagawa, Y., and Racker, E. (1971). *J. Biol. Chem.* **246**, 5477–5487.

Ke, B., and Shaw, E. R. (1972). *Biochim. Biophys. Acta* **275**, 192–198.

Ke, B., Vernon, L. P., and Chaney, T. H. (1972). *Biochim. Biophys. Acta* **256**, 345–357.

Ke, B., Sahu, S., Shaw, E., and Beinert, H. (1974). *Biochim. Biophys. Acta* **347**, 36–48.

Ke, B., Sugahara, K., and Shaw, E. R. (1975). *Biochim. Biophys. Acta* **408**, 12–25.

Keck, R. W., Dilley, R. A., Allen, C. F., and Biggs, S. (1970). *Plant Physiol.* **46**, 692–698.

Kirk, J. T. O. (1971). *Annu. Rev. Biochem.* **40**, 161–196.

Kirk, J. T. O. (1972). *Proc. Int. Congr. Photosynth. Res., 2nd, Stresa, 1971* **3**, 2333–2347.

Kirk, J. T. O., and Goodchild, D. J. (1972). *Aust. J. Biol. Sci.* **25**, 215–241.

Klein, S. M., and Vernon, L. P. (1974). *Photochem. Photobiol.* **19**, 43–49.

Kok, B. (1975). *Annu. Rev. Biochem.* **44**, 409–429.

Levine, R. P., Burton, W. G., and Duram, H. A. (1972). *Nature (London), New Biol.* **237**, 176–177.

Lin, D. C., and Nobel, P. S. (1971). *Arch. Biochem. Biophys.* **145**, 622–632.

Lockau, W., and Selman, B. R. (1976). *Z. Naturforsch. C* **31**, 48–54.

Machold, O., Meister, H., Sagronsky, H., Hoeyer-Hansen, G., and Von Wettstein, D. (1977). *Photosynthetica* **11**, 200–206.

Marsho, T. V., and Kok, B. (1974). *Biochim. Biophys. Acta* **333**, 353–365.

Melis, A., and Homann, P. H. (1976). *Photochem. Photobiol.* **23**, 343–350.

Miller, K. R., and Staehelin, L. A. (1976). *J. Cell Biol.* **68**, 30–47.

Miller, K. R., and Staehelin, L. A. (1973). *Protoplasma* **77**, 55–78.

Miller, K. R., Miller, G. J., and McIntyre, K. R. (1976). *J. Cell Biol.* **71**, 624–638.

Miller, M. M., and Nobel, P. S. (1972). *Plant Physiol.* **49**, 535–541.

Mitchell, P. (1965). *Biol. Rev. Cambridge Philos. Soc.* **41**, 445–502.

Moya, I. (1974). *Biochim. Biophys. Acta* **368**, 214–227.

Murakami, S., and Packer, L. (1969). *Biochim. Biophys. Acta* **180**, 420–423.

Murakami, S., and Packer, L. (1970). *J. Cell Biol.* **47**, 332–351.

Murakami, S., Torres-Pereira, J., and Packer, L. (1975). *In* "Bioenergetics of Photosynthesis" (Govindjee, ed.), pp. 555–618. Academic Press, New York.

Murata, N. (1969). *Biochim. Biophys. Acta* **189**, 171–181.

Murata, N. (1971). *Biochim. Biophys. Acta* **226**, 422–432.

Murata, N., Troughton, J. H., and Fork, D. C. (1975). *Plant Physiol.* **56**, 508–517.

Myers, J. (1971). *Annu. Rev. Plant Physiol.* **22**, 289–312.

Nelson, N., and Neumann, J. (1972). *J. Biol. Chem.* **247**, 1817–1824.

Nelson, N., and Racker, E. (1972). *J. Biol. Chem.* **247**, 3848–3853.

Nobel, P. S. (1968). *Biochim. Biophys. Acta* **153**, 170–182.

Nobel, P. S. (1970). *Plant Cell Physiol.* **11**, 467–474.

Nolan, W. G., and Park, R. B. (1975). *Biochim. Biophys. Acta* **375**, 406–421.

Ojakian, G. K., and Satir, P. (1974). *Proc. Natl. Acad. Sci. U.S.A.* **21**, 2052–2056.

Oleszko, S., and Moudrianakis, E. N. (1974). *J. Cell Biol.* **63**, 936–947.

Ophir, I., and Ben-Shaul, Y. (1974). *Protoplasma* **80**, 109–127.

Park, R. B. (1965). *In* "Plant Biochemistry" (J. Bonner and J. E. Varner, eds.), pp. 124–150. Academic Press, New York.

Park, R. B., and Sane, P. V. (1971). *Annu. Rev. Plant Physiol.* **22**, 395–430.

Philipson, K. D., and Sauer, K. (1973). *Biochemistry* **12**, 3454–3458.

Phung-Nhu-Hung, S., Lacourly, A., and Sarda, C. (1970). *Z. Pflanzenphysiol.* **62**, 1–16.

Racker, E. (1970). *In* "Membranes of Mitochondria and Chloroplasts" (E. Racker, ed.), pp. 127–171. Van Nostrand, New York.

Radunz, A., Schmid, G. H., and Menke, W. (1971). *Z. Naturforsch. B* **26**, 435–446.

Reimer, S., and Trebst, A. (1975). *Biochem. Physiol. Pflanz.* **168**, 225–232.

Renger, G. (1976). *Biochim. Biophys. Acta* **440**, 287–300.

Ruben, G. C., and Telford, J. N. (1975). *Annu. Proc. Electron Microsc. Soc. Am., 33rd* pp. 282–283.

Schmid, G. H. (1972). *Proc. Int. Congr. Photosynth. Res. 2nd, Stresa, 1971a* **2**, 1603–1612.

Schmid, G. H., Radunz, A., and Menke, W. (1975). *Z. Naturforsch. C* **30**, 201–212.

Schmid, G. H., Menke, W., Koenig, F., and Radunz, A. (1976a). *Z. Naturforsch. C* **31**, 304–311.

Schmid, G. H., Renger, G., Graser, M., Koenig, F., Radunz, A., and Menke, W. (1976b). *Z. Naturforsch. C* **31**, 594–600.

Schooley, R. E., and Govindjee (1976). *FEBS Lett.* **65**, 123–125.

Seely, J. R. (1973). *J. Theor. Biol.* **40**, 173–187.

Seidow, J. N., Curtis, V. A., and San Pietro, A. (1973). *Arch. Biochem. Biophys.* **158**, 889–897.

Senger, H., Bishop, N. I., Wehrmeyer, W., and Kulandaivelu, J. (1975). *Proc. Int. Congr. Photosynth. 3rd, Rehovot, 1974* **3**, 1913–1923.

Serrano, R., Kanner, B. I., and Racker, E. (1976). *J. Biol. Chem.* **251**, 2453–2461.

Shibata, K. (1971). *In* "Photosynthesis and Nitrogen Fixation," Part A (A. San Pietro, ed.), Methods in Enzymology, Vol. 23, pp. 296–302. Academic Press, New York.

Shneyour, A., Raison, J. K., and Smillie, R. M. (1973). *Biochim. Biophys. Acta* **292**, 152–161.

Siggel, U., Regner, G., Stiehl, H. H., and Rumberg, B. (1972). *Biochim. Biophys. Acta* **256**, 328–335.

Singer, S. J. (1974). *Annu. Rev. Biochem.* **43**, 805–833.

Singer, S. J., and Nicolson, G. (1972). *Science* **175**, 720–731.

Staehelin, L. A. (1975). *Biochim. Biophys. Acta* **408**, 1–11.

Staehelin, L. A. (1976). *J. Cell Biol.* **71**, 136–158.

Staehelin, L. A., Armond, P. A., and Miller, K. R. (1976). *Brookhaven Symp. Biol.* **28**, 278–315.

Strotmann, H., Hesse, H., and Edelmann, K. (1973). *Biochim. Biophys. Acta* **314**, 202–210.

Stuart, A. L., and Wasserman, A. R. (1973). *Biochim. Biophys. Acta* **314**, 284–297.

Telfer, A., Nicolson, J., and Barber, J. (1976). *FEBS Lett.* **65**, 77–83.

Thornber, J. P. (1975). *Annu. Rev. Plant Physiol.* **26**, 127–158.

Trebst, A. (1974). *Annu. Rev. Plant Physiol.* **25**, 423–458.

Trebst, A., Reimer, S., and Hauska, G. (1975). *In* "Electron Transfer Chains and Oxidative Phosphorylation" (E. Quagliariello, S. Papa, F. Palmieri, E. C. Slater, and N. Siliprandi, eds.), pp. 343–350. North-Holland Publ., Amsterdam.

Vernon, L. P., and Klein, S. M. (1975). *Ann. N. Y. Acad. Sci.* **244**, 281–295.

Vernon, L. P., and Shaw, E. R. (1971). *In* "Photosynthesis and Nitrogen Fixation," Part A (A. San Pietro, ed.), Methods in Enzymology, Vol. 23, pp. 277–289. Academic Press, New York.

Vernon, L. P., Shaw, E. R., Ogawa, T., and Raveed, D. (1971). *Photochem. Photobiol.* **14**, 343–357.

Vernotte, C., Briantais, J.-M., Armond, P., and Arntzen, C. J., (1975). *Plant Sci. Lett.* **4**, 115–123.

Vernotte, C., Briantais, J.-M. and Remy, R. (1976). *Plant Sci. Lett.* **6**, 135–141.

Wang, A. Y.-I., and Packer, L. (1973). *Biochim. Biophys. Acta* **305**, 488–492.

Wang, R. T., and Myers, J. (1974). *Biochim. Biophys. Acta* **347**, 134–140.

Wessels, J. S. C., and Borchert, M. T. (1975). *Proc. Int. Congr. Photosynth. 3rd, Rehovot, 1974* **1**, 473–484.

Wessels, J. S. C., and van Leeuwen, M. J. F. (1971). *In* "Energy Transduction in Respiration and Photosynthesis" (E. Quagliariello, S. Papa, and C. S. Rossi, eds.), pp. 537–550. Adriatica Editrice, Bari.

Wessels, J. S. C., and Voorn, G. (1972). *Proc. Int. Congr. Photosynth. Res. 2nd, Stresa, 1971* **1**, pp. 833–845.

Wessels, J. S. C., Van Alphen-Van Waveren, O., and Voorn, G. (1973). *Biochim. Biophys. Acta* **292**, 741–752.

Witt, H. T. (1975). *In* "Bioenergetics of Photosynthesis" (Govindjee, ed.), pp. 493–554. Academic Press, New York.

Younis, H. M., and Winget, C. D. (1977). *Biochem. Biophys. Res. Commun.* **77**, 168–174.

Zilinskas, B. A., and Govindjee, R. (1976). *Z. Pflanzenphysiol.* **77**, 302–314.

Structure and Development of the Membrane System of Photosynthetic Bacteria

GERHART DREWS

Lehrstuhl für Mikrobiologie
Institut für Biologie 2 der Albert Ludwigs-Universität
Freiburg, West-Germany

I. Introduction

Two groups of prokaryotic organisms, the Cyanobacteria and the phototrophic bacteria (Rhodospirillales) (Buchanan and Gibbons, 1974), have developed a photosynthetic apparatus during evolution.

The photosynthetic apparatus (PSA)* of all phototrophic prokaryotes is membrane bound but not enclosed in a compartment comparable with chloroplasts of green plants. The PSA of Cyanobacteria, in terms of composition and function is similar to that of some algae (Chrysomonadales, Rhodophyceae) and will not be discussed in this chapter. Cyano-

---

* List of nonstandard abbreviations: PSA, photosynthetic apparatus; BChl, bacteriochlorophyll; ICM, intracytoplasmic membrane; CM, cytoplasmic membrane; LH, light harvesting (antenna); RC, reaction center; *R.*, *Rhodospirillum*; *Rps.*, *Rhodopseudomonas*; B870, LH-BChl with main IR absorption band at 870 nm; B850, LH-BChl with main IR absorption bands at 800 + 850 nm; IR, infrared.

bacteria perform an oxygenic photosynthesis with two photosystems. Chlorophyll a, phycobiliproteins, and the two most common carotenoids, β-carotene and zeaxanthin, are the principal photosynthetic pigments (Stanier, 1974).

The phototrophic green and purple bacteria, in contrast, carry out an anoxygenic photosynthesis with one photosystem functioning mainly in membrane energization and ATP production. They use electron donors, such as reduced sulfur compounds, molecular hydrogen, or organic compounds. Instead of oxygen, the corresponding oxidation products from photosynthesis are sulfate, protons, more oxidized organic compounds, and CO_2. Photosynthetic pigments are the bacteriochlorophylls a, b, c, d, or e and a great variety of aliphatic, mono-, or bicyclic carotenoids (Pfennig, 1978). In most photosynthetic bacteria (Rhodospirillaceae, Chromatiaceae) the PSA is localized on intracytoplasmic membranes (Oelze and Drews, 1972) known as chromatophores (Schachmann et al., 1952) or thylakoids (Drews and Giesbrecht, 1963). Small amounts of PSA are also detectable in the cytoplasmic membrane (CM) under special growth conditions. In green bacteria (Chlorobiaceae) the reaction center (RC) seems to be localized in the CM, whereas the light-harvesting (LH) complex is in the chlorobium vesicles, which are attached to the CM (Olson et al., 1976c). The terminology of PSA-supporting structures has been discussed recently (Oelze and Drews, 1972).

Some members of Rhodospirillaceae are able to produce ATP coupled either to respiratory or photochemical electron flow. The amount of membranes per cell, its composition, and functional patterns were found to be altered in response to growth conditions. Thus, these species seemed to be suitable systems for study of cell differentiation in a prokaryotic organism. Structure, function, and development of the bacterial PSA have been recently reviewed (Oelze and Drews, 1972; Baltscheffsky and Baltscheffsky, 1974; Dutton and Wilson, 1974; Parson, 1974; Parson and Cogdell, 1975; Clayton and Sistrom, 1978; Kaplan, 1978).

II. Structure and Composition of Membranes of Photosynthetic Bacteria

A. MORPHOLOGY, SUBSTRUCTURE, AND PHYSICAL PROPERTIES OF INTRACYTOPLASMIC MEMBRANES

1. Morphology and Substructure

Under conditions of photosynthetic growth, all members of Chromatiaceae and Rhodospirillaceae (Pfennig and Trüper, 1974) develop intracytoplasmic membranes (ICM) that form species-specific patterns (Oelze and Drews, 1972; Pfennig and Trüper, 1974). The basic structures are vesicles, tubules, and double membranes, which are connected to each other and/or to the CM, at least during morphogenesis. They are clustered to form granalike stacks or bundles of tubules, and in some cases they form branched membrane or vesicle aggregates or irregular and small membrane structures (Oelze and Drews, 1972; Pfennig and Trüper, 1974).

In cross sections of fixed, dehydrated, and embedded cells ICM and CM exhibited the typical double-track structure of the unit membrane with a total thickness of 70–80 Å. The surface of most of the negative-stained membranes appeared to be smooth. However, negative-stained and freeze-etch preparations of *Rhodospirillum* (*R.*) *rubrum* ICM vesicles showed a pattern of regularly distributed particles (Ketchum and Holt, 1970; Oelze and Golecki, 1975). On the surface of *Rhodopseudomonas* (*Rps.*) *viridis* ICM linearly arranged particles, forming a periodic pattern with an average distance of 100 Å, were detected (Giesbrecht and Drews, 1966). Stalked knoblike structures were observed on ICM of *Rps. capsulata* which have about the same distance as the surface particles on *Rps. viridis* ICM and a diameter of about 80 Å (Lampe *et al.*, 1972). The knoblike particles were found in many photosynthetic bacteria on the cytoplasm side of membranes (Löw and Afzelius, 1964; Lampe *et al.*, 1972). It has been suggested that these surface particles are coupling factors (Reed and Raveed, 1972a).

By the freeze-fracturing process, membranes were split within the hydrophobic inner regions (Branton, 1966). The exposed inner fracture faces of membranes from photosynthetic bacteria exhibited "rough," particle covered, concave surfaces and smooth convex surfaces (Takacs and Holt, 1971; Reed and Raveed, 1972a; Golecki and Drews, unpublished observations) (Fig. 1). An asymmetric distribution of particles on inner fracture faces was also observed in other electron-transport membranes (Altendorf and Staehelin, 1974). The diameter of the particles was 60–80 Å in membranes of *Thiocapsa roseopersicina,* 95 Å in membranes of *Rps. capsulata*, 85 Å in membranes of *Rps. palustris*, and 100 Å in membranes of *R. rubrum* (Takacs and Holt, 1971; Golecki and Drews, unpublished observations). The particle distribution varied depending on growth conditions. In *Rps. palustris* 4220 particles/μm^2 were observed in CM of aerobically grown cells, 5235/μm^2 in CM of semiaerobically grown cells, and 6250 in ICM of cells grown anaerobically in the light (Golecki and Drews, unpublished observations). A double ICM (thylakoid) of *Rps. viridis* showed in a perpendicular cross-fracture a total thickness of 190–200 Å. The intrinsic particle pattern had a period of about 100 Å (Giesbrecht and Drews, 1966). More than one type of particle seemed to be present.

FIG. 1. Inner fracture faces of intracytoplasmic membranes isolated from *Rhodopseudomonas capsulata*. The concave faces are covered with particles; the convex faces are smooth. The marker bar represents 100 nm. Micrograph by courtesy of J. Golecki.

Although the particles found in bacterial membranes are not functionally identified, from the structural point of view it seemed to be clear that they are different in size and distribution from the particles found in chloroplast membranes (Staehelin, 1975).

2. Isolation of Membranes; Physical Properties of Membrane Fractions

The methods of isolation and purification of membranes from photosynthetic bacteria have been recently discussed (Niederman, 1978). Crude membrane fractions were separated and purified by density gradient centrifugation (Worden and Sistrom, 1964; Gibson, 1965; Holt and Marr, 1965b; Oelze *et al.*, 1969a,b; Fraker and Kaplan, 1971; Niederman and Gibson, 1971; Lampe *et al.*, 1972; Niederman *et al.*, 1972; Oelze *et al.*, 1975; Collins and Niederman, 1976a). The fractions of CM and ICM differ in their buoyant densities as shown in Table I. CM has always a lower density in comparison with ICM. The so-called heavy fraction in sucrose density gradients was identified as an ICM fraction contaminated with cell wall material. This fraction has a density of about 1.303 gm/cm^3 in CsCl (Collins and Niederman, 1976b) and 1.22 or 1.24 in sucrose density gradients (Oelze *et al.*, 1975; Collins and Niederman, 1976b). The different density values result from the substances forming the gradient, which have different osmotic effects (Table I). A fraction of intermediate density was isolated from aerobically grown cells of *Rps. capsulata* 37b4 consisting of tubular membranes with respiratory and photosynthetic activities (Lampe, 1972). A heavy fraction with reversed orientation of intramembrane particles and reversed H$^+$ flow in comparison with intracytoplasmic vesicles was described by Hochman *et al.* (1975).

As already discussed, a continuous membrane system was demonstrated in ultrathin sections of whole cells (Boatman and Douglas, 1961; Drews and Giesbrecht, 1963, 1965; Holt and Marr, 1965a,b; Tauschel and Drews, 1967; Weckesser *et al.*, 1969) and in preparations of spheroplasts and ghosts (Giesbrecht and Drews, 1962; Boatman, 1964; Hurlbert *et al.*, 1974). The continuity of the whole-membrane system is presumably a consequence of morphogenetic events (Oelze and Drews, 1972). The dynamic state of a membrane system in a living cell, however, includes the possibility that some parts of the whole membrane system become reversibly detached or fused by membrane flow (Singer, 1974). This process has been demonstrated not only in living cells, but also in *in vitro* systems by formation of vesicles from CM of ghosts (Kaback, 1974; Hellingwerf *et al.*, 1975).

Good evidence that membrane components are arranged in the same asymmetric manner in CM and ICM comes from measurement of H$^+$ transport, localization of ATPase and electron-transport components, and the particle distribution in inner fracture faces. In consequence, in whole cells proton translocation, associated with light-dependent electron flow or respiration, is directed outward. In isolated vesicular ICM fragments the proton translocation is directed into vesicles (Chance *et al.*, 1966; von Stedingk and Baltscheffsky, 1966; Scholes *et al.*, 1969; Hochman *et al.*, 1975). ATPase is always localized on the cytoplasmic side of membranes (Reed and Raveed, 1972b). Cytochrome c$_2$ was found in the periplasmatic space or in the intractyoplasmic vesicles (Prince *et al.*, 1974b, 1975; Hochman *et al.*, 1975) (Fig. 2).

TABLE I

BUOYANT DENSITIES OF MEMBRANE FRACTIONS

| Strain | Material for preparation of gradients | Buoyant densities (gm/cm³)[a] | | | References |
		CM	ICM	Cell wall	
Rhodospirillum rubrum	Sucrose	1.145	1.165	1.246	Oelze *et al.* (1975); Oelze (1976)
R. rubrum	Sucrose	—	1.17	—	Ketchum and Holt (1970)
R. rubrum	Sucrose	1.12	—	1.22	Collins and Niederman (1976b)
R. rubrum	Ficoll	—	1.07	—	Ketchum and Holt (1970)
R. rubrum	NaBr	—	1.20	—	Ketchum and Holt (1970)
R. rubrum	CsCl	1.166	1.91	1.303	Collins and Niederman (1976b)
Rhodopseudomonas sphaeroides	CsCl	1.181	—	1.240	Ding and Kaplan (1976)
Rps. sphaeroides	—	—	1.17–1.18	—	Gibson (1965)

[a] All values are from equilibrium density gradient centrifugations. CM, cytoplasmic membrane; ICM, intracytoplasmic membranes.

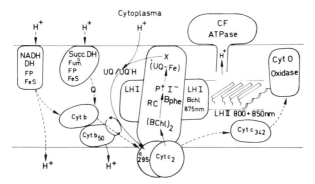

FIG. 2. Hypothetical scheme of the localization of electron-transport components from *Rhodopseudomonas sphaeroides* and *Rps. capsulata*. As far as known, experimental evidence is given in the text.

B. ORGANIZATION OF THE PHOTOSYNTHETIC APPARATUS

1. Components

All members of Rhodospirillaceae and Chromatiaceae have only one type of BChl, alternatively BChl a or b (Pfennig and Trüper, 1974). BChl a and b are the only bacteriochlorophylls photochemically active in the reaction center. The major esterifying alcohol of BChl a and BChl b is phytol (a_p). However, BChl a (a_{Gg}) of R. rubrum contains all-*trans*-geranylgeraniol instead of phytol (Katz et al., 1972; Brockmann et al., 1973; Künzler and Pfennig, 1973). A few species of purple bacteria contain small amounts of BChl b ($\leq 5\%$), which is esterified with all-*trans*-geranylgeraniol (Gloe and Pfennig, 1974). It is unknown whether BChl a_p and BChl a_{Gg} in those strains that have both derivatives are in different pigment-protein complexes. In green and brown Chlorobiaceae BChl c, d, or e occurs as the major BChls in addition to BChl a (Pfennig and Trüper, 1974; Gloe et al., 1975). Gloe et al. (1975) summarized the structures of all BChls.

Bchls bound to membranes of living cells show numerous spectral forms (Thornber et al., 1978). The strong red shift of infrared (IR) maxima and formation of two to five special absorption bands in the IR region are the chief characteristics. The variations in spectra were thought to be indicative of alterations in the molecular interactions of BChls or interactions of BChl with the environment. The absorption spectra of the monomeric form of BChl in organic solution differ greatly, particularly in the near infrared region, from absorption spectra of BChl molecules that are in the aggregated state *in vitro* (Litvin and Krasnovsky, 1957, 1958; Katz and Norris, 1973) or *in vivo* associated to membranes (Wassink *et al.*, 1939). It was originally proposed that each absorption band or spectral form of pigment corresponded to a particulate BChl–protein complex (Katz and Wassink, 1939; Wassink *et al.*, 1939; French and Young, 1956). Later, when data on aggregated BChl molecules *in vitro* were published (Krasnovsky *et al.*, 1952; Litvin and Krasnovsky, 1957, 1958; Katz *et al.*, 1966; Krasnovsky, 1969; Katz and Norris, 1973), the individual absorption bands *in vivo* were attributed to various states of aggregation of BChl molecules (Clayton, 1965). However, strong BChl–protein interactions *in vivo* have been proved by X-ray diffraction studies on the native BChl a–protein complex from *Chlorobium limicola* strain 2 K (Fenna and Matthews, 1975). The observation that addition of the proteolytic enzyme, α-chymotrypsin, rapidly shifted the 850 nm peak to 780 nm supports the idea that protein plays the major role in maintenance of the special environment for the BChl in the membrane (Reed *et al.*, 1970). Electron nuclear double resonance (ENDOR) experiments have proved that in reaction-center preparations the odd electron produced in the primary photochemical reaction is shared between two BChl molecules (Feher *et al.*, 1973, 1975; Norris *et al.*, 1973, 1975). This indicates that the primary donor is a specialized BChl dimer (Norris *et al.*, 1971). In conclusion, BChl–BChl and BChl–protein interactions and possibly BChl–lipid interactions contribute in varying extents to the different spectral forms of BChl.

Five groups of carotenoids were found in purple bacteria: (1) normal spirilloxanthin series, (2) rhodopinal branch of spirilloxanthin series, (3) alternative spirilloxanthin series, (4) okenone series, (5) isorenieratene series (Pfennig and Trüper, 1974; Pfennig, 1978). The distribution of these carotenoids within the photosynthetic bacteria is characteristic (Pfennig, 1967) and has some taxonomic relevance (Pfennig and Trüper, 1974). The comparative chemistry of these carotenoids has been reviewed recently (Liaanen Jensen, 1967; Schmidt, 1978). The carotenoids are responsible for the red, purple, or brownish color of bacterial suspensions. They protect cells from photooxidative killing (Sistrom *et al.*, 1956) and transfer the adsorbed light energy very effectively to the reaction center (Cogdell *et al.*, 1976).

Energy conservation coupled to light-induced electron and proton transport is dependent on electron and proton transfer carriers including cytochromes, nonheme iron proteins, quinones, flavins, and pterins (Horio and Kamen, 1970; Kamen and Horio, 1970; Kennel and Kamen, 1971; Oelze and Drews, 1972; Shanmugam *et al.*, 1972; Meyer *et al.*, 1973; Hall *et al.*, 1974, 1975; Yoch *et al.*, 1975; Jennings and Evans, 1976; Bartsch, 1978; San Pietro, 1978). Redox potentiometry (Dutton and Wilson, 1974), measurement of flash-induced kinetics in the picosecond and nanosecond range (Parson and Cogdell, 1975; Prince and Dutton, 1976) and electron paramagnetic resonance (EPR) spectroscopy (Okamura *et al.*, 1975) of these redox carriers have greatly increased our knowledge on the functions of these compounds. It is beyond the scope of this article to review the recent literature on chemistry and functions of these compounds, which are active in the different electron-transport system of photosynthetic bacteria.

2. *Arrangement of Polypeptides, Pigments, and Redoxcomponents in the Photosynthetic Apparatus and in the Membrane.*

Three main events occur in the photosynthetic apparatus: (1) absorption of light quanta in the antenna pigment complexes; (2) channeling of excitation energy from antenna to reaction centers where, through a charge separation, an electric field is produced across the membrane; (3) electron and proton transport and coupled phosphorylation in consequence of the primary events.

Spectroscopic studies suggest that BChl forms different complexes in membranes (Fig. 2). Some pigment-protein complexes have been isolated and characterized (Bril, 1958, 1960; Clayton, 1962; Olson and Romano, 1962; Garcia *et al.*, 1966). The proteins always have a high content of apolar residues, a very low cysteine content, and weights in the range of 10 to 40 kdaltons (Schmitz, 1967; Smith *et al.*, 1972; Feick and Drews, 1978). Since reaction centers are the best characterized BChl–protein complexes, they will be discussed first, followed by the antenna pigment complexes.

a. Reaction-Center (RC) Preparations. The discovery of a specialized BChl that shows reversible photoinduced absorbance changes in the near-infrared region (Duysens, 1952; Clayton, 1963), a characteristic EPR signal (Feher and Okamura, 1976), and resistance to photochemical or chemical oxidation (Clayton, 1963; Loach *et al.*, 1963) led to isolation and characterization of RC particles (Garcia *et al.*, 1968; Reed and Clayton, 1968). The first isolated preparations were found to be heterogeneous in size and composition (reviewed in Oelze and Drews, 1972). Further efforts have led to highly purified RCs (Reed, 1969; Clayton and Wang, 1971; Feher, 1971; Thornber, 1971; Noël *et al.*, 1972; Smith *et al.*, 1972; Wang and Clayton, 1973; Okamura *et al.*, 1974; Lin and Thornber, 1975; Nieth *et al.*, 1975; Oelze and Golecki, 1975; Oelze and Mechler, 1976; Pucheau *et al.*, 1976). A survey of various preparations has been published recently (Sauer, 1975). In contrast to earlier results, recent studies on the pigment content demonstrate that each functional RC of *Rps. sphaeroides*, i.e., each unit that is functional in translocation of one electron with the use of one quantum, contains four BChls, two bacteriopheophytin molecules (Straley *et al.*, 1973; Parson and Cogdell,

1975), two molecules of ubiquinone, and one Fe^{2+} (Okamura et al., 1975). BChls of the RC appear to be linked in dimers. The high stability of BChl in RC preparations, the spectral properties, and the reactivity indicate strong BChl–protein as well as BChl–BChl interactions. Ke (in chapter 3 of Volume 7 of this series) and Dutton et al. (1976) present a more detailed discussion of the photochemistry of bacterial RC.

Absorption spectra of RC preparations from R. rubrum, Rps. capsulata, Rps. sphaeroides, and Chromatium show major peaks at about 755, 800, and 865 nm and minor peaks at 590 and 1250 nm (Okamura et al., 1974; van der Rest and Gingras, 1974; Lin and Thornber, 1975; Nieth et al., 1975). The characteristic reversible bleaching at 590, 865, and 1250 nm and the blue shift of the 800-nm peak were thought to be due to the primary events of oxidation of BChl, not to reaction of different BChl species (Parson, 1974). RC preparations of Rps. viridis (Pucheau et al., 1976) show main absorption maxima at 835 and 975 nm. Extinction coefficients for BChl measurements in RC preparations or membrane preparations were calculated on the 870-nm values. They vary according to the strain and the type of preparation (Straley et al., 1973; van der Rest and Gingras, 1974; Dutton et al., 1975). RC BChl molecules are a small portion of the total BChl population.

RCs of wild-type strains contain carotenoids (Thornber et al., 1969; Thornber, 1970; Noël et al., 1972; van der Rest and Gingras, 1974; Jolchine and Reiss-Husson, 1974, 1975; Cogdell et al., 1975; Lin and Thornber, 1975). RCs of Rps. sphaeroides contain 1 mole of a specific carotenoid per RC, which differs from the predominant carotenoid of the light-harvesting (LH) pigment, i.e, spheroidene in strain 2.4.1 and chloroxanthin in strain Ga (Cogdell et al., 1976). RCs of R. rubrum contain 1 mole of spirilloxanthin per equivalent of P870 (van der Rest and Gingras, 1974).

RCs of Rps. sphaeroides and Rps. capsulata contain three polypeptides with molecular weights of 21,000, 24,000, and 28,000 with a stoichiometry of 1:1:1 (Okamura et al., 1974; Nieth et al., 1975). The two smaller subunits are especially hydrophobic in composition (Steiner et al., 1974). The large subunit can be separated from the two smaller subunits by a mild treatment with detergents (Okamura et al., 1974; Nieth et al., 1975). The two smaller polypeptides remain associated in a complex with BChl that is photochemically active at room temperature but is less stable in comparison with the complex of three proteins and inactive at cryogenic temperatures (Okamura et al., 1974; Nieth et al., 1975). At present it is open to discussion whether the four BChl and the two bacteriopheophytin molecules are associated with two smaller subunits (1 × 24 + 1 × 21 kdaltons) or four (2 × 24 + 2 × 21 kdaltons) as was suggested by measurements of the BChl:protein ratio (Noël et *al.*, 1972; Nieth *et al.*, 1975; Sauer, 1975). RC preparations from other purple bacteria differ in their polypeptid and pigment patterns (Pucheau *et al.*, 1976; van der Rest *et al.*, 1974; Gingras, 1978). The large subunit (28 kdaltons) does not seem to be bound to RC-BChl. However, the strict stoichiometry of the three RC subunits observed under various culture conditions (Drews *et al.*, 1975; Takemoto, 1974; Nieth and Drews, 1975) suggested that even in the membrane all three subunits are associated. The sum of the polar amino acids is 38% in the largest, or H, subunit, but only 29 and 30% in the L (21 kdaltons) and M (24 kdaltons) units (Steiner *et al.*, 1974). The L and H subunits, but not the M subunit, contain 1 mole of half-cystine. Antibodies against RCs of *Rps. sphaeroides* do not cross-react with RCs of *R. rubrum* or *Chromatium*. The amino acid composition of *R. rubrum* G-9 RC was different from that of *Rps. sphaeroides* R-26, notably in the basic residues (Steiner *et al.*, 1974). The mobilities of the three RC proteins of *R. rubrum* are a few percent lower than those of the subunits of *Rps. sphaeroides* (Okamura *et al.*, 1974). The carotenoid does not reside on a separate protein in the RC but instead is bound to the same pair of proteins (L+M) as are the BChl and bacteriopheophytin of the RC (Cogdell *et al.*, 1976). A strong circular dichroism in absorption bands of spheroidene was observed. The long axis of the carotenoid molecule seems to lie approximately parallel to the direction of the 600-nm transition vector of the BChl complex. This was shown by examination of the fluorescence polarization spectrum (Cogdell *et al.*, 1976). The efficiency with which energy was transferred from carotenoid to BChl was found to be approximately 80% (Göbel, 1978). This was determined by comparing the fluorescence yield elicited from BChl when the carotenoid was excited at 470 or 510 nm with the fluorescence yield when the BChl was excited directly at 600 nm. The binding of one specific carotenoid molecule in the right orientation to the RC BChl suggests that the measured high efficiency of energy transfer is also valid for the living cell (Göbel, 1978). The RC of *Rps. sphaeroides* R-26 binds two molecules of ubiquinone. One molecule is loosely bound and can easily be removed with LDAO and *o*-phenanthroline (Okamura *et al.*, 1975). It probably represents the secondary electron acceptor (Feher *et al.*, 1972; Slooten, 1972; Clayton, 1973; Ke *et al.*, 1973; Knaff *et al.*, 1973; Halsey and Parson, 1974). Ubiquinone was also found in *R. rubrum* chromatophores in a loosely bound and a tightly bound form. One-half ubiquinone per phototrap was tightly bound. Removal of the tightly bound ubiquinone was related to the loss of primary photochemical capacity (Morrison *et al.*, 1977; Okamura *et al.*, 1976).

It has been suggested that the tightly bound ubiquinone plays an obligatory role in the primary photochemistry of *Rps. sphaeroides*

(Okamura *et al.*, 1975; Morrison *et al.*, 1977; Dutton *et al.*, 1976). In addition to ubiquinone a nonheme iron protein detected by EPR signals (g = 1.82, 1.68) was considered to be a primary electron acceptor (Clayton, 1973; Dutton *et al.*, 1973; Dutton and Leigh, 1973). The chemical evidence, i.e., removal of iron and its replacement with manganese (Loach and Hall, 1972; Feher *et al.*, 1974) and the unchanged valence of iron upon reduction of the acceptor (Feher *et al.*, 1974), indicates that iron plays a relatively minor role in the primary events. Complexes of Fe^{2+} and ubiquinone have been proposed as primary acceptor (Bolton *et al.*, 1969; Feher *et al.*, 1972; Loach and Hall, 1972; Bolton and Cost, 1973). These data are reviewed by Ke in Volume 7 (Chapter 3), who concludes that an Fe–ubiquinone complex is the most likely candidate for the primary electron acceptor. The midpoint potential of the primary acceptor has been shown to be pH dependent (Parson and Cogdell, 1975). At acid values of pH the equilibrium reduction of the primary acceptor involves both an electron and a proton.

During light-induced electron flow, reduction of the primary acceptor is a very rapid reaction ($t_{1/2}$ = 200 psec) and the subsequent oxidation and reduction of ubiquinone occurs in the 50–400 μsec range. The primary acceptor is not protonated if secondary electron flow to ubiquinones occurs. Consequently under these conditions the reduction of the primary acceptor involves only one electron (X/X^-), implying that the kinetically operational midpoint potential is the equilibrium value measured above the pK of the reduced form. In *Chromatium vinosum* this is -160 mV, in *Rps. sphaeroides* it is -180 mV, and in *R. rubrum* it is -200 mV (Prince and Dutton, 1976). In RC preparations from *Chromatium* (strain D), ubiquinone is replaced by menaquinone (Okamura *et al.*, 1975; Romijn, 1976).

Subsequent to the oxidation of RC BChl, the RC is re-reduced by Cyt c_2 or another high-potential cytochrome of the c type (Horio and Kamen, 1962; Reed and Clayton, 1968; Ke *et al.*, 1970; Dutton and Jackson, 1972; Evans and Crofts, 1974). The rate of reaction varies with ionic strength, but not with pH over the range 4.5–11.0 (Prince *et al.*, 1974a). The half-reduction potential of Cyt c_2 in a RC-protein-associated state has a value of about $+295$ mV, which is pH independent in the range of pH 5–9. This shows that a proton is not involved in the oxidation and reduction of Cyt c_2 in the physiological pH range (Dutton *et al.*, 1975). In living cells of *Rps. sphaeroides*, 2 moles of Cyt c_2 seem to be associated with one RC (Dutton *et al.*, 1975). A similar ratio was found for Cyt c_{555} in *Chromatium* (Case and Parson, 1971). Experiments dealing with the effects of monospecific antibodies for Cyt c_2 from *Rps. sphaeroides* and *Rps. capsulata* have shown that cytochrome c_2 is located *in vivo* in the periplasmic space between cell wall and cytoplasmic membrane. When ICM vesicles are prepared from whole cells, the Cyt c_2 becomes trapped inside these vesicles (Prince et al., 1975). As a consequence these vesicles are still active in photophosphorylation, but sphaeroplasts which are washed free of Cyt c_2 have lost this activity. At least one b-type cytochrome is involved between Cyt c_2 and ubiquinone. Of the three b cytochromes with E_m of $+155$, $+50$, and -90 mV, respectively, the Cyt b_{50} with a pK of about pH 7.4 associated with the reduced form participates in cyclic electron flow (Petty and Dutton, 1976). This Cyt b_{50} shows a reduced α - band prominent at 560 nm (Dutton and Jackson, 1972). In the uncoupled state the kinetics of reoxidation of Cyt b are the same as those for the re-reduction of Cyt c_2 (Prince and Dutton, 1976). A proton coming from ubisemiquinone (Q·⁻/ Q·H) is involved in the oxidation and reduction of Cyt b_{50} below the pK (Petty and Dutton, 1976).

RC seem to be embedded in the hydrophobic zone of the ICM. This conclusion is based on the high hydrophobicity of the RC proteins, especially the L and M porteins (Steiner et al., 1974), and the insolubility of RC in any buffers or solutions of chaotropic salts. However, after removal of the coupling factor, RCs in membranes are accessible to antibodies against RC (Reed et al., 1975; Drews et al., unpublished observations). Antibodies against chromatophores cross-react with RC (van der Rest et al., 1974). The L and M proteins of RC were accessible to ferritin-labeled antibodies both on the outer and the inner surface of the chromatophore vesicles: however, the H subunit seems to be accessible only on the cytoplasm side of the membrane (Feher and Okamura, 1976). The H subunit of RC is heavily labeled by iodination (Phillips and Morrison, 1970) on the cytoplasm side of ICM (Oelze, 1977; Zürrer et al., 1977). These data indicate that RC particles span across the membrane. RCs are also present in the CM, but the density of ferritin label was found to be much lower than in the ICM (Drews et al., unpublished observations), which is in accordance with the low BCh1 content of the CM (Oelze et al., 1969a,b; Lampe and Drews, 1972).

b.   Reaction-Center Preparation from Green Photosynthetic Bacteria. The PSA of green bacteria (Chlorobiaceae) differs in its organization and composition from that of purple bacteria (Fig. 3). Because of its complexity and the high light-harvesting BChl to RC BChl ratio, the isolation of RC preparations is more difficult than in purple bacteria. Photochemically active preparations were isolated from cell-free extracts by density gradient centrifugation (Fowler et al., 1971; Fowler, 1974; Boyce et al., 1976; Prince and Olson, 1976). Incubation of the BChl a–RC complex I of Chlorobium limicola (M_r 1.5 $\times$ 10⁶, Fowler et al., 1971) with 2 M guanidine HC1 strips off most of the BChl a from the RC complex, which can be separated by chromatography on Sepharose

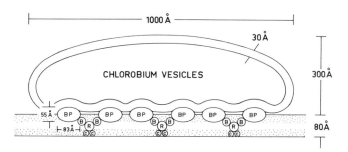

Fig. 3. Organization of the photosynthetic apparatus of green bacteria. BP, bacterio-chlorophyll (BChl) a–protein trimers; R, reaction center; B, BChl a; C, cytochrome c. From Olson *et al.*, (1976c.)

4B (Olson *et al.*, 1976a,c). The *in vivo* absorption spectrum of the enriched RC preparation (complex II) shows maxima at 341, (375), 603, 813, and (835) nm due to BChl a. The peaks at 414 and 674 nm are attributed to bacteriopheophytin c, and the peak at 744 nm is attributed to BChl c. Carotenoid peaks are found at 504 and 560 nm (Olson *et al.*, 1976a). Complex II apparently contains no polypeptides larger than approximately 15 kdaltons (Olson *et al.*, 1976b). This result differs from the data on purple bacteria, where proteins of 20–30 kdaltons contribute to RC complexes. Evidence obtained using biochemical markers for the subcellular components shows that RC (P840) and BChl a reside on the CM (Boyce *et al.*, 1976). Membranes were detected in enriched RC preparations (Boyce *et al.*, 1976; Olson *et al.*, 1976a,c). ATPase activity of *Chlorobium thiosulfatophilum* is bound to the CM (Burns and Midgley, 1976).

Light minus dark difference spectra of complex I showed troughs at 790, 830, and 842 nm and a peak at 1157 nm (Olson *et al.*, 1976a,c). The troughs at 830 and 842 nm are different from the typical pattern seen in purple bacteria and suggest different arrangements of RC BChl a molecules in purple and green bacteria, respectively. Upon illumination, approximately two Cyt c_{553} molecules were oxidized for every P840 oxidized (Prince and Olson, 1976). That is consistent with the finding that the RC in complex I is associated with two functional Cyt c molecules (Fowler *et al.*, 1971; Olson *et al.*, 1976a,c). However, the total Cyt c content of complex I is approximately eight times the P840 content. In complex II approximately six Cyt c molecules were found for every 100 BChl a molecules (Olson *et al.*, 1976b). Concomitant with the oxidation of Cyt c, a light-induced reduction of Cyt b was observed (Fowler, 1974; Olson *et al.*, 1976b). In *Chlorobium limicola* strain 2-K, 0.63 nmole of Cyt $c_{551.5}$ and 1.06 nmoles of Cyt c_{555} per milligram of protein were found, but no Cyt c_{553} or b-type cytochromes (Shioi *et al.*,

1974). The primary electron acceptor of the RC has an equilibrium midpoint potential lower than -450 mV at pH 6.8 (Prince and Olson, 1976). It has been proposed that the primary electron acceptor for the RC is an iron–sulfur protein ($g = 1.94$) with a potential of $E_m = -550$ mV (Jennings and Evans, 1976; Olson et al., 1976a). The reversible light-induced ESR signal in complex I is centered at $g = 2.003$ with a peak-to-peak linewidth of 9.2 ± 0.5 (Olson et al., 1976b). These values are characteristic for RCs from purple bacteria (Feher, 1971; Okamura et al., 1975) and are consistent with the primary electron donor being a BChl dimer (Norris et al., 1971; Katz and Norris, 1973).

c. *Light-Harvesting (Antenna) Pigment Complexes.* Bulk BChl is organized in light-harvesting (antenna) BChl complexes, which absorb light quanta. The absorbed excitation energy is transferred to RC or delivered as fluorescence emission (Litvin and Sineshchekov, 1975; Sauer, 1975; Monger and Parson, 1977).

IR-absorption spectra of membranes from purple bacteria appear to be due to different BChl-protein complexes. Numerous data on the different spectroscopic forms of BChl have been recently collected (Biebl and Drews, 1969; Pfennig and Trüper, 1974; Amesz, 1978; Thornber et al., 1978). LH–BChl complexes have been isolated from membranes of some photosynthetic bacteria and described in the next paragraph. Another approach to study BChl–protein interrelationships comes from comparative studies on the different BChl spectral types and protein patterns of membranes from cells grown under various conditions (Dierstein and Drews, 1975; Oelze and Pahlke, 1976; Niederman et al., 1976). Results and conclusions of experiments of this type are presented in Section III.

All proteins associated with antenna pigments fall into the category of "integral membrane proteins" (Singer, 1974; Singer and Nicolson, 1972). They can be dissociated from the membrane only by detergents and organic solvents. Chaotropic agents or salts in a high ionic strength solubilize the pigment complexes only in the presence of detergents. The pigment protein complexes reaggregate or precipitate in a neutral aqueous buffer. Thus, particles of quite different sizes have been isolated (Reed et al., 1970; Rivas et al., 1970; Thornber, 1970; Biedermann, 1971; Fraker and Kaplan, 1971, 1972).

BChl is thought to be associated with the protein by ionic or hydrophobic interaction, but not by covalent linkage. Therefore, the possibility that an artificial complex is formed by association of BChl and a protein from a detergent-treated membrane extract has to be considered. The following criteria were used in the literature to prove that the pigment complexes exist *in vivo*: (1) The IR-absorption band, characteristic for the antenna BChl species, is still present after isolation of the complex from the membrane (Reed et al., 1970). (2) A strong affinity between antenna BChl and one major membrane protein was observed (Fraker and Kaplan, 1971, 1972; Clayton and Clayton, 1972). (3) LH BChl and the protein belonging to the LH complex were synthesized stoichiometrically under different growth conditions (Dierstein and Drews, 1975). (4) Genetic experiments have shown that lack or modification of an "antenna pigment protein" by mutation inhibited the formation of the respective LH complex, even in the presence of BChl (Drews et al., 1976a).

The absorption spectrum of R. rubrum (Rhodospirillaceae) represents a simple spectral form. The photoreceptor particle of R. rubrum membranes (AUT treatment), with an average particle size of 100 kdaltons, contains all membrane pigments: 40 BChl molecules and 20 carotenoid molecules (Loach et al., 1970). The near-IR absorption spectrum is identical to that of whole membranes. It seems probable that the total LH BChl in R. rubrum is represented by B870.

From chromatophore membranes or photoreceptor complex preparations of R. rubrum, one protein was extracted with chloroform–methanol 1:1 (v/v) and purified by column chromatography using Sephadex (Tonn et al., 1977). Chemical analysis supports a molecular weight of 18,000. SDS–polyacrylamide gel electrophoresis resulted in a molecular weight of 12,000. The polarity of this protein was found to be 42% (Tonn et al., 1977). This is higher than the polarity of the RC proteins isolated from R. rubrum (Steiner et al., 1974). Cysteine and tyrosine were absent from the preparations. It has been suggested that this protein functions in the organization of the LH BChl complex. A calculation shows that there are between 3 and 7 BChl molecules and 1 to 2 carotenoid molecules per polypeptide (Tonn et al., 1977).

A LH–BChl protein complex from the blue-green strain VI of R. rubrum (FR1) was found to be enriched in two proteins (8.5 and 9.5 kdaltons) and B870. A BChl content of about 2 moles per mole of protein has been suggested. This preparation retains the typical vesicular shape and the double-track structure characteristic for untreated ICM (Oelze and Golecki, 1975). A similar protein pattern of LDAO-extracted membranes was obtained from wild-type cells of R. rubrum (van der Rest and Gingras, 1974). The two-dimensional arrangement of the B870-complex seems to indicate that the complex is a "morphogenetic" component of the membrane and stabilizes the BChl molecules. The organization seems to be similar to the B850 complex of Rps. sphaeroides. (Fig. 2)

Some nonsulfur purple bacteria grown photosynthetically appear to contain two or more species of LH BChl complexes. LH BChl I of Rps. capsulata has a single major IR-absorbance peak at approximately 875–880 nm and LH BChl II has two absorbance maxima at 802 and 855 nm

(Lien et al., 1973). Photobleaching experiments indicated that membrane-bound RC BChl in this bacterium has an absorbancy maximum at 880–890 nm (Lien et al., 1971; Dierstein and Drews, 1975). In Rps. sphaeroides one LH BChl component absorbs at 875 nm and the other at 800 nm and 850 nm (Aagaard and Sistrom, 1972). The RC BChl of Rps. sphaeroides has an absorbancy maximum at 870 nm. The conclusion that in both species the maxima at 800 and 850 nm belong to the same BChl component is based on the observation that the absorption at 800 nm relative to that at 850 nm is constant (Sistrom, 1964; Drews et al., 1976b). Additional proof has come from the observation that the mutant Y 9 of Rps. capsulata, which has no RC BChl and no LH BChl I, but LH BChl II shows two IR maxima at 800 and 858 nm (Drews et al., 1976a; Feick and Drews, 1978).

The LH BChl II (B850)–protein complex from Rps. sphaeroides, consists of BChl and protein in the molar ratio of 1.4:1 (Fraker and Kaplan, 1972) or 1.6:1 (Clayton and Clayton, 1972). The weight of the polypeptide monomer was estimated by SDS–polyacrylamide gel electrophoresis to be 9 or 9.9 dalton (Clayton and Clayton, 1972; Fraker and Kaplan, 1972). The peptide and amino acid analysis, which gives molecular weights of 7000–11,000, seems to indicate that only one polypeptide is involved in complex formation (Fraker and Kaplan, 1972; Huang and Kaplan, 1973). The complex (band 15) consists of 59% protein, 35% phospholipid, and 6% BChl (Fraker and Kaplan, 1972). Small amounts of carotenoids seem to be present (Clayton and Clayton, 1972; Fraker and Kaplan, 1972). The isolated protein contains about 60% nonpolar amino acids. Histidine and cysteine were absent. A C-terminal sequence, NH_2 . . . Tyr-Ser-Glu-Glu-(Leu, Ala, Ala, Val, Ala, Ala)-Gly COOH was proposed (Huang and Kaplan, 1973). The complex tends to form high-molecular-weight aggregates under mild conditions of treatment and low detergent concentrations in aqueous solution (Clayton and Clayton, 1972; Fraker and Kaplan, 1971). More than 60% of the total BChl of ICM was recovered as a complex with the 9-kdalton protein (Clayton and Clayton, 1972). The B850-BChl species of Rps. sphaeroides, which seems to be identical with the BChl-protein complex of this bacterium, has absorption maxima at 800 and 850 nm. It is the variable portion of the photosynthetic unit (Aagaard and Sistrom, 1972). A polypeptide of apparent M_r 12,000 has been isolated from phototrap particles of Rps. sphaeroides. The preparation retains 24 LH BChl and 8 carotenoid molecules per phototrap. The molecular weight on the basis of one histidine was calculated to be 18,800, and in SDS gels the M_r was 12,000. The polypeptide contains no cysteine and no tyrosine (Loach, 1976).

The other LH-BChl component, B875, has not been isolated in a pure complex. The absence of B875 from the solubilized membrane was explained as an effect of LDAO transforming the B875 form of BChl into monomeric BChl, which absorbs at 770 nm (Clayton and Clayton, 1972). Spectrophotometric studies on membranes of *Rps. sphaeroides* grown under different conditions have shown that 25–35 B875 molecules are associated with each RC BChl in a fixed stoichiometry (Aagaard and Sistrom, 1972).

Similarly, in *Rps. capsulata* a close biosynthetic association between RC BChl and LH BChl I (875 nm) was observed, but with a relatively independent regulation of LH BChl II (800 + 850 nm) (Lien *et al.*, 1971, 1973; Drews *et al.*, 1976b). The LH BChl I–protein complex was isolated from the carotenoidless mutant Ala$^+$ of *Rps. capsulata* by SDS treatment of membranes, hydroxyapatite column chromatography, and preparative polyacrylamide gel electrophoresis (Drews, 1976). After hydroxyapatite chromatography, the enriched fraction showed a major absorption peak at 872 nm and minor bands at 800, 776 and 690 nm. In polyacrylamide gel electrophoresis (10–15%) one polypeptide was detected with an M_r about 12,000 (Drews, 1976). This band was in the same position as band 2 of "LH proteins" from membranes of wild-type strains (Drews *et al.*, 1976a). The molar ratio BChl per protein in the LH I BChl–protein complex was found to be approximately two in the enriched complex, but 0.5 in the purified fraction. After dissociation of the complex into the monomeric subunit, the IR maximum of BChl was shifted to 772 nm.

The protein bands 1, 3, and 4 (M_r approximately = 14,000, 10,000, and 8000) are consistent with the presence of LH BChl II (800 + 850 nm) in *Rps. capsulata* (Drews *et al.*, 1976 a,b). The LH II BChl–protein complex was isolated from the mutant strain Y 5 of *Rps. capsulata* which is free of RC and LH I BChl. The purified complex contains three polypeptides with apparent M_r of 14,000, 10,000, and 8000. It contains 200 μg BChl per milligram of protein and carotenoids. The absorption spectrum of the complex showed IR maxima at 802 and 855 nm (Feick and Drews, 1978). Data on isolated and purified BChl–protein complexes from Rhodospirillaceae are summarized in Table II.

LH BChl II is the dominating BChl component in the PSA from wild-type cells of *Rps. capsulata* and *Rps. sphaeroides* grown photosynthetically (Aagaard and Sistrom, 1972; Lien *et al.*, 1971, 1973; Schumacher and Drews, 1978).

The present data on composition of the photosynthetic units are summarized in Fig. 2 and Table III. The weight of a unit consisting of RC plus LH complexes (B870) in *Rps. sphaeroides* is about 360 kdaltons. The amino acid composition and the solubility of this particle indicate that it is mainly localized in the hydrophobic zone of the

TABLE II

LIGHT-HARVESTING (LH) BACTERIOCHLOROPHYLL–PROTEIN COMPLEXES ISOLATED FROM RHODOSPIRILLACEAE

Strain	Type of complex	IR maxima in vivo (nm)	Polypeptides M_r	BChl:protein molar ratio	Remarks	References
Rhodospirillum rubrum	LH I	870	18,800[a]	3–7:1	No Cys and Tyr, 42% polarity, soluble in pyridine–water	Tonn *et al.* (1977)
R. rubrum VI	LH I	870	8,500[b] 9,500[b]	2:1	Membranelike structure	Oelze and Golecki (1975)
Rhodopseudomonas sphaeroides	LH II	800 + 850	9,000[b]	1.6:1	—	Clayton and Clayton (1972)
Rps. sphaeroides	LH II	800 + 850	10,000[b]	1.4:1	5% protein, 35% P-lipids, 6% BChl; no His and Cys	Fraker and Kaplan (1972)
Rps. sphaeroides	LH II		18,000[a]		No Cys, 37% polarity	Tonn (unpublished 1976)
Rps. capsulata	LH I	872	12,000	2:1	—	Drews (1976)
Rps. capsulata	LH II	802 + 855	14,000[b] 10,000[b] 8,000[b]	4:1	1.0 µg lipid P per mg membrane protein, 42% polarity of amino acids	Feick and Drews (1978)

[a] Estimated by chemical and enzymic cleavages.
[b] Estimated by polyacrylamide gel electrophoresis.

membrane. A calculation of the weight from electron microscopical observations on the basis of the formula 0.315 × (diameter of the particles in the fracture faces of the ICM = 100 Å in *Rps. capsulata*)^3 × 1.2 gives a value of about 370 kdaltons, which is in agreement with the value calculated from the weights of BCh1 and proteins. The LH–BChl complex II (B850) changes the size of the photosynthetic unit by a factor of about three (Table III). The tendency for a two-dimensional aggregation of that fraction and the observation that the mass of BChl molecules is regularly oriented in the plane of the membrane support the assumption that LH BChl complex II is attached to the RC + LH I particles but is extended sheetlike in the hydrophobic zone of the membrane. Up to 80% of the total ICM protein can be proteins of the photosynthetic apparatus (Table III).

Chromatium vinosum has more near-IR absorption bands than most other photosynthetic bacteria: 795, 805, 825, 850, 888 nm (Goedheer, 1972). Membrane subfractions enriched in 890-nm absorbing material and 800- and 850-nm material, respectively, have been described (Clayton, 1962; Garcia *et al.*, 1966). Hydroxyapatite chromatography of SDS-dissociated membranes resulted in three fractions enriched in (1) RC plus B890, (2) 803 and 846 nm, and (3) 800- and 820-nm bands (Thornber, 1970). All fractions contained carotenoids. The carotenoid pattern of fraction 1 was different from those of fractions 2 and 3. It appears that one 12-kdalton protein may be involved in a B890 BChl–protein complex (Thornber, 1970). Two polypeptides of 9.8 and 7.6 kdaltons occur in the B800 + B850 BChl fraction (Erokhin and Moskalenko, 1973). The polypeptide composition of fraction 3, containing 800- and 820-nm bands appears to be identical with that of fraction 2 (Thornber *et al.*, 1978).

   d. The Antenna Complexes in Green Bacteria. In contrast to purple bacteria, the antenna BChl molecules of green bacteria (Chlorobiaceae) are not organized within the hydrophobic zone of ICM. They are either attached to the CM (BChl *a* complex) or are contained in the *Chlorobium* vesicles (BChl *c*, *d*, or *e*) (Fig. 3). The ratio of antenna BChl to RC BChl in green bacteria is an order of magnitude greater than in purple bacteria: 1000–2000 molecules of BChl *c*, *d*, or *e* per RC (Fowler *et al.*, 1971). One reason for the large amount of LH BChl in the PSA of green bacteria might be that some species of the brown-green bacteria (*Pelodictyon, Anacalachloris*) grow deep in lakes at very low light intensities (Trüper and Genovese, 1968; Fenchel, 1969; Kusnezow and Gorlenko, 1973; Pfennig, 1978). However, this argument is not valid for those green bacteria growing at moderate or high light intensities and still having high BChl contents. The BChls *c*, *d*, or *e* are localized in *Chlorobium* vesicles, which are 30–40 nm wide and 100–140 nm long and lie immediately under the CM (Cohen-Bazire *et al.*, 1964; Holt *et al.*, 1966;

TABLE III
SIZE OF THE PHOTOSYNTHETIC UNIT

Strain	Growth conditions	Nanomoles BChl per 100 μg membrane protein		Moles total BChl per RC	Weight of the PSU (kdaltons)[a]	Micrograms protein of PSU per 100 μg membrane protein
		Total BChl	RC BChl			
Rhodopseudomonas sphaeroides Ga[a]	Anaerobic light, 8000–50 ft-c	1.86–12.2	0.024–0.036	77–333	620–2130	13.5–78
Rps. capsulata 37b4[b]	Semiaerobic, dark, or anaerobic, saturating light, chemostate, NH$_3$-limited	2.09–5.7	0.019	106–300	780–1930	13.3–37
Rhodospirillum rubrum FR1[c]	Anaerobic, light, 6000–100 ft-c	3.5–7.3	0.15–0.30	23	285	37–76
R. rubrum 1.1.1.[a]		4.8–9.4	0.13–0.27	35	356	43–85

[a] Calculated from data of Aagaard and Sistrom (1972).
[b] Calculated from data of Dierstein and Drews (1975).
[c] Calculated from data of Irschik and Oelze (1973).
[d] The weight of the photosynthetic unit (PSU) is calculated on the following basis: One of reaction center (RC) (4 moles of RC BChl + 2 moles of bacteriopheophytin bound to polypeptides with total weight of 146 kdaltons) expels one electron when excited by one quantum. It is assumed that in Rps. capsulata and Rps. sphaeroides 30 moles of LH BChl I are bound to 15 moles of LH protein (12 kdaltons) and associated with one RC. LH BChl II = total BChl − (LH BChl I + RC BChl). LH BChl II is assumed to be bound to 10-kdalton polypeptides in the molar ratio 2:1. In R. rubrum only one LH complex is present, which is assumed to be composed of 2 moles of BChl per mole of protein (10 kdaltons). The ratio of LH BChl to RC BChl is constant (Aagaard and Sistrom, 1972; Oelze and Pahlke, 1976).

Pfennig and Cohen-Bazire, 1967). They are surrounded by a (single-layered) 2–3 nm thick membrane (not a unit membrane). The vesicles stick tightly to the CM (Holt *et al.*, 1966). The isolated chlorobium vesicles have to be stabilized by glutaraldehyde treatment; otherwise they disintegrate rapidly (Cruden and Stanier, 1970). In such preparations regular intravesicular structures, 9–10 nm wide with a small central hole were observed, which were still present in glutaraldehyde-fixed vesicles after extraction of BChl (*Chlorobium thiosulfatophilum, C. limicola*; Cruden and Stanier, 1970). The vesicle fraction of *C. thiosulfatophilum* strain 6130 contained 845 mg of BChl *c* per milligram of protein. This BChl:protein ratio is 10,000 times higher than the BChl:protein ratio in ICM of purple bacteria. It is clear that the organization of BChl in the vesicles must be completely different from that of antenna pigment complexes in membranes. Presumably most of the protein contributed to the layer which surrounds the vesicles. A significant amount of carbohydrate and lipids was present (about 15% each on a dry weight basis). Monogalactosyl diglyceride was the major lipid in the vesicle fraction. The phospholipid content was low (Cruden and Stanier, 1970). The high BChl:protein ratio seems to indicate that mainly BChl–BChl interactions are responsible for the red shift of the IR peaks of *Chlorobium* chlorophylls *in vivo*. Light energy absorbed by *Chlorobium* chlorophyll is transferred to the RC via BChl *a* (Sybesma and Olson, 1963; Sybesma and Vredensberg, 1963, 1964). Thus, the *Chlorobium* vesicles have a similar function in the photosynthetic apparatus of green bacteria as the phycobilisomes in Cyanobacteria.

    e. Bacteriochlorophyll a Proteins. Bacteriochlorophyll *a*-protein, discovered by Olson and Romano (1962) and Sybesma and Olson (1963), is a LH complex, accepting excitation energy from BChl *c, d,* or *e*, and transferring this excitation energy to the RC BChl P840 (Sybesma and Vredenberg, 1963, 1964). It has been suggested that the BChl *a*-protein is located between the *Chlorobium* vesicles and the CM (Fig. 3) (Olson *et al.*, 1976c). Large subcellular fractions (complex I, M_r about 1.5 × 10^6) contain about 80 BChl *a* molecules per RC (Fowler *et al.*, 1971; Olson *et al.*, 1976b). BChl *a*-protein has been obtained from three strains of *Chlorobium limicola* in a highly purified and crystallized state (Olson *et al.*, 1969; Fenna *et al.*, 1974; Fenna and Matthews, 1975). The BChl *a*-protein from *Chlorobium limicola*, strain 2 K, is a trimer. The calculated subunit weight is 42 kdaltons for both BChl *a* and protein together. The corresponding trimer weight is 127 kdaltons, calculated from its chemical composition (Olson, 1978). The complex contains 7 BChl *a* molecules and at least 327 amino acid residues per subunit (Fenna *et al.*, 1974). The amino acid composition was studied by Thornber and Olson

(1968) and Olson *et al*. (1976a,b). The "polarity" is 45%. The isoelectric point of BChl *a*-protein of strain 2 K was centered at pH 6.0. The structure of the protein has been studied by X-ray diffraction of single hexagonal crystals of native BChl *a*-protein to a resolution of 2.8 Å (Fenna and Matthews, 1975). The polypeptide of the monomer subunit forms a distorted hollow bag with 7 BChl *a* molecules inside. The BChl *a* molecules occupy the space within an ellipsoid of axial dimensions 45–35 × 15 Å. They interact by hydrophobic interaction. Each BChl molecule is apparently anchored to the protein through extensive hydrogen bonding and liganding to the Mg atom in addition to hydrophobic interactions through the phytyl tail (Fenna and Matthews, 1975). The BChl *a*-protein from *Chlorobium limicola* strain 2 K is most stable in aqueous solutions in the presence of salt, buffered at pH 7–8, and in the presence of Triton X-100 (Ghosh and Olson, 1968) or 8 *M* urea (Kim and Ke, 1970). The BChl *a*-protein has main absorption bands at 370, 602, and 808 nm (Olson, 1978). The fluorescence emission peak at room temperature is at 818 nm (Sybesma and Olson, 1963).

3. The Respiratory Electron-Transport System

Eight species of *Rhodopseudomonas* and *Rhodospirillum* can grow aerobically in the dark, producing ATP by oxidative phosphorylation. Five other species of Rhodospirillaceae and *Thiocapsa roseopersicina* are able to grow slowly in the dark under microaerophilic conditions. Some Chromatiaceae can respire but are unable to grow aerobically in the dark (Pfennig, 1970; Pfennig and Trüper, 1974; Keister, 1978). Thus, most of the purple bacteria have a respiratory chain in their membrane system, which, however, has been described in only a few species (Taniguchi and Kamen, 1965; Boll, 1968; Thore *et al*., 1969; Baltscheffsky *et al*., 1971; Baccarini-Melandri *et al*., 1973; King and Drews, 1973, 1975, 1976; Nisimoto *et al*., 1973; Melandri *et al*., 1975; Saunders and Jones, 1974; La Monica and Marrs, 1976; Zannoni *et al*., 1976).

The enzymes and electron-transport carrier of the respiratory chains contribute to the organization of membrane components. Few of the enzymes (Hatefi *et al*., 1972; King and Drews, 1976) but many cytochromes (Horio and Kamen, 1970; Kamen and Horio, 1970; Meyer and Jones, 1973; Bartsch, 1978) and nonheme iron proteins (Shanmugam *et al*., 1972; Jennings and Evans, 1976) have been isolated and described. As far as is known, at least in those species that grow aerobically in the dark, respiratory-chain components dominate in the CM of these organisms (Oelze *et al*., 1969a,b; Oelze and Drews, 1970b, 1972; Throm *et al*., 1970; Lampe and Drews, 1972; Niederman, 1974). Light inhibits the respiratory-chain electron transport (Clayton, 1953, 1955; Ramirez and

Smith, 1968; Oelze and Weaver, 1971; Keister, 1978). Electron flow between the cyclic light-driven and the respiratory chains seems to be possible but is not well understood (Keister, 1978).

4. Coupling Factor

The coupling factor is a macromolecular enzyme complex, endowed with ATPase activity, which is involved in the synthesis of ATP coupled to electron transport and membrane energization. It consists of a knoblike particle of 12-nm size, which is localized on the cytoplasmic face of CM and ICM, and a hydrophobic component embedded in the membrane, which might be involved in transmembrane proton translocation. The isolated multicomponent complex from Rhodospirillaceae has a molecular weight of about 300 kdaltons (Melandri et al., 1971a; Johansson et al., 1973). The apparent weights of the five subunits estimated by SDS–gel electrophoresis are 54(α), 50(β), 32 (γ), 13(δ), and 7.5(ϵ) kdaltons. It was shown that in photosynthetic bacteria, which are able to produce ATP either by photophosphorylation or oxidative phosphorylation, the same coupling factor can catalyze both processes (Melandri et al., 1971b; Lien and Gest, 1973). The literature on coupling factors of photosynthetic bacteria was recently reviewed (Baccarini-Melandri and Melandri, 1978).

III. Biosynthesis and Differentiation of the Membrane System of Photosynthetic Bacteria

A. THE INFLUENCE OF EXOGENOUS FACTORS ON THE STEADY-STATE CONCENTRATIONS OF THE PHOTOSYNTHETIC APPARATUS

Biosynthesis and differentiation of the bacterial photosynthetic apparatus were reviewed recently (Lascelles, 1968; Oelze and Drews, 1972; Kaplan, 1978). Thus, the discussion in this chapter will be restricted to newer results and some actual problems.

1. The Influence of the Oxygen Partial Pressure

The actual oxygen partial pressure in the culture, resulting from oxygen consumption by cells and diffusion of oxygen from the bubbles in the aerated medium into the solution (Aiba et al., 1965) is the main factor in regulation of BChl formation (Cohen-Bazire et al., 1957; Lascelles, 1959; Biedermann et al., 1967; Oelze and Drews, 1972; Dierstein and Drews, 1974). In continuous dark cultures of Rps. capsulata strain 37b4, a marked increase of the BChl concentration per dry weight of cell mass was observed when the oxygen partial pressure was decreased below 10 mmHg. (Dierstein and Drews, 1974). However, BChl was still formed above the threshold level. Even at 400 Torr (pO_2)

10 pmoles of BChl per milligram of membrane protein were formed (Drews et al., 1976b). In contrast, R. rubrum, strain FR1, does not form measurable amounts of BChl above 50 mmHg (pO_2) in dark-batch cultures. The cellular BChl level strongly increased below 5 mmHg, pO_2 (Biedermann et al., 1967). Thus, the threshold level of pO_2 is more pronounced in this species. In cultures of both strains, Rps. capsulata 37b4 and R. rubrum FR1, the formation of the PSA is induced when the oxygen tension is lowered in the dark cultures to 5 mmHg (pO_2), although the growth rate is unimpaired (Drews et al., 1969a; Dierstein and Drews, 1974, 1975). When the oxygen tension is decreased below 2 mmHg, the formation of the PSA is still induced, but limited by the energy metabolism (Schumacher and Drews, 1978). In R. rubrum the formation of BChl and ICMs was observed even in strictly anaerobic dark cultures (Schön and Drews, 1966; Schön and Ladwig, 1970; Uffen and Wolfe, 1970). Other strains of photosynthetic bacteria may have slightly different oxygen thresholds and characteristic ratios of BChl steady-state concentrations to oxygen tensions, but the tendency is always the same. The rate of pigment synthesis increases as the oxygen partial pressure decreases (Lascelles, 1959; Cohen-Bazire and Kunisawa, 1960; Drews and Giesbrecht, 1963). BChl synthesis can be regulated independently from the energy metabolism.

BChl synthesis was shown to be inhibited by inhibitors of protein synthesis (Lascelles, 1959; Sistrom, 1962; Bull and Lascelles, 1963; Drews, 1965; Higuchi et al., 1965). Later, a correlation between specific proteins of ICM and BChl was demonstrated (Biedermann and Drews, 1968; Oelze et al., 1969a,b; Oelze and Drews, 1970a,b; Takemoto and Lascelles, 1973). It has been shown recently that the membrane bound BChl–protein complexes; i.e., RC and LH complexes were formed by coordinated synthesis of specific proteins and BChl (Lampe and Drews, 1972; Takemoto and Lascelles, 1974; Takemoto, 1974; Nieth and Drews, 1975; Dierstein and Drews, 1975; Niederman et al., 1976; Oelze and Pahlke, 1976). In addition, the isolation of BChl–protein complexes as described in Section II, and the isolation of mutants disturbed in biosynthesis of BChl–protein complexes (Oelze et al., 1970; Drews et al., 1976a) demonstrate clearly the strict interdependence between synthesis of BChl and specific membrane proteins.

Threshold levels of the partial pressure of oxygen regulate both the synthesis of different BChl species and of proteins forming complexes with the respective BChl. It was proposed that the oxygen partial pressure has an indirect influence on an "effector" molecule (a) by variation of the redox balance of components of the electron-transport system including pyridine nucleotides (Cohen-Bazire et al., 1957), or (b) by variation of the ATP (Fanica-Gaignier et al., 1971) or other nucleotide concentrations or by changing the energy charge (Sojka and Gest, 1968), or (c) because the oxygen partial pressure directly affects an effector molecule (Marrs and Gest, 1973) or a step of protein synthesis (reviewed in Oelze and Drews, 1972). At present we have no additional experimental data on the processes which translate the intracellular oxygen partial pressure into processes of regulation. Kaplan (1978) has developed a working hypothesis to explain regulation on different levels. He proposed that a direct interaction of oxygen with an effector molecule (Marrs and Gest, 1973) is active in corepressor synthesis. The corepressor bound to an aporepressor would induce or inhibit mRNA formation specifically for the PSA or interfere with biosynthetic processes on other levels.

A decrease in oxygen partial pressure below a threshold level not only induces the formation of the PSA, but also stimulates the process of membrane formation, resulting in invaginations of the CM or enlargement of the ICM (Cohen-Bazire and Kunisawa, 1963; Drews and Giesbrecht, 1963; Tauschel and Drews, 1967; Golecki and Oelze, 1975). An increase of oxygen tension inhibits both formation of the photosynthetic units and membrane formation. The concentrations of different components of the PSA in the membrane vary in correlation to the oxygen partial pressure (Drews et al., 1967; Lampe and Drews, 1972; Dierstein and Drews, 1975; Niederman et al., 1976). The observation that formation of the PSA and the ICM are not strictly correlated can be explained by the presence of other membrane components in the ICM, i.e., components of the respiratory chain, transport systems, etc. The synthesis of those components is partly regulated in the opposite direction (Keister and Minton, 1969; Klemme and Schlegel, 1969; Oelze and Drews, 1970b; Throm et al., 1970; Lampe and Drews, 1972; King and Drews, 1975). For example, NADH oxidase and cytochrome oxidase are still formed in dark cultures at pO_2 of 5 mmHg, but the rate of LH BChl synthesis is much higher (Lampe and Drews, 1972; King and Drews, 1975). Therefore formation of the PSA is the dominating event in the biosynthesis of the ICM.

Assuming that membrane formation is the sum of partial processes leading to the incorporation of all components into the membrane, synthesis of phospholipids should be coregulated to membrane protein synthesis. An increase of phospholipids and fatty acids parallel to membrane proteins per cell dry weight was observed (Lascelles and Szilágyi, 1965; Cohen-Bazire and Sistrom, 1966; Gorchein, 1968b; Gorchein et al., 1968; Schröder and Drews, 1968). The lipid composition of ICM differs significantly from that of cells lacking ICMs (Steiner et al., 1970; Gorchein, 1968b). Some of the key enzymes for phospholipid synthesis seem to be membrane bound (Salton, 1974). Coregulation of protein and lipid synthesis in the two membranes (CM and ICM), which differ in their composition, can be attributed either to a stoichiometric binding of lipids and proteins in the membrane or by a coregulation between the membrane protein and lipid synthesis. Thus, coregulation may be due to incorporation of lipids into the membrane from pools in the same amounts as proteins are incorporated (see Weiss, 1973). Alternatively, the synthesis of proteins of the PSA would be correlated to synthesis of specific enzymes of phospholipid metabolism. Studies on *Escherichia coli* and other nonphotosynthetic bacteria seem to indicate that phospholipid formation is regulated by synthesis of the enzymes envolved in that synthesis (Overath *et al.*, 1971; Cronan, 1974; Salton, 1974; Machtiger and Fox, 1973). On the basis of generation time, the turnover rates of membrane lipids (Oelze and Drews, 1970a; Gorchein *et al.*, 1968) and of membrane proteins (Takemoto, 1974) are low. Therefore, differentiation of the membrane system takes place almost exclusively by incorporation of new components, proteins, pigments, and lipids, into the membrane. Consequently, during differentiation (Oelze and Drews, 1972) concentrations of those compounds that are incorporated into membranes increase relative to those compounds that are not incorporated or are incorporated at a lower rate.

Mutant strains of *Rps. sphaeroides* have been isolated which contain 5–50 times more BChl and carotenoids than the wild type when grown under highly aerobic conditions in the dark (Lascelles and Wertlieb, 1971). In contrast to the wild-type strains, magnesium protoporphyrin–S–adenosylmethionine methyltransferase activity in membrane preparations from the mutants was not repressed by growth under aerobic conditions in light or dark (Lascelles and Wertlieb, 1971). This type of mutant seems to be suitable for studies on the regulatory system controlling pigment and ICM synthesis.

2. The Influence of Light Intensity on Differentiation

The development of the bacterial PSA is light independent. A functional PSA and a fully developed ICM system are synthesized under low oxygen partial pressure in the dark (Cohen-Bazire and Kunisawa, 1960; Biedermann *et al.*, 1967; Drews *et al.*, 1969a). However, light is needed in anaerobic cultures to provide cells with energy for metabolic processes. The intensity of light modifies the morphogenetic process of formation of ICMs (Cohen-Bazire and Sistrom, 1966; Kaplan, 1978). Light intensity has an influence on the growth rate via photophosphorylation. This effect is difficult to separate from the "morphogenetic effect" of light. Unfortunately, the action spectrum for BChl formation is identical to the absorption spectrum of photopigments (Drews and Jaeger, 1963). Thus, light controls the synthesis of BChl via the photosynthetic apparatus or one of its products. Preillumination with low light intensities at different wavelength does not reveal any influence of pigments other than BChl and carotenoids on the formation of the photosynthetic apparatus (Drews and Jaeger, 1963). The relationship between radiation intensity, growth rate, and BChl formation is discussed in Section III,A,4.

Light intensity controls the total BChl concentration per cell and per membrane protein (Cohen-Bazire et al., 1957; Lascelles, 1959; Uemura et al., 1961; Drews and Giesbrecht, 1963; Gorchein, 1968a; Drews et al., 1969b; Irschik and Oelze, 1973, 1976; Göbel, 1978) and the amount of ICMs per cell (Cohen-Bazire and Kunisawa, 1963; Gorchein, 1968a; Drews et al., 1967; Biedermann et al., 1967; Irschik and Oelze, 1976). At low light intensity, cells of R. rubrum are filled with chromatophores which have a high concentration of BChl (for example, 100 μg of BChl per milligram of membrane protein and 20 μg of BChl per milligram of cell protein (Oelze et al., 1969a; Oelze and Drews, 1970a). In contrast, with strong light only a peripheral layer of chromatophores with a low concentration of BChl was observed [for example, 80 μg of BChl per milligram of membrane protein and 10 μg of BChl per milligram of cell protein (Cohen-Bazire and Kunisawa, 1963; Holt and Marr, 1965c; Oelze et al., 1969b; Oelze and Drews, 1970a,b)]. Similar observations were obtained from other photosynthetic bacteria (Cohen-Bazire and Sistrom, 1966; Gorchein, 1968a). At a specific growth rate of 0.1 hr^{-1}, a maximum concentration of 25.1 μg of BChl per milligram dry weight on irradiation with 1.8 nEinsteins/sec cm^2 and a minimum concentration of 1.4 μg of BChl per milligram dry weight on irradiation with 55.0 nEinsteins sec cm^2 were found in continuous cultures of R. rubrum (Göbel, 1978). Other values are summarized in Table IV. Light intensity also regulates the ratios of steady-state concentrations of the different pigment species (RC BChl:LH BChl(s):carotenoids) (Firsow and Drews, 1977; Schumacher and Drews, 1978). The photochemical activities of membranes measured as rate of photophosphorylation (micromoles of ATP per micromole of total BChl or per micromole of RC BChl per minute) were found to be dependent on the culture conditions of the bacteria (Cohen-Bazire and Kunisawa, 1960; Lien et al., 1973; Dierstein and Drews, 1975; Nieth and Drews, 1975). Rates of photophosphorylation measured at saturating light intensity increased with decreasing BChl concentrations in membranes. Thus, membranes of bacteria, which were cultivated using high light intensities, have higher photophosphorylation rates (per micromole of BChl) than membranes from cells grown at low light intensities (Cohen-Bazire and Kunisawa, 1960; Lien et al., 1973; Irschik and Oelze, 1973, 1976).

TABLE IV

BACTERIOCHLOROPHYLL CONTENTS OF PHOTOTROPICALLY GROWN CELLS OF
RHODOSPIRILLACEAE[a]

| Strains | Bacteriochlorophyll content (μg BChl per mg dry weight) at mean irradiation[b] of | | |
	2 nEinsteins per sec cm^2	40 nEinsteins per sec cm^2	Ratio max:min
Rhodospirillum rubrum Ha	12.6	1.9	6.6
Rhodospirillum tenue 3661	12.3	3.3	3.7
Rhodopseudomonas acidophila 7050	21.5	4.3	5.0
Rhodopseudomonas capsulata Kb1	20.1	13.0	1.5

[a] From Göbel (1978).
[b] The continuous cultures were illuminated with monochromatic light of wavelength corresponding to maximal *in vivo* absorption of BChl.

Phototrophically grown cells subjected to a sudden shift in light intensity vary not only in the content of BChl and ICM per cell, but also in the content of cytochromes and enzymes of the respiratory chain. Cells of *R. rubrum*, transferred from low to high light intensities did not synthesize BChl for about two cell mass doublings. The BChl content of chromatophores decreased during this period from 117 to 70 μg of BChl per milligram of protein (Irschik and Oelze, 1976). The NADH-oxidizing electron-transport systems (NADH dehydrogenase and NADH–Cyt c oxidoreductase) were incorporated during this period into both CM and ICM. Succinate oxidizing system and cytochrome oxidase, however, were incorporated exclusively into ICM (Irschik and Oelze, 1976). Activities of photophosphorylation and succinate-dependent NAD$^+$ reduction in the light, as well as fast light-induced "on" reactions at 422 nm and soluble Cyt c_2 levels increased on the basis of BChl (Irschik and Oelze, 1976).

Cells of *Rps. sphaeroides* grown under low light intensities (400 ft-c) assembled low-molecular-weight polypeptides (supposedly light-harvesting) 5–7 times faster than they assembled RC polypeptides when compared to cells grown under higher light intensities (4500 ft-c). Cells adapting from higher to lower light intensities assembled LH polypeptides at higher rates than cells adapting from lower to higher light intensities. In contrast, the relative amounts of RC polypeptides were approximately the same with varying incident light levels (Takemoto and Huang Kao, 1977).

Growing cells subjected to an increase in light intensity reduce the differential rate of ICM synthesis and BChl formation. This response is referred to as repression. Conversely, a decrease in light intensity, resulting in an increase in the differential rate of membrane formation is referred to as derepression (Kaplan, 1978). All recent data, especially the careful measurements of Göbel (1978) on continuous cultures, suggest that the stimulus of a sudden change in light intensity is transmitted via BChl to the biosynthetic machinery of the cell.

It has been proposed that the light specifically interacting with RC BChl is effective in producing a corepressor. Light energy reaching RC BChl by way of LH BChl was assumed to be not only ineffective in generating a corepressor, but actually counterproductive for corepressor formation (Kaplan, 1978). In steady-state cells growing anaerobically in high light, corepressor concentrations should be high and cells would be restricted to the synthesis of small amounts of ICM. These membranes will be enriched in RC BChl relative to LH BChl (B850) owing to the greater affinity of the repressor for those regulatory sites involved in LH BChl synthesis (Kaplan, 1978). Conversely, steady-state cells growing in dim light were assumed to have lower corepressor levels owing to the greatly decreased interaction of the limiting light with LH BChl (Kaplan, 1978). Clearly, this model was developed on the basis of observations with *Rps. sphaeroides* and *Rps. capsulata* (see Section III,A,5) and has to be modified when used for *R. rubrum*. Earlier hypotheses on the regulation of BChl synthesis by light intensity have been recently discussed (Oelze and Drews, 1972).

3. The Influence of Other Factors

Temperature and carbon and nitrogen sources influence steady-state concentrations of pigments and other membrane components (Pfennig, 1967; Biebl and Drews, 1969; Schön and Ladwig, 1970; Dierstein and Drews, 1974, 1975; Göbel, 1978). However, no data are available which demonstrate unequivocally a direct influence of one of these factors on membrane morphogenesis. The effects could not be separated from influences on the growth rate and are discussed in the next section.

4. Growth Rate and Morphogenesis

Studies on the relationship between the specific cellular BChl content and the growth rate in batch or continuous phototrophic cultures of different species have shown that with increasing growth rate the BChl content decreases linearly (Sistrom, 1962; Holt and Marr, 1965c; Cohen-Bazire and Sistrom, 1966). The chemostat was limited by succinate (Cohen-Bazire and Sistrom, 1966) and the batch cultures by light energy.

In a recent study with ammonium-limited continuous cultures, contrary results were obtained. In cultures growing in the dark at low oxygen partial pressure, BChl content and the ratio of membrane protein to total cell protein increased nonlinearly when the growth rate was increased (Dierstein and Drews, 1975). On the other hand, in cells growing anaerobically under saturating light intensity, only a small increase of both LH BChl and LH proteins was observed when the growth rate was increased (Dierstein and Drews, 1975).

These few selected data clearly show that simple relationships do not always exist between BChl content and growth rate. When growth is limited by light, the light dependency of the specific growth rate resembles a saturation function similar to a substrate-limited growth rate (Monod, 1950; Göbel, 1978). At a fixed light intensity, cells form a species-specific pattern of ICM with RC and LH complexes, which adapt to an optimal quantum uptake rate under the experimental light regime. The quantum yield of light-limited cultures possibly varies in relation to the growth rate because of the variable portions of maintenance energy (Göbel, 1978). Thus, under conditions of energy-limited growth, light intensity is the dominating factor which regulates the formation of the photosynthetic apparatus. When the energy metabolism is light saturated and the growth rate is limited by ammonium concentration, light intensity also seems to be the dominating factor in the regulation of BChl synthesis, keeping the photosynthetic apparatus optimally adapted to quanta uptake. In cells growing in a succinate-limited chemostat at 400 ft-c light intensity, the cellular BChl concentration decreases about 20% while the growth rate increases 50% (Cohen-Bazire and Sistrom, 1966). The discrepancy in results from nitrogen- and carbon-limited chemostates cannot be resolved as yet. However, the variation in the C:N ratio and differences in fermentation rates and production rates of slime or storage material might influence growth rate and BChl formation (Dierstein and Drews, 1974; Göbel, 1978).

In dark-grown cultures the photosynthetic apparatus is formed but not used for energy production. Therefore a regulatory device might be active that is independent of the photosynthetic apparatus and regulated by the respiratory chain.

It has been suggested that the level of the corepressor for the synthesis of BChl and protein is affected by the specific interaction of RC BChl with light or interaction of an effector molecule with oxygen. In addition, the metabolism of a carbon (and nitrogen?) source influenced the level of the corepressor (Kaplan, 1978). Thus, the rate of biosynthesis of the PSA is influenced by all factors that determine the growth rate. Light intensity and oxygen partial pressure, however, can specifically effect the morphogenesis of the photosynthetic apparatus. This specific effect can be separated from the influence of the growth rate when the energy metabolism is not limited by light or oxygen. The general problem of how differentiation relates to growth arises from the experiments discussed above in this section. As shown, BChl and membrane formation are more repressed in NH_4-limited cultures when the photosynthetic apparatus is induced but inactive (dark cultures) and less repressed when the photosynthetic apparatus is induced and active (light cultures). This seems to indicate the presence of a feedback regulatory system that reflects whether the induced complex is active or not.

5. Two Types of Differentiation

It was shown in Section II,B2 that the ICM of *R. rubrum* contain two BChl complexes, RC and B870, which are formed in a fixed stoichiometric ratio (Aagaard and Sistrom, 1972; Oelze and Pahlke, 1976). Thus, the size of the photosynthetic unit in *R. rubrum* is relatively constant. However, the cellular BChl content can vary between 0 and more than 20 μg of BChl per milligram of cell protein. *R. rubrum* adapts to various oxygen tensions or light intensities by variation of (1) BChl content of ICM (variation of number of photosynthetic units per membrane area) and (2) of membrane area per cell (Cohen-Bazire and Kunisawa, 1960, 1963; Oelze *et al.*, 1969a).

Rps. capsulata and *Rps. sphaeroides* were shown to have RC BChl– LH BChl I– and LH BChl II–protein complexes (Aaagaard and Sistrom, 1972; Lien *et al.*, 1973; Feick and Drews, 1978; Drews *et al.*, 1976a,b). Cells of both species adapt to various oxygen tensions or light intensities by (1) variation of size of the photosynthetic unit (ratio LH II (B850) BChl to RC + LH I (B875) BChl), (2) variation of number of photosynthetic units per membrane area, and (3) variation of membrane area per cell (Lien *et al.*, 1971, 1973; Aagaard and Sistrom, 1972; Dierstein and Drews, 1974, 1975; Takemoto, 1974; Drews *et al.*, 1976a; Firsow and Drews, 1977; Schumacher and Drews, 1978). The ratio of RC BChl to LH I BChl seems to be constant under a steady-state condition (Aagaard and Sistrom, 1972).

There is some evidence that the ratio between BChl and the accompanying proteins in the different complexes is also nearly constant (Oelze and Drews, 1972; Biedermann and Drews, 1968; Dierstein and Drews, 1975; Nieth and Drews, 1975; Takemoto, 1974; Oelze and Pahlke, 1976). Thus, the main difference between type A (*R. rubrum*) and type B (*Rps. capsulata* and *Rps. sphaeroides*) is that in type A the number of photosynthetic units per membrane area varies, keeping the size of the photosynthetic unit constant. Type B, however, varies the size of the photosynthetic unit and the number of photosynthetic units per membrane area. It is still necessary to learn whether all purple bacteria can be assigned to one of these types or have developed special mechanisms to adapt their photosynthetic apparatus to the environment. In cells of *Chromatium vinosum* the ratio RC to B880 BChl is constant. B850 BChl is the variable spectral form (Oelze and Mechler, 1976).

6. Summary on Membrane Differentiation

After induction of formation of the PSA by changing oxygen partial pressure, light intensity, or other regulator levels, RC– and LH BChl–protein complexes, carotenoids, electron carriers and other compounds are incorporated into the membrane by a multistep process. During the transient state of morphogenesis the various components of the PSA are synthesized at different and variable rates until a new steady state is reached. Growth and differentiation are independent processes.

Morphogenesis begins in *R. rubrum* and other Rhodospirillaceae by incorporation of the PSA into the CM. Afterward an increasing number of invaginations of CM is formed, which is the beginning of ICM formation (Golecki and Oelze, 1975). The cells adapt to different regulator levels (pO_2, etc.) in the culture by reaching specific steady-state concentrations of the PSA per cell.

Under various steady states the parameters are as listed below:

R. rubrum	*Rps. capsulata, Rps. sphaeroides,* and *Rps. palustris*
CONSTANT	
Ratio RC:B870 (size of the photosynthetic unit)	Ratio RC:LHI (B870)
VARIABLE	
1. Numer of photosynthetic units per ICM area	1. Size of the photosynthetic unit (LH II (B850):RC+LH I)
2. ICM area per cell	2. ICM area per cell
	3. Number of photosynthetic units per ICM area

B. THE PROCESS OF DIFFERENTIATION

1. Regulation of Bacteriochlorophyll Synthesis

The regulation of BChl synthesis has been reviewed recently (Lascelles, 1968; Oelze and Drews, 1972). The key enzymes for synthesis of

BChl are 5-aminolevulinate synthetase, the first enzyme in the biosynthetic pathway, Mg-chelatase and methyltransferase, the first enzymes behind the branching point leading to Mg-tetrapyrrols. The 5-aminolevulinate synthetase can exist in low-activity (b) and high-activity (a) forms *in vitro*. Oxygenation of a semiaerobic or anaerobic culture results in the disappearance of high-activity enzyme and the accumulation of low-activity enzyme in the cell (Sandy *et al.*, 1975). The active and the inactive enzyme have been separated and purified (Tuboi *et al.*, 1970; Sandy *et al.*, 1975). The low-activity enzyme could be reactivated *in vitro* by addition of cysteine, cystine, cystine trisulfide, glutathione trisulfide, and mixed trisulfides of glutathione and cystine (Neuberger *et al.*, 1973; Hayasaka and Tuboi, 1974; Sandy *et al.*, 1975). It has been proposed that the cellular content of cystine is involved in the regulation of the dynamic equilibrium between the *b* form and the *a* form in the cell (Sandy *et al.*, 1975). Magnesium-protophorphyrin chelatase activity is also inhibited by oxygen in whole cells (Gorchein, 1972, 1973). The mechanism of activation and inactivation of both enzymes needs further study.

2. The Sequential Process of Formation of the Photosynthetic Apparatus

Early events of morphogenesis of the PSA were studied in shift experiments. Photosynthetic bacteria able to grow chemotrophically as well as phototrophically were cultivated under high oxygen tension in the dark. These cells, which are completely or nearly devoid of BChl and photochemically inactive, were resuspended in a fresh medium and cultivated at very low oxygen tension in the dark to induce the morphogenetic process. The low aeration led to cessation of growth, but PSA was formed. It was studied whether BChl and the respective protein moieties were simultaneously synthesized and whether the assembly of the PSA proceeded through a multistep or a single step mechanism. The early formation of the PSA in *R. rubrum* will be described first, then the early events in the cells of *Rps. sphaeroides* and *Rps. capsulata* will follow.

BChl synthesis starts very soon after transferring a culture of *R. rubrum* to low aeration. The BChl content rises from zero (not detectable) to 20 nmoles of BChl per milligram of cell protein in 4 hours. The ratio of RC BChl to LH BChl does not show any significant variation during this period. This was also demonstrated by the IR *in vivo* spectrum of membranes isolated from cells after different incubation periods: the 800- to 875- nm peak ratio and the shape of the main peak did not change significantly (Oelze and Pahlke, 1976).

Concomitantly with BChl, the proteins of the PSA are incorporated into the membrane. However, contrary to the results from the BChl species, the different proteins of the PSA were not synthesized in constant proportions. During the initial phase, incorporation of the LH protein dominated over incorporation of RC proteins. After 4 hours a coordinated formation of the different proteins belonging to the LH and RC complexes was observed (Oelze and Pahlke, 1976). Neither Cyt c_2 levels nor incorporation of carotenoids into the membrane were found to be correlated with BChl synthesis (Oelze and Pahlke, 1976). This indicates that the PSA is assembled in *R. rubrum* through a multistep mechanism, at least during the initial phase. Immediately after the induction of morphogenesis, BChl was incorporated predominantly into the CM. With increasing pigment content the newly formed ICM became the site of preferential BChl incorporation (Oelze, 1976). During this period the IR maximum shifted from 874 to 880 nm. The BChl content increased from 0.46 to 12.0 μg of BChl per milligram of protein. Proteins incorporated into the CM became constituents of ICM during the same initial period (Oelze, 1976). These experiments are in accordance with the hypothesis that early ICM arise by invagination of CM (Oelze and Drews, 1972).

In cells of *Rps. capsulata* and *Rps. sphaeroides,* representatives of type B of differentiation, the mode of assembly of the PSA is different from the sequences observed in cells of *R. rubrum*, representative of type A. In low aerated cultures of *Rps. capsulata* the BChl content rises from 0.1 to 2.8 nmoles of BChl per milligram of membrane protein within 150 min (Drews *et al.*, 1976b; Schumacher and Drews, 1978). RC BChl and LH BChl I are the first detectable BChl species (Nieth and Drews, 1975; Drews *et al.*, 1976b; Schumacher and Drews, 1978). LH BChl II (B800 + B850) was observed in low-temperature spectra of membranes after 90 minutes of induction and dominated after 130 minutes of incubation (Drews *et al.*, 1976b; Schumacher and Drews, 1978). In both species, *Rps. capsulata* and *Rps. sphaeroides*, the size of the photosynthetic unit (molar ratio of total BChl to RC-BChl) decreased during the very early period of incubation at low aeration and increased during the following period (Takemoto, 1974; Nieth and Drews, 1975).

During the first 40–60 minutes of incubation at low oxygen partial pressure, radioactive labeled amino acids were incorporated at a higher rate into membrane proteins associated with RC BChl as compared to the membrane protein fraction associated with LH BChl (Takemoto, 1974; Nieth and Drews, 1975). Preliminary results have shown that in *Rps. capsulata* formation of RC proteins is accompanied by formation of LH BChl I protein (band 2 of LH proteins), whereas protein bands 3 and 4 together with LH BChl II were incorporated into the membrane fraction in the following step of morphogenesis (Drews *et al.*, 1976b; Schumacher and Drews, 1978). In *Rps. capsulata* and *Rps. sphaeroides* LH BChl complex II (B850) becomes the dominating fraction of the PSA shortly (60–120 minutes) after induction (Takemoto, 1974; Nieth and Drews, 1975; Schumacher and Drews, 1978; Niederman *et al.*, 1976). Although the diffferent BChl species were incorporated concomitantly with their respective proteins, a stoichiometry between BChl and the respective proteins on a molar basis has not yet been proved.

The activity of photophosphorylation increases with incubation time on basis of membrane protein (Takemoto, 1974), but decreases on the basis of total BChl and remains constant on the basis of RC BChl (Nieth and Drews, 1975). The increase of carotenoids in the membrane shows the same tendency as total BChl (Takemoto, 1974; Niederman *et al.*, 1976). However, the level of both pigments increases differently (Takemoto, 1974). In conclusion, at present the data indicate that the bacterial PSA is assembled through a multistep mechanism. The proposed types A and B of differentiation differ in the events of morphogenesis.

The synthesis of ICM has been examined in cell populations of *Rps. sphaeroides* that divide synchronously (Kaplan, 1977). Total cell number and DNA increased stepwise in synchronized cells of cultures grown anaerobically in the light. Total cellular protein, BChl, carotenoids, RC BChl, and Cyt *c* increased exponentially. In contrast, discontinuous increases in the levels of succinic dehydrogenase and NADH oxidase activities were observed throughout the period of synchronous growth. Results from density-shift experiments (method of Kosakowski and Kaplan, 1974) suggest that a rapid burst of ICM biosynthesis occurs prior to cell division. Incorporation of protein and lipid phosphorus is also a discontinuous process (Kaplan, 1977).

The observation that the incorporation of pigment and protein into the membrane is coordinated suggests that not single components, but macromolecular complexes, are the membrane precursors (see discussions in Oelze and Drews, 1972; Kaplan, 1978; Shaw and Richards, 1971, 1972). If such precursors do exist, the pool size should be small and a separation of such precursors from membrane fragments and artifically formed aggregates would be extremely difficult. Mutant strains blocked in BChl and/or membrane protein synthesis excrete protein-bound BChl precursors into the medium (Oelze and Drews, 1972; Drews, 1974; Richards *et al.*, 1975). The protein in the excreted pigment complex of *Rps. capsulata* strain Ala⁻ is not identical in SDS–gel electrophoresis with any protein of the membrane-bound PSA (Drews, 1974), but runs to the same position, as an early labeled 40-kdalton polypeptide in the wild-type strain (Drews *et al.*, unpublished).

It seems to be possible that precursor proteins were modified before the incorporation into the membrane. At present, however, we have no proof that these excreted pigment complexes are precursors that are enriched in a mutant strain.

A subunit fraction isolated from wild-type cells of *Rps. sphaeroides* in the early stages of adaptation contains a major polypeptide with an apparent weight of 38–42 kdaltons. The molecular weight is similar to that of the pigment–protein complex excreted from poorly adapting cells. The so-called prephore subunit fraction was also isolated from adapting cultures and from cultures able to form BChl but not ICM. It has been suggested that aggregates of these pigment–protein complexes are modified during adaption and incorporated into invaginations of the CM during morphogenesis (Shaw and Richards, 1971, 1972, and unpublished results). Results from pulse-chase experiments with *Rps. sphaeroides* indicate that ICM are formed from precursor structures. The proteins of the small membrane fragments have the behavior expected of precursors (Gibson *et al.*, 1972).

The processes of biosynthesis and assembly leading to the PSA of green bacteria are unknown and no data are available. However, it seems clear that these bacteria can adapt their BChl content to different light intensities.

3. Correlation between Photosynthetic and Respiratory Apparatus in Development

Photosynthetic bacteria, able to grow chemoheterotrophically or phototrophically, do not lose the capacity of oxidative phosphorylation, even when cultivated anaerobically in the light for many generations. Components of the respiratory chain are incorporated into both CM and ICM grown phototrophically (Lampe and Drews, 1972; Throm *et al.*, 1970; Irschik and Oelze, 1976; Collins and Niederman, 1976a). The rate of incorporation is influenced by oxygen tension and light intensity (Thore *et al.*, 1969; Lampe and Drews, 1972; King and Drews, 1975). The differences between low and high levels of enzymes and cytochromes on the basis of cell and membrane protein are smaller than the corresponding differences in BChl content.

It has been discussed that PSA and respiratory chain have a close contact in the membrane (Oelze and Drews, 1972). Electrons can flow from respiratory chain to the PSA and vice versa. The coupling factor is active for both oxidative and cyclic photophosphorylation processes (Lien and Gest, 1973). A cyclic light-driven electron flow (Jones and Plewis, 1974; Garcia *et al.*, 1975) and photophosphorylation were reconstituted by recombining the coupling factor and RC from a photosynthetically active carotenoidless strain with uncoupled membranes from a carotenoidless and BChl-less but respiratory active mutant strain (Garcia *et al.*, 1974).

Light inhibits respiration in strains that are active in both photosynthetic and respiratory electron transport (Clayton, 1953; Keister, 1978). However, most of the effects can be explained by an energization of the membrane. The energized state of the membrane can drive energy-dependent reactions. Thus, active transport of amino acids in whole cells and membrane vesicles from *Rps. sphaeroides* is coupled to electron flow in the respiratory chain and in the cyclic light-driven electron-transport system as well (Helligwerf *et al.*, 1975). Clearly, both electron-transport chains must be arranged so as to result in a unidirectional electron flow into isolated chromatophores or outward in whole cells. The regulation of membrane differentiation becomes more complicated when both electron-transport systems are considered. Induction of the formation of the PSA does not always inhibit synthesis of respiratory chain components and vice versa.

IV. The Regulation of Differentiation on a Molecular Basis

The architecture and development of the membrane system of photosynthetic bacteria have been described in the preceding sections. Differentiation of the membrane system is known to be regulated by some external factors. The most important factors are oxygen tension and light intensity. The model proposed for the formation of ICM in *R. rubrum* (Oelze and Drews, 1972) has been confirmed by many experimental data during the past few years.

Kaplan's (1978) model will be helpful in leading to experiments on the molecular basis of coregulation of BChl and protein synthesis in formation of specific pigment complexes. All strains that are fully inducible in the synthesis of the PSA and assembly of ICM are able to synthesize both BChl and the respective proteins. Mutants blocked in the BChl synthesis or formation of one of its respective proteins cannot form active subunits. Some data on spectroscopic forms of BChl and membrane protein patterns (Drews *et al.*, 1976a,b) seem to indicate that genes for enzymes of the BChl synthesis and genes for RC plus LH I proteins are clustered together on the bacterial chromosome or are coregulated on the transcriptional level. The transcription of both might be switched on and off by a common holorepressor. It was shown that genes of BChl and carotenoids are on adjacent regions of the chromosome of *Rps. capsulata* (Yen and Marrs, 1976).

Another explanation for the observed coregulation of BChl and proteins might be a mutual control: a precursor of BChl synthesis controls the synthesis of the LH proteins (Lascelles, 1968), and a peptide split from the nascent LH protein controls the BChl synthesis.

It was shown that ribosomes bind specifically to membranes (Chua *et al.*, 1973). Proteins of the PSA might be synthesized on those ribosomes. Small peptides can be split from nascent polypeptides (Blobel and Dobberstein, 1975). It is proposed that such a peptide could control the assembly of the protein with the synthesized BChl in the hydrophobic zone of the membrane. This hypothesis does not need the involvement of subchromatophore precursors, and the whole process of BChl–protein complex formation would be a membrane-bound process. To test this hypothesis a cell-free system of membrane formation is required.

ACKNOWLEDGMENTS

The author is grateful to Dr. Jürgen Oelze for constructive criticism of the manuscript. The work of the author was supported by grants of the Deutsche Forschungsgemeinschaft.

REFERENCES

Aagaard, J., and Sistrom, W. R. (1972). *Photochem. Photobiol.* **15**, 209–225.
Aiba, S., Humphrey, A. E., and Millis, N. F. (1965). "Biochemical Engineering." Academic Press, New York.
Altendorf, K. H., and Staehelin, L. A. (1974). *J. Bacteriol.* **117**, 888–899.
Amesz, J. (1978). *In* "The Photosynthetic Bacteria" (R. K. Clayton and W. R. Sistrom, eds.) Plenum, New York, in press.
Baccarini-Melandri, A., and Melandri, B. A. (1978). *In* "The Photosynthetic Bacteria" (R. K. Clayton and W. R. Sistrom, eds.) Plenum, New York, in press.
Baccarini-Melandri, A., Zannoni, D., and Melandri, B. A. (1973). *Biochim. Biophys. Acta* **314**, 298–311.
Bachofen, R., Hanselmann, K. W., and Snozzi, M. (1976). *Brookhaven Symp. Biol.* **28**, Abstr. G2.
Baltscheffsky, H., and Baltscheffsky, M. (1974). *Annu. Rev. Biochem.* **43**, 871–897.
Baltscheffsky, H., Baltscheffsky, M., and Thore, A. (1971). *Curr. Top. Bioenerg.* **4**, 273.
Bartsch, R. G. (1978). *In* "The Photosynthetic Bacteria" (R. K. Clayton and W. R. Sistrom, eds.) Plenum, New York, in press.
Biebl, H., and Drews, G. (1969). *Zentralbl. Bakteriol. Parasitenkd. Infektionskr. Hyg., Abt. 2* **123**, 425–452.
Biedermann, M. (1971). *Hoppe Seyler's Z. Physiol. Chem.* **352**, 567–574.
Biedermann, M., and Drews, G. (1968). *Arch. Mikrobiol.* **61**, 48–58.
Biedermann, M., Drews, G., Marx, R., and Schröder, J. (1967). *Arch. Mikrobiol.* **56**, 133–147.
Blobel, G., and Dobberstein, B. (1975). *J. Cell Biol.* **67**, 835–851.
Boatman, E. S. (1964). *J. Cell Biol.* **20**, 297–311.
Boatman, E. S., and Douglas, H. C. (1961). *J. Biophys. Biochem. Cytol.* **11**, 469–483.
Boll, M. (1968). *Arch. Mikrobiol.* **64**, 85–102.
Bolton, J. R., and Cost, K. (1973). *Photochem. Photobiol.* **18**, 417–421.
Bolton, J. R., Clayton, R. K., and Reed, D. W. (1969). *Photochem. Photobiol.* **9**, 209–218.
Boyce, C. O., Oyewole, S. H., and Fuller, R. C. (1976). *Brookhaven Symp. Biol.* **28**, Abstr. G1, p. 365.
Branton, D. (1966). *Proc. Natl. Acad. Sci. U.S.A.* **55**, 1048–1056.
Bril, C. (1958). *Biochim. Biophys. Acta* **29**, 458.
Bril, C. (1960). *Biochim. Biophys. Acta* **39**, 296–303.

200 GERHART DREWS

Brockmann, H., Knobloch, G., Schweer, I., and Trowitzsch, W. (1973). *Arch. Microbiol.* **90**, 161–164.

Buchanan, R. E., and Gibbons, N. E., eds. (1974). "Bergey's Manual of Determinative Bacteriology," 8th Ed. Williams & Wilkins, Baltimore, Maryland.

Bull, M. J., and Lascelles, J. (1963). *Biochem. J.* **87**, 15–28.

Burns, D. D., and Midgley, M. (1976). *Eur. J. Biochem.* **67**, 323–333.

Case, G. D., and Parson, W. W. (1971). *Biochim. Biophys. Acta* **253**, 187–202.

Chance, B., Nishimura, M., Avron, M., and Baltscheffsky, H. (1966). *Arch. Biochem. Biophys.* **117**, 158–166.

Chua, N.-H., Blobel, G., Siekevitz, P., and Palade, G. E. (1973). *Proc. Natl. Acad. Sci. U.S.A.* **70**, 1554–1558.

Clayton, R. K. (1953). *Arch. Mikrobiol.* **19**, 107–124.

Clayton, R. K. (1955). *Arch. Mikrobiol.* **22**, 180–194.

Clayton, R. K. (1962). *Photochem. Photobiol.* **1**, 201–210.

Clayton, R. K. (1963). *Biochim. Biophys. Acta* **75**, 312–323.

Clayton, R. K. (1965). "Molecular Physics in Photosynthesis," pp. 149–156. Ginn (Blaisdell), Boston, Massachusetts.

Clayton, R. K. (1973). *Annu. Rev. Biophys. Bioeng.* **2**, 131–156.

Clayton, R. K., and Clayton, B. J. (1972). *Biochim. Biophys. Acta* **283**, 492–504.

Clayton, R. K., and Sistrom, W. R., eds. (1978). "The Photosynthetic Bacteria." Plenum, New York.

Clayton, R. K., and Wang, R. T. (1971). *In* "Photosynthesis" Part A (A. San Pietro, ed.), Methods in Enzymology, Vol. 23, pp. 696–704. Academic Press, New York.

Cogdell, R. J., Monger, T. G., and Parson, W. W. (1975). *Biochim. Biophys. Acta* **408**, 189–199.

Cogdell, R. J., Parson, W. W., and Kerr, M. A. (1976). *Biochim. Biophys. Acta* **430**, 83–93.

Cohen-Bazire, G., and Kunisawa, R. (1960). *Proc. Natl. Acad. Sci. U.S.A.* **46**, 1543–1553.

Cohen-Bazire, G., and Kunisawa, R. (1963). *J. Cell Biol.* **16**, 401–419.

Cohen-Bazire, G., and Sistrom, W. R. (1966). *In* "The Chlorophylls" (L. P. Vernon and G. R. Seely, eds.), pp. 313–341. Academic Press, New York.

Cohen-Bazire, G., Sistrom, W. R., and Stanier, R. Y. (1957). *J. Cell Comp. Physiol.* **49**, 25, 68.

Cohen-Bazire, G., Pfennig, N., and Kunisawa, R. (1964). *J. Cell Biol.* **22**, 207–225.

Collins, M. L. P., and Niederman, R. A. (1976a). *J. Bacteriol.* **126**, 1316–1325.

Collins, M. L. P., and Niederman, R. A. (1976b). *J. Bacteriol.* **126**, 1326–1338.

Cronan, J. E. (1974). *Proc. Natl. Acad. Sci. U.S.A.* **71**, 3758–3762.

Cruden, D. L., and Stanier, R. Y. (1970). *Arch. Mikrobiol.* **72**, 115–134.

Dierstein, R., and Drews, G. (1974). *Arch. Microbiol.* **99**, 117–128.

Dierstein, R., and Drews, G. (1975). *Arch. Microbiol.* **106**, 227–235.

Ding, D. H., and Kaplan, S. (1976). *Prep. Biochem.* **6**, 61–79.

Drews, G. (1965). *Arch. Mikrobiol.* **51**, 186–198.

Drews, G. (1974). *Arch. Microbiol.* **100**, 397–407.

Drews, G. (1976). *Brookhaven Symp. Biol.* **28**, G3.

Drews, G., and Giesbrecht, P. (1963). *Zentralbl. Bakteriol., Parasitenkd., Infektionskr. Hyg., Abt. I: Orig.* **190**, 508–536.

Drews, G., and Giesbrecht, P. (1965). *Arch. Mikrobiol.* **52**, 242–250.

Drews, G., and Jaeger, K. (1963). *Nature (London)* **199**, 1112–1113.

Drews, G., Biedermann, M., and Schön, G. (1967). *Zentralbl. Bakteriol., Parasitenkd., Infektionskr. Hyg. Abt. I: Orig.* **205**, 38–41.

Drews, G., Lampe, H.-H., and Ladwig, R. (1969a). *Arch. Mikrobiol.* **65**, 12–28.

Drews, G., Biedermann, M., and Oelze, J. (1969b). *Prog. Photosynth. Res.* **1**, 204–208.

Drews, G., Dierstein, R., and Nieth, K.-F. (1975). *Proc. Int. Congr. Photosynth. 3rd, Rehovot,* Vol. III (M. Avron, ed.), pp. 2139–2146.

Drews, G., Dierstein, R., and Schumacher, A. (1976a). *FEBS Lett.* **68,** 132–136.

Drews, G., Schumacher, A., and Dierstein, R. (1976b). *Proc. Int. Symp. Photosynthetic Prokaryotes, 2nd, Dundee* (G. A. Codd and W. D. P. Stewart, eds.), pp. 61–63.

Dutton, P. L., and Jackson, J. B. (1972). *Eur. J. Biochem.* **30,** 495–510.

Dutton, P. L., and Leigh, J. S. (1973). *Biochim. Biophys. Acta* **314,** 178–190.

Dutton, P. L., and Wilson, D. F. (1974). *Biochim. Biophys. Acta* **346,** 165–212.

Dutton, P. L., Leigh, J. S., and Reed, D. W. (1973). *Biochim. Biophys. Acta* **292,** 654–664.

Dutton, P. L., Petty, K. M., Bonner, H. S., and Morse, S. D. (1975). *Biochim. Biophys. Acta* **387,** 536–556.

Dutton, P. L., Prince, R. C., Tiede, D. M., Petty, K. M., Kaufmann, K. L., Netzel, T. L., and Rentzepis, P. M. (1976). *Brookhaven Symp. Biol.* **28,** 213–237.

Duysens, L. N. M. (1952). Ph.D. Thesis, Univ. Utrecht.

Erokhin, Y. E., and Moskalenko, A. A. (1973). *Dokl. Biol. Sci.* **212,** 429.

Evans, E. H., and Crofts, A. R. (1974). *Biochim. Biophys. Acta* **357,** 78–88.

Fanica-Gaignier, M., Clement-Mentral, J., and Kamen, M. D. (1971). *Biochim. Biophys. Acta* **226,** 135–143.

Feher, G. (1971). *Photochem. Photobiol.* **14,** 373–387.

Feher, G., and Okamura, M. Y. (1976). *Brookhaven Symp. Biol.* **28,** 183–194.

Feher, G., Okamura, M. Y., and McElroy, J. D. (1972). *Biochim. Biophys. Acta* **267,** 222–226.

Feher, G., Hoff, A. J., Isaacson, R. A., and Mc Elroy, J. D. (1973). *Biophys. Soc. Abstr.* **13,** 61.

Feher, G., Isaacson, R. A. Mc Elroy, J. D., Ackerson, L. C., and Okamura, M. Y. (1974). *Biochim. Biophys. Acta* **368,** 135–139.

Feher, G., Hoff, A. J., Isaacson, R. A., and Ackerson, L. C. (1975). *Ann. N.Y. Acad. Sci.* **244,** 239.

Feick, R., and Drews, G. (1978). *Biochim. Biophys. Acta,* in press.

Fenchel, T. (1969). *Ophelia* **6,** 1–182.

Fenna, R. E., and Matthews, B. W. (1975). *Nature (London)* **258,** 573–577.

Fenna, R. E., Matthews, B. W., Olson, J. M., and Shaw, E. K. (1974). *J. Mol. Biol.* **84,** 231–240.

Firsow, N. N., and Drews, G. (1977). *Arch. Microbiol.* **115,** 299–306.

Fowler, C. F. (1974). *Biochim. Biophys. Acta* **357,** 327–331.

Fowler, C. F., Nugent, N. A., and Fuller, R. C. (1971). *Proc. Natl. Acad. Sci. U.S.A.* **68,** 2278–2282.

Fraker, P. J., and Kaplan, S. (1971). *J. Bacteriol.* **108,** 465–473.

Fraker, P. J., and Kaplan, S. (1972). *J. Biol. Chem.* **247,** 2732–2737.

French, C. S., and Young, U. M. K. (1956). *In* "Radiation Biology III" (A. Hollaender, ed.), pp. 343–390. McGraw-Hill, New York.

Garcia, A. F., Vernon, L. P., and Mollenhauer, H. (1966). *Biochemistry* **5,** 2399–2407.

Garcia, A. F., Vernon, L. P., Ke, B., and Mollenhauer, H. (1968). *Biochemistry* **7,** 319–332.

Garcia, A. F., Drews, G., and Kamen, M. D. (1974). *Proc. Natl. Acad. Sci. U.S.A.* **71,** 4213–4216.

Garcia, A. F., Drews, G., and Kamen, M. D. (1975). *Biochim. Biophys. Acta* **387,** 129–134.

Ghosh, A. K., and Olson, J. M. (1968). *Biochim. Biophys. Acta* **162,** 135–148.

Gibson, K. D. (1965), *Biochemistry* **4,** 2042–2059.

Gibson, K. D., Segen, B. J., and Niederman, R. A. (1972). *Arch. Biochem. Biophys.* **152,** 561–568.

Giesbrecht, P., and Drews, G. (1962). *Arch. Mikrobiol.* **43**, 152–161.

Giesbrecht, P., and Drews, G. (1966). *Arch. Mikrobiol.* **54**, 297–330.

Gingras, G. (1978). *In* "The Photosynthetic Bacteria" (R. K. Clayton and W. R. Sistrom, eds.) Plenum, New York, in press.

Gloe, A., and Pfennig, N. (1974). *Arch. Microbiol.* **96**, 93–101.

Gloe, A., Pfennig, N., Brockmann, H., Jr., and Trowitzsch, W. (1975). *Arch. Microbiol.* **102**, 103–109.

Göbel, F. (1978). *In* "The Photosynthetic Bacteria" (R. K. Clayton and W. R. Sistrom, eds.) Plenum, New York, in press.

Goedheer, J. C. (1972). *Biochim. Biophys. Acta* **275**, 169–176.

Golecki, J. R., and Oelze, J. (1975). *J. Gen. Microbiol.* **88**, 253–258.

Gorchein, A. (1968a). *Proc. R. Soc. Ser. B* **170**, 247–254.

Gorchein, A. (1968b). *Proc. R. Soc. Ser. B.* **170**, 265–278.

Gorchein, A. (1972). *Biochem. J.* **128**, 1159–1169.

Gorchein, A. (1973). *Biochem. J.* **134**, 833–845.

Gorchein, A., Neuberger, A., and Tait, G. H. (1968). *Proc. R. Soc., Ser. B* **170**, 311–318.

Hall, D. O., Cammack, R., and Rao, K. K. (1974). *In* "Iron in Biochemistry and Medicine" (A. Jacob and M. Worwood, eds.), pp. 279–334. Academic Press, New York.

Hall, D. O., Rao, K. K., and Cammack, R. (1975). *Sci. Prog. (London)* **62**, 285–316.

Halsey, Y. D., and Parson, W. W. (1974). *Biochim. Biophys. Acta* **34**, 404–416.

Hatefi, Y., Davis, K. A., Baltscheffsky, H. Baltscheffsky, M., and Johansson, B. C. (1972). *Arch. Biochem. Biophys.* **152**, 613–618.

Hayasaka, S., and Tuboi, S. (1974). *J. Biochem. (Tokyo)* **76**, 157–168.

Hellingwerf, K. J., Michels, P. A. M., Dorpema, J. W., and Konings, W. N. (1975). *Eur. J. Biochem.* **55**, 397–406.

Higuchi, M. Goto, K., Fujimoto, M., Namiki, O., and Kikuchi, G. (1965). *Biochim. Biophys. Acta* **95**, 94–110.

Hochman, A., Fridberg, I., and Carmeli, C. (1975). *Eur. J. Biochem.* **58**, 65–72.

Holt, S. C., and Marr, A. G. (1965a). *J. Bacteriol.* **89**, 1402–1412.

Holt, S. C., and Marr, A. G. (1965b). *J. Bacteriol.* **89**, 1413–1420.

Holt, S. C., and Marr, A. G. (1965c). *J. Bacteriol.* **89**, 1421–1429.

Holt, S. C., Conti, S. F., and Fuller, R. C. (1966). *J. Bacteriol.* **91**, 311–323.

Horio, T., and Kamen, M. D. (1962). *Biochemistry* **1**, 144–145.

Horio, T., and Kamen, M. D. (1970). *Annu. Rev. Microbiol.* **24**, 399–428.

Huang, J. W., and Kaplan, S. (1973). *Biochim. Biophys. Acta* **307**, 317–342.

Hurlbert, R. E., Golecki, J. R., and Drews, G. (1974). *Arch. Microbiol.* **101**, 169–186.

Irschik, H., and Oelze, J. (1973). *Biochim. Biophys. Acta* **330**, 30–89.

Irschik, H., and Oelze, J. (1976). *Arch. Microbiol.* **109**, 307–313.

Jennings, J. V., and Evans, M. C. V. (1976). *Abstr. Int. Conf. Primary Electron Transp. Energy Transduction Photosynthetic Bact.* Brussels.

Johansson, B. C., Baltscheffsky, M., Baltscheffsky, H., Baccarini-Melandri, A., and Mealandri, B. A. (1973). *Eur. J. Biochem.* **40**, 109–117.

Jolchine, G., and Reiss-Husson, F. (1974). *FEBS Lett.* **40**, 5–8.

Jolchine, G., and Reiss-Husson, F. (1975). *FEBS Lett.* **52**, 33–36.

Jones, O. T. G., and Plewis, K. M. (1974). *Biochim. Biophys. Acta* **357**, 204–214.

Kaback, H. R. (1974). *Science* **186**, 882–892.

Kamen, M. D., and Horio, T. (1970). *Annu. Rev. Biochem.* **39**, 673–700.

Kaplan, S. (1977). *Abstr. Annu. Meeting Am. Soc. Microbiol., 77th,* New Orleans, p. 210, K 147.

Kaplan, S. (1978). *In* "The Photosynthetic Bacteria" (R. K. Clayton and W. R. Sistrom, eds.) Plenum, New York, in press.

Katz, E., and Wassink, E. C. (1939). *Enzymologia* **7**, 97.

Katz, J. J., and Norris, J. (1973). *Curr. Top. Bioenerg.* **5**, 41–75.

Katz, J. J., Dougherty, R. C., and Boucher, L. J. (1966). *In* "The Chlorophylls" (L. P. Vernon and G. R. Seely, eds.), p. 185. Academic Press, New York.

Katz, J. J., Strain, H. H., Harkness, A. L., Studier, M. H., Svec, W. A., Janson, T. R., and Cope, B. T. (1972). *J. Am. Chem. Soc.* **94**, 7938–7939.

Ke, B., Chaney, T. H., and Reed, D. W. (1970). *Biochim. Biophys. Acta* **216**, 373–383.

Ke, B., Garcia, A. F., and Vernon, L. P. (1973). *Biochim. Biophys. Acta* **292**, 226–236.

Keister, D. L. (1978). *In* "The Photosynthetic Bacteria" (R. K. Clayton and W. R. Sistrom, eds.) Plenum, New York, in press.

Keister, D. L., and Minton, N. J. (1969). *Progr. Photosynth. Res.* **3**, 1299–1305.

Kennel, S. J., and Kamen, M. D. (1971). *Biochim. Biophys. Acta* **253**, 153–166.

Ketchum, P. A., and Holt, S. C. (1970). *Biochim. Biophys. Acta* **196**, 141–161.

Kim, Y. D., and Ke, B. (1970). *Arch. Biochem. Biophys.* **140**, 341–353.

King, M.-T., and Drews, G. (1973). *Biochim. Biophys. Acta* **305**, 230–248.

King, M.-T., and Drews, G. (1975). *Arch. Microbiol.* **102**, 219–231.

King, M.-T., and Drews, G. (1976). *Eur. J. Biochem.* **68**, 5–12.

Klemme, J.-H., and Schlegel, H. G. (1969). *Arch. Mikrobiol.* **68**, 326–354.

Knaff, D. B., Buchanan, B., and Malin, R. (1973). *Biochim. Biophys. Acta* **325**, 94–101.

Kosakowski, M. H., and Kaplan, S. (1974). *J. Bacteriol.* **118**, 1144–1157.

Krasnovsky, A. A. (1969). *Progr. Photosynth. Res.* **2**, 709–727.

Krasnovsky, A. A., Voinovskaya, K. K., and Kosobutskaya, L. M. (1952). *Dokl. Akad. Nauk SSSR* **85**, 389.

Künzler, A., and Pfennig, N. (1973). *Arch. Mikrobiol.* **91**, 83–86.

Kusnezow, S. I., and Gorlenko, W. M. (1973). *Arch. Hydrobiol.* **71**, 475–486.

La Monica, R. F., and Marrs, B. L. (1976). *Biochim. Biophys. Acta* **423**, 432–439.

Lampe, H.-H. (1972). Ph.D. Thesis, Univ. Freiburg, Freiburg.

Lampe, H.-H., and Drews, G. (1972). *Arch. Mikrobiol.* **84**, 1–19.

Lampe, H.-H., Oelze, J., and Drews, G. (1972). *Arch. Mikrobiol.* **83**, 78–94.

Lascelles, J. (1959). *Biochem. J.* **72**, 508–518.

Lascelles, J. (1968). *Adv. Microb. Physiol.* **2**, 1–42.

Lascelles, J., and Szilágyi, J. F. (1965). *J. Gen. Microbiol.* **38**, 55–64.

Lascelles, J., and Wertlieb, D. (1971). *Biochim. Biophys. Acta* **226**, 328–340.

Liaanen Jensen, S. (1967). *Pure Appl. Chem.* **14**, 227–244.

Lien, S., and Gest, H. (1973). *Arch. Biochem. Biophys.* **159**, 730–737.

Lien, S., San Pietro, A., and Gest, H. (1971). *Proc. Natl. Acad. Sci. U.S.A.* **68**, 1912–1915.

Lien, S., Gest, H., and San Pietro, A. (1973). *Bioenergetics,* **4**, 423–434.

Lin, L., and Thornber, J. P. (1975). *Photochem. Photobiol.* **22**, 37–40.

Litvin, F. F., and Krasnovsky, A. A. (1957). *Dokl. Akad. Nauk SSSR* **117**, 106. Cited in Litvin and Sineshchekov (1975).

Litvin, F. F., and Krasnovsky, A. A. (1958). *Dokl. Akad. Nauk SSSR* **120**, 764. Cited in Litvin and Sineshchekov (1975).

Litvin, F. F., and Sineshchekov, V. A. (1975). *In* "Bioenergetics of Photosynthesis" (Govindjee, ed.), pp. 619–661. Academic Press, New York.

Loach, P. (1976). *Abstr. Int. Conf. Primary Electron Transp. Energy Transduction Photosynth. Bact. Brussels.*

Loach, P. A., and Hall, R. L. (1972). *Proc. Natl. Acad Sci. U.S.A.* **69**, 786–790.

Loach, P. A., Androes, G. M., Maksim, A. F., and Calvin, M. (1963). *Photochem. Photobiol.* **2**, 443–454.

Loach, P. A., Hadsell, R. M., Sekura, D. L., and Stemer, A. (1970). *Biochemistry* **9**, 3127–3135.

Löw, H., and Alfzelius, B. A. (1964). *Exp. Cell Res.* **35**, 431–434.
Machtinger, N. A., and Fox, C. F. (1973). *Annu. Rev. Biochem.* **42**, 575–600.
Marrs, B. (1974). *Proc. Natl. Acad. Sci. U.S.A.* **71**, 971–973.
Marrs, B., and Gest, H. (1973). *J. Bacteriol.* **114**, 1052–1057.
Melandri, B. A., Baccarini-Melandri, A., Gest, H., and San Pietro, A. (1971a). *In* "Energy Transduction in Respiration and Photosynthesis" (E. Quagliariello, S. Papa, and C. S. Rossi, eds.), p. 593. Adriatica Editrice, Bari.
Melandri, B. A., Baccarini-Melandri, A., San Pietro, A., and Gest, H. (1971b). *Science* **174**, 514–516.
Melandri, B. A., Zannoni, D., Casadio, R., and Baccarini-Melandri, A. (1975). *Proc. Int. Congr. Photosynth. 3rd, Rehovot, 1974* **2**, 1147–1162.
Meyer, D. J., and Jones, C. W. (1973). *Int. J. Syst. Bacteriol.* **23**, 459–467.
Meyer, T. E., Kennel, S. J., Tedro, S. M., and Kamen, M. D. (1973). *Biochim. Biophys. Acta* **292**, 634–643.
Monger, T. G., and Parson, W. W. (1977). *Biochim. Biophys. Acta* **460**, 393–407.
Monod, J. (1950). *Ann. Inst. Pasteur, Paris* **79**, 390–410.
Morrison, L., Runquist, J., and Loach, P. (1977). *Photochem. Photobiol.* **25**, 73–84.
Neuberger, A., Sandy, J. D., and Tait, G. H. (1973). *Biochem. J.* **136**, 477–490.
Niederman, R. (1974). *J. Bacteriol.* **117**, 19–28.
Niederman, R. A. (1978). *In* "The Photosynthetic Bacteria" (R. K. Clayton and W. R. Sistrom, eds.). Plenum, New York, in press.
Niederman, R. A. and Gibson, K. D. (1971). *Prep. Biochemistry* **1**, 141–150.
Niederman, R. A., Segen, B. J., and Gibson, K. D. (1972). *Arch. Biochem. Biophys.* **152**, 547–560.
Niederman, R. A., Mallon, D. E., and Langan, J. J. (1976). *Biochim. Biophys. Acta* **440**, 429–447.
Nieth, K.-F., and Drews, G. (1975). *Arch. Microbiol.* **104**, 77–82.
Nieth, K.-F., Drews, G., and Feick, R. (1975). *Arch. Microbiol.* **105**, 43–45.
Nisimoto, Y., Kakuno, T., Yamashita, J., and Horio, T. (1973). *J. Biochem. (Tokyo)* **74**, 1205–1216.
Noël, H., van der Rest, M., and Gingras, G. (1972). *Biochim. Biophys. Acta* **275**, 219–230.
Norris, J. R., Uphaus, R. A., Crespi, H. L., and Katz, J. J. (1971). *Proc. Natl. Acad. Sci. U.S.A.* **68**, 625–628.
Norris, J. R., Druysan, M. E., and Katz, J. J. (1973). *J. Am. Chem. Soc.* **95**, 1680.
Norris, J. R., Scheer, H., and Katz, J. J. (1975). *Ann. N.Y. Acad Sci.* **244**, 260.
Oelze, J. (1976). *Biochim. Biophys. Acta* **436**, 95–100.
Oelze, J. (1978). *Biochim. Biophys. Acta,* in press.
Oelze, J., and Drews, G. (1970a). *Biochim. Biophys. Acta* **203**, 189–198.
Oelze, J., and Drews, G. (1970b). *Biochim. Biophys. Acta* **219**, 131–140.
Oelze, J., and Drews, G. (1972). *Biochim. Biophys. Acta* **265**, 209–239.
Oelze, J., and Golecki, J. R. (1975). *Arch. Microbiol.* **102**, 59–64.
Oelze, J., and Mechler, B. (1976). *Proc. Int. Symp. Photosynth. Prokaryotes, 2nd, Dundee* (G. A. Codd and W. D. P. Stewart, eds.), pp. 54–56.
Oelze, J., and Pahlke, B. (1976). *Arch. Microbiol.* **108**, 218–285.
Oelze, J., and Weaver, P. (1971). *Arch. Mikrobiol.* **79**, 108–121.
Oelze, J., Biedermann, M., Freund-Mölbert, E., and Drews, G. (1969a). *Arch. Mikrobiol.* **66**, 154–165.
Oelze, J., Biedermann, M., and Drews, G. (1969b). *Biochim. Biophys. Acta* **173**, 436–447.
Oelze, J., Schröder, J., and Drews, G. (1970). *J. Bacteriol.* **101**, 669–674.
Oelze, J., Golecki, J. R., Kleinig, H., and Weckesser, J. (1975). *Antonie van Leeuwenhoek; J. Microbiol. Serol.* **41**, 273–286.

Okamura, M. Y., Steiner, L. A., and Feher, G. (1974). *Biochemistry* **13**, 1394–1402.
Okamura, M. Y., Isaacson, R. A., and Feher, G. (1975). *Proc. Natl. Acad. Sci. U.S.A.* **72**, 3491–3495.
Okamura, M. Y., Ackerson, L. C., Isaacson, R. A., Parson, W. W., and Feher, G. (1976). *Biophys. J.* **16**, 223.
Olson, J. M. (1978). *In* "The Photosynthetic Bacteria" (R. K. Clayton and W. R. Sistrom, eds.). Plenum, New York, in press.
Olson, J. M., and Romano, C. A. (1962). *Biochim. Biophys. Acta* **59**, 726–728.
Olson, J. M. Thornber, J. P., Koenig, D. F., Ledbetter, M. C., Olson, R. A., and Jennings, W. H. (1969). *Prog. Photosynth. Res.* **1**, 217–225.
Olson, J. M., Giddings, T. H., and Shaw, E. K. (1976a). *Biochim. Biophys. Acta* **449**, 197–208.
Olson, J. M., Ke, B., and Thompson, K. H. (1976b). *Biochim. Biophys. Acta* **430**, 524–537.
Olson, J. M., Prince, R. C., and Brune, D. C. (1976c). *Brookhaven Symp. Biol.* **28**, 238–245.
Overath, P., Hill, F., and Lamnek-Hirsch, T. (1971). *Nature (London), New Biol.* **239**, 264–267.
Parson, W. W. (1974). *Annu. Rev. Microbiol.* **28**, 41–59.
Parson, W. W., and Cogdell, R. J. (1975). *Biochim. Biophys. Acta* **416**, 105–149.
Petty, K. M., and Dutton, P. L. (1976). *Arch. Biochem. Biophys.* **172**, 346–353.
Pfennig, N. (1967). *Annu. Rev. Microbiol.* **21**, 285–324.
Pfennig, N. (1970). *J. Gen. Microbiol.* **61**, ii.
Pfennig, N. (1978). *In* "The Photosynthetic Bacteria" (R. K. Clayton and W. R. Sistrom, eds.). Plenum, New York, in press.
Pfennig, N., and Cohen-Bazire, G. (1967). *Arch. Mikrobiol.* **59**, 226–236.
Pfennig, N., and Trüper, H. G. (1974). *In* "Bergey's Manual of Determinative Bacteriology" (R. E. Buchanan and N. E. Gibbons, eds.), 8th Ed., pp. 24–64. Williams & Wilkins, Baltimore, Maryland.
Phillips, D. R., and Morrison, M. (1970). *Biochem. Biophys. Res. Commun.* **40**, 284–289.
Prince, R. C., and Dutton, P. L. (1976). *Arch. Biochem. Biophys.* **172**, 329–334.
Prince, R. C., and Olson, J. M. (1976). *Biochim. Biophys. Acta* **423**, 357–362.
Prince, R. C., Cogdell, R. J., and Crofts, A. R. (1974a). *Biochim. Biophys. Acta* **347**, 1–13.
Prince, R. C., Hauska, G. A., Crofts, A. R., Melandri, A., and Melandri, B. A. (1974b). *Proc. Int. Congr. Photosynth. 3rd, Rehovot* **1**, 769–774.
Prince, R. C., Baccarini-Melandri, A., Hauska, G. A., Melandri, B. A., and Crofts, A. R. (1975). *Biochim. Biophys. Acta* **387**, 212–227.
Pucheau, N. L., Kerber, N. L., and Garcia, A. F. (1976). *Arch. Microbiol.* **109**, 301–305.
Ramirez, J., and Smith, L. (1968). *Biochim. Biophys. Acta* **153**, 466–475.
Reed, D. W. (1969). *J. Biol. Chem.* **244**, 4936–4941.
Reed, D. W., and Clayton, R. K. (1968). *Biochem. Biophys. Res. Commun.* **30**, 471–475.
Reed, D. W., and Raveed, D. (1972a). *Proc. Int. Congr. Photosynth. Res., 2nd, Stresa, 1971* **2**, 1441–1452.
Reed, D. W., and Raveed, D. (1972b). *Biochim. Biophys. Acta* **283**, 79–91.
Reed, D. W., Raveed, D., and Israel, H. (1970). *Biochim. Biophys. Acta* **223**, 281–291.
Reed, D. W., Raveed, D., and Reporter, M. (1975). *Biochim. Biophys. Acta* **387**, 368–378.
Richards, W. R., Wallace, R. B., Tsao, M. S., and Ho, E. (1975). *Biochemistry* **14**, 5554–5561.
Rivas, E. A., Kerber, N. L., Viale, A. A., and Garcia, A. F. (1970). *FEBS Lett.* **11**, 37–40.
Romijn, J. C. (1976). *Abstr. Int. Conf. Primary Electron Transp. Energy Transduction Photosynth. Bact. Brussels.*

Salton, M. R. J. (1974). Adv. Microbiol. Physiol. 11, 213–283.

Sandy, J. D., Davies, R. C., and Neuberger, A. (1975). Biochem. J. 150, 245–257.

San Pietro, A. (1978). In "The Photosynthetic Bacteria" (R. K. Clayton and W. R. Sistrom, eds.). Plenum, New York, in press.

Sauer, K. (1975). In "Bioenergetics of Photosynthesis" (Govindjee, ed.), p. 142. Academic Press, New York.

Saunders, V. A., and Jones, O. T. G. (1974). Biochim. Biophys. Acta 333, 439–445.

Schachman, H. K., Pardee, A. B., and Stanier, R. Y. (1952). Arch. Biochem. Biophys. 38, 245–260.

Schmidt, K. (1978). In "The Photosynthetic Bacteria" (R. K. Clayton and W. R. Sistrom, eds.). Plenum, New York, in press.

Schmitz, R. (1967). Z. Naturforsch. B 22, 645–648.

Schön, G., and Drews, G. (1966). Arch. Mikrobiol. 54, 109–214.

Schön, G., and Ladwig, R. (1970). Arch. Mikrobiol. 74, 356–371.

Scholes, P., Mitchell, P., and Moyle, J. (1969). Eur. J. Biochem. 8, 450–454.

Schröder, J., and Drews, G. (1968). Arch. Mikrobiol. 64, 59–70.

Schumacher, A., and Drews, G. (1978). Biochim. Biophys. Acta 501, 183–194.

Shanmugam, K. T., Buchanan, B. B., and Arnon, D. I. (1972). Biochim. Biophys. Acta 256, 477–486.

Shaw, M. A., and Richards, W. R. (1971). Biochem. Biophys. Res. Commun. 45, 863–869.

Shaw, M. A., and Richards, W. R. (1972). Proc. Int. Congr. Photosynth., 2nd, 3, 2733–2745.

Shioi, Y., Takamiya, J., and Nishimura, M. (1974). J. Biochem. (Tokyo) 76, 241–250.

Singer, S. J. (1974). Annu. Rev. Biochem. 43, 805–833.

Singer, S. J., and Nicolson, G. L. (1972). Science 175, 720–731.

Sistrom, W. R. (1962). J. Gen. Microbiol. 28, 599–605.

Sistrom, W. R. (1964). Biochim. Biophys. Acta 79, 419–421.

Sistrom, W. R., Griffiths, M., and Stanier, R. Y. (1956). J. Cell. Comp. Physiol. 48, 473–515.

Slooten, L. (1972). Biochim. Biophys. Acta 275, 208–218.

Smith, W. R., Sybesma, C., and Dus, K. (1972). Biochim. Biophys. Acta 267, 609–615.

Sojka, G. A., and Gest, H. (1968). Proc. Natl. Acad. Sci. U.S.A. 61, 1486–1493.

Staehelin, A. (1975). Biochim. Biophys. Acta 408, 1–11.

Stanier, R. Y. (1974). Symp. Soc. Gen. Microbiol. 24, 219–240. Cambridge Univ. Press.

Steiner, L. A., Okamura, M. Y., Lopes, A. D., Moskowitz, E., and Feher, G. (1974). Biochemistry 13, 1403–1410.

Steiner, S., Sojka, G. A., Conti, S. F., Gest, H., and Lester, R. L. (1970). Biochim. Biophys. Acta 203, 571–574.

Straley, S. C., Parson, W. W., Mauzerall, D., and Clayton, R. K. (1973). Biochim. Biophys. Acta 305, 597–609.

Sybesma, C., and Olson, J. M. (1963). Proc. Natl. Acad. Sci. U.S.A. 49, 248–253.

Sybesma, C., and Vredenberg, W. J. (1963). Biochim. Biophys. Acta 75, 439–441.

Sybesma, C., and Vredenberg, W. J. (1964). Biochim. Biophys. Acta 88, 205–207.

Takacs, B. J., and Holt, S. C. (1971). Biochim. Biophys. Acta 233, 258–295.

Takemoto, J. (1974). Arch. Biochem. Biophys. 163, 515–520.

Takemoto, J., and Huang Kao, M. Y. C. (1977). J. Bacteriol. 129, 1102–1109.

Takemoto, J., and Lascelles, J. (1973). Proc. Natl. Acad. Sci. U.S.A. 70, 799–803.

Takemoto, J., and Lascelles, J. (1974). Arch. Biochem. Biophys. 163, 507–514.

Taniguchi, S., and Kamen, M. D. (1965). Biochim. Biophys. Acta 96, 395–428.

Tauschel, H.-D., and Drews, G. (1967). Arch. Mikrobiol. 59, 381–404.

Thore, A., Keister, D. L., and San Pietro, A. (1969). Arch. Mikrobiol. 67, 378–396.

Thornber, J. P. (1970). *Biochemistry* **9**, 2688–2698.
Thornber, J. P. (1971). *In* "Photosynthesis," Part A (A. San Pietro, ed.), Methods in Enzymology, Vol. 23, pp. 688–691. Academic Press, New York.
Thornber, J. P.,and Olson, J. (1968). *Biochemistry* **7**, 2242–2250.
Thornber, J. P., Olson, J. M., Williams, D. M., and Clayton, M. L. (1969). *Biochim. Biophys. Acta* **172**, 351–354.
Thornber, J. P., Trosper, T. L., and Strouse, C. E. (1978). *In* "The Photosynthetic Bacteria" (R. K. Clayton and W. R. Sistrom, eds.). Plenum, New York, in press.
Throm, E., Oelze, J., and Drews, G. (1970). *Arch. Mikrobiol.* **72**, 361–370.
Tonn, S. J., Gogel, G. E., and Loach, P. A. (1977). *Biochemistry* **16**, 877–885.
Trüper, H. G., and Genovese, S. (1968). *Limnol. Oceanogr.* **13**, 225–232.
Tuboi, S., Kim, H. J., and Kikuchi, G. (1970). *Arch. Biochem. Biophys.* **138**, 147–154.
Uemura, T., Suzuki, K., Nagano, K., and Morita, S. (1961). *Plant Cell Phyiol.* **2**, 451–461.
Uffen, R. L., and Wolfe, R. S. (1970). *J. Bacteriol.* **104**, 462–472.
van der Rest, M., and Gingras, G. (1974). *J. Biol. Chem.* **249**, 6446–6453.
van der Rest, M., Noël, H., and Gingras, G. (1974). *Arch. Biochem. Biophys.* **164**, 285–292.
von Stedingk, L.-V., and Baltscheffsky, H. (1966). *Arch. Biochem. Biophys.* **117**, 400–404.
Wang, R. T., and Clayton, R. K. (1973). *Photochem. Photobiol.* **17**, 57–61.
Wassink, E. C., Katz, E., and Dorrestein, R. (1939). *Enzymologia* **7**, 113.
Weckesser, J., Drews, G., and Tauschel, H.-D. (1969). *Arch. Mikrobiol.* **65**, 346–368.
Weiss, D. E. (1973). *Sub-Cell Biochem.* **2**, 201–235.
Worden, P. B., and Sistrom, W. R. (1964). *J. Cell Biol.* **23**, 135–150.
Yen, H.-C., and Marrs, B. (1976). *J. Bacteriol.* **126**, 619–629.
Yoch, D. C., Arnon, D. I., and Sweeney, W. V. (1975). *J. Biol. Chem.* **250**, 8330–8336.
Zannoni, D., Melandri, B. A., and Baccarini-Melandri, A. (1976). *Biochim. Biophys. Acta* **423**, 413–430.
Zürrer, H., Snozzi, M., Hanselmann, K., and Bachofen, R. (1977). *Biochim. Biophys. Acta* **460**, 273–279.

Genetic Control of the Photosynthetic Membrane

Genetic Control of Chloroplast Proteins

N. W. Gillham and J. E. Boynton
Departments of Zoology and Botany
Duke University
Durham, North Carolina

N.-H. Chua
The Rockefeller University
New York, New York

I. Introduction

This review considers both the genetic control of chloroplast proteins and the sites of synthesis of these proteins. Perhaps the most obvious approach to defining the roles of individual genes in the biogenesis of an organelle is the isolation and characterization of mutations that affect the structure and function of that organelle. In both green algae and higher plants, many mutations blocking the normal biogenesis or function of the chloroplast are known (cf. reviews by Kirk and Tilney-Bassett, 1967; Kirk, 1972; cf. also Birky *et al.*, 1975; Bücher *et al.*, 1976; Nasyrov and Šesták, 1975). The pigment deficiency in many of these mutants is but one of several pleiotropic consequences of the failure to make specific gene products necessary for normal biogenesis of the chloroplast and usually does not result from a block in the biosynthetic pathway of chlorophyll per se. Such mutations have been characterized physiologically and ultrastructurally in a number of higher plants including maize (Bachmann et al., 1969, 1973; Miles and Daniel, 1974; Shumway and Weier, 1967), tomato (Lefort, 1959; Boynton, 1966), Arabidopsis (Röbbelen, 1959; Velemínský and Röbbelen, 1966; Rédei, 1973; Rédei and Plurad, 1973), pea (Highkin et al., 1969), cotton (Benedict and Kohel, 1968), barley (von Wettstein et al., 1971, 1974; Gough, 1972; Henningsen et al., 1973; Nielsen, 1974), tobacco (Schmid et al., 1966; Schmid, 1967, 1971, Shumway and Kleinhofs, 1973), peanut (Alberte, et al., 1976), and sunflower (Walles, 1965). Mutants affecting chloroplast structure and function have also been isolated and characterized in a number of algal species including Euglena (Gibor and Herron, 1968; Schiff et al., 1971; Schmidt and Lyman 1976), Scenedesmus (Bishop, 1971a,b; Senger and Bishop, 1972a,b), Chlorella (Allen, 1971; Granick, 1971), and Chlamydomonas (Levine, 1968, 1969, 1971; Levine and Goodenough, 1970; Goodenough and Levine, 1969, 1970; Goodenough and Staehelin, 1971; Ohad et al., 1967a,b; Boynton et al., 1972). However, only Chlamydomonas has a sufficiently well-defined sexual cycle to permit genetic analysis of these algal mutants.

The primary genetic defect or gene product involved has been identified in only a very few of the many mutations known in both algae and higher plants. In certain cases, the genetic defect has been localized to a specific metabolic pathway, e.g., chlorophyll biosynthesis (Gough, 1972; von Wettstein et al., 1974; Wang et al., 1974, 1975), or to a specific portion of the photosynthetic electron-transport chain (Levine, 1968; Bishop, 1971a). Most of these mutants in higher plants and Chlamydomonas show Mendelian inheritance in crosses, a characteristic of genes located on chromosomes in the nucleus.

However, certain mutants of higher plants possessing variegated chlorophyll deficiencies have long been known to show non-Mendelian inheritance (Baur, 1909; Correns, 1909; cf. recent reviews by Kirk and Tilney-Bassett, 1967; Sager, 1972; Tilney-Bassett, 1975). These have been termed plastome mutants since they obviously affect the chloroplast, are inherited in a fashion distinct from nuclear gene mutations, and segregate somatically. No fully convincing proof exists that any plastome mutation in higher plants actually results from an alteration in chloroplast DNA. Perhaps the best evidence is that of Wong-Staal and Wildman (1973), who have described a white plastome mutant in tobacco in which the chloroplast DNA has about 1% higher G+C content in thermal denaturation experiments than that of wild type. Analyses of heteroduplexes formed in mixtures of chloroplast DNA from the mutant and wild type suggest that the mutant contains a nonhomologous region consisting of 500–1000 base pairs. Wildman *et al.* (1973) also showed that this plastome mutant was maternally inherited in reciprocal crosses. These two sets of results support the contention made by Chan and Wildman (1972) that maternal inheritance in *Nicotiana* reflects the inheritance of chloroplast genes.

While plastome mutations show maternal inheritance in many species of higher plants, such as tobacco and barley, they may be inherited in a non-Mendelian biparental fashion in other plants, notably in *Oenothera* and *Pelargonium* (Kirk and Tilney-Bassett, 1967; Tilney-Bassett, 1975). In general, plastome mutations segregate somatically and hence produce the variegated leaves and shoots observed. No formal genetic analysis or mapping of these plastome mutants has been possible to date.

Non-Mendelian, uniparentally inherited mutations affecting chloroplast functions are also known in the green alga *Chlamydomonas reinhardtii* (Sager, 1972; Adams *et al.*, 1976). On the basis of several pieces of indirect evidence, these mutations are also thought to be localized in chloroplast DNA (Adams *et al.*, 1976). The putative chloroplast mutants of *Chlamydomonas* also segregate somatically, and their low spontaneous frequency of biparental transmission can be raised experimentally to a level that permits formal genetic analysis (Sager, 1972; Adams *et al.*, 1976). All chloroplast mutants appear to map in a single linkage group, and certain of these are known to alter specific chloroplast components, as discussed in subsequent sections of this chapter.

A second approach to determining whether a chloroplast component is coded in the nuclear or organelle genome is to identify closely related species or strains of a given species that differ with respect to a given component and can be hybridized. One then makes reciprocal crosses and determines the phenotype of the F_1 hybrids with respect to this component. If the component shows only maternal inheritance in the two reciprocal F_1 hybrids, then the component is presumed to be coded by chloroplast DNA, whereas if it shows biparental inheritance in the two F_1 hybrids, then it is thought to be coded by nuclear DNA. In cases in which plastids are inherited biparentally, one would expect to see somatic segregation within and between individual F_1 plants for the components coded by chloroplast genes.

Finally, in maize and spinach restriction enzyme analysis of chloroplast DNA is now yielding physical maps of the chloroplast genome and the rRNA cistrons are being localized by molecular hybridization experiments (Bedbrook and Bogorad, 1976a,b; Herrmann *et al.*, 1976). We may soon expect that the mRNAs for specific chloroplast proteins will be isolated and localized to individual restriction fragments using techniques like those employed by Howell *et al.* (1976) and Gelvin *et al.* (1977) for the localization of the messenger for the large subunit of fraction I protein in *Chlamydomonas* (see Section II,A).

The fact that at present we know of many more nuclear than chloroplast mutations that affect chloroplast function probably has a 3-fold basis. First, the chloroplast depends on the cytoplasm of the cell for many of its metabolites, and the pathways by which these are produced are under nuclear gene control. Second, analyses of purified chloroplast DNA from several higher plants (Herrmann *et al.*, 1975; Kolodner and Tewari, 1975) and from *Chlamydomonas reinhardtii* (Behn and Herrmann, 1977; Lambowitz *et al.*, 1976; Rochaix, 1976) give estimates of 0.9 to $1.3 \times 10^8 \ M_r$, which would provide sufficient genetic information to code for 300–400 proteins of molecular weight 20,000. Only a small percentage of this information would be required to code for the ribosomal and transfer RNAs. In contrast, the nuclear genome has manyfold greater informational content. Third, chloroplasts appear to be highly polyploid, e.g., 50–75 copies of the genome are contained in the single chloroplast of a *Chlamydomonas* cell, and in addition most algae and higher plant cells contain several chloroplasts per cell. Since most mutations are recessive and involve changes in only a single DNA molecule, the chances of a single mutant DNA copy segregating to homozygosity in one or more plastids and thus expressing the mutant phenotype will be low (Kirk, 1972).

Two general approaches have been taken to establish the sites of synthesis of organelle proteins that become assembled into complex macromolecular structures, such as membranes. The *in vitro* approach asks what polypeptides are made within isolated organelles and incorporated into the membrane systems. The *in vivo* approach makes use of inhibitors specific for cytoplasmic and chloroplast protein synthesis to determine where a given protein is translated. Ideally, proteins made in the chloroplast will continue to be synthesized in the presence of an inhibitor of cytoplasmic protein synthesis, whereas proteins whose messages are translated on cytoplasmic ribosomes should be made in the presence of inhibitors of chloroplast protein synthesis. In practice both methodologies have certain drawbacks. These have been considered at length in the excellent review by Schatz and Mason (1974) on the biosynthesis of mitochondrial proteins.

In this chapter we restrict our attention to those chloroplast components that are either affected only by chloroplast genes or that appear to involve the joint participation of nuclear and chloroplast genes. Furthermore, not only will the site of genetic information be examined for each, but also the site of translation of this information. The specific chloroplast components to be considered include the CO_2-fixing enzyme ribulose-1,5-bisphosphate carboxylase, proteins comprising the chloroplast envelope and thylakoid membranes including those associated with photosystems I and II (PS I, PS II); the chloroplast ribosomal proteins; and the aminoacyl-tRNA synthetases of the chloroplast.

II. Ribulose-1,5-Bisphosphate Carboxylase (Fraction I Protein)

The enzyme ribulose-1,5-bisphosphate carboxylase (RubPCase, EC 4.1.1.39) was first identified by Wildman and Bonner (1947) as the major soluble protein in green-plant leaves. They named this protein fraction I because it had a high molecular weight and it separated clearly from the low-molecular-weight proteins upon analytical ultracentrifugation. Fraction I protein from all higher plants and green algal species examined to date is similar in structure and function. This protein has a sedimentation coefficient ($s_{20,w}$) of about 18 and a molecular weight (M_r) of ~560,000. It consists of an aggregate of two types of polypeptide subunits of M_r ~55,000 and ~12,000, respectively (cf. Kung, 1976; Chen et al., 1976). Each molecule of fraction I protein is presently viewed as consisting of 8 large and 8 small subunits (Chen et al., 1976). The primary function of fraction I protein has long been recognized to catalyze the first CO_2 fixation step in photosynthesis, using ribulose 1,5-bisphosphate and CO_2 as substrates to produce two molecules of 3-phosphoglyceric acid (Kawashima and Wildman, 1970). Recently, however, fraction I protein has been shown also to catalyze the oxygenation of ribulose 1,5-bisphosphate to form one molecule each of 3-phosphoglyceric acid and phosphoglycolic acid, the first intermediate in the glycolate pathway for photorespiration (Kung, 1976). Furthermore, both the carboxylase and oxygenase activities of this enzyme are activated by preincubation of the enzyme with CO_2, but not with ribulose 1,5-bisphosphate (Lorimer et al., 1976). The catalytic site is thought to reside on the large subunits of the enzyme, whereas the small subunits serve to regulate the activity of the enzyme (Kung, 1976). In addition, O_2 and CO_2 appear to compete for the same catalytic site (Badger and Andrews, 1974).

Kung et al. (1974) have resolved 55 peptides from the large subunit and 32 from the small subunit of fraction I protein of Nicotiana tabacum following trypsin digestion and two-dimensional electrophoresis. In barley Strøbaek and Gibbons (1976) observed 42 and 25 peptides, respectively, following trypsin digestion of the large and small subunits of fraction I protein. In five higher plant species and Chlamydomonas, fraction I protein shows generally similar amino acid composition overall with certain noticeable differences (Strøbaek and Gibbons, 1976). The ratio of tryosine to tryptophan in the large and small subunits of fraction

I protein from barley is 1.0 and 1.2, respectively (Strøbaek and Gibbons, 1976). On the basis of amino acid compositions, tryptic peptide analysis, and immunological comparisons, Kung (1976) concluded that in a given species the large subunit of fraction I protein shares no similarities with the small subunit. Large subunits from different species are very similar in these properties whereas the small subunits vary considerably. The amino acid sequence of the small subunit of fraction I protein is being determined in several species, and differences are already found between barley, pea, bean, and tobacco in nine positions out of 25 examined (Poulsen et al., 1976).

Wildman and co-workers (1975), using isoelectric focusing of carboxymethylated fraction I protein, have been able to resolve the large subunits of fraction I protein from all Nicotiana species into a cluster of three polypeptides, with a total of four clusters representing all 63 species examined. Isoelectric focusing of the small subunit of fraction I protein from these same 63 species resolves from 1 to 4 polypeptides per species, with a total of 13 different combinations (Fig. 1). Kung (1976) raised the obvious question of whether the three large-subunit polypeptides seen upon isoelectric focusing actually represent three distinct

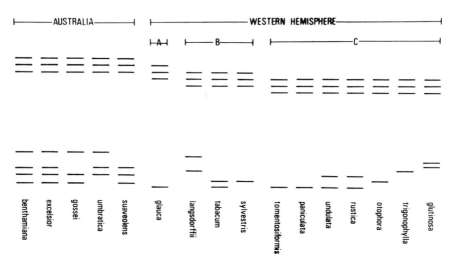

FIG. 1. Polypeptide composition of the large and small subunits of fraction I protein from different Australian and Western Hemisphere species of Nicotiana resolved by isoelectric focusing of carboxymethylated material. From "Evolution of Ferredoxin and Fraction I Protein in the Genus Nicotiana," by S. G. Wildman, K. Chen, J. C. Gray, S. D. Kung, P. Kwanyuen, and K. Sakano, in Genetics and Biogenesis of Mitochondria and Chloroplasts, edited by C. William Birky, Jr., Philip S. Perlman, and Thomas J. Byers. Ohio State University Biosciences Colloquia, no. 1. Columbus: Ohio State University Press, 1975.

polypeptides coded by three separate chloroplast genes or result from posttranslational modifications of a single gene product producing three polypeptides of different charge. He calculated that, based on the known arginine–lysine composition, if each subunit contained 3 different polypeptides having the same molecular weight one would expect more than the 55 peptides seen upon trypsin digestion. Therefore the nature of the 3 polypeptides obtained following isoelectric focusing of carboxymethylated material remains to be resolved. In spite of this, the polypeptides seen upon isoelectric focusing of the carboxymethylated material are characteristic of individual species and serve as reliable genetic markers for differences between species.

A. LOCATION OF GENES CODING FOR FRACTION I POLYPEPTIDES

Kawashima *et al*. (1971) prepared tryptic peptide maps for the large and small subunits of fraction I protein from five American species of *Nicotiana*. Several differences were found that served as the basis for determining the pattern of inheritance of the small subunit in reciprocal interspecific hybrids, but no differences were found in the large subunit. Sephadex chromatography of the tryptic peptides of the small subunit revealed in *N. tabacum* four peaks that absorbed at 280 nm, whereas *N. glauca* and *N. glutinosa* produced only three peaks (Kawashima and Wildman, 1972). Peak 3, which was unique to *N. tabacum,* resulted because the peptides in this region of the column contained tyrosine, which absorbs 280 nm UV light, whereas similar peptides in *N. glauca* and *N. glutinosa* contained an amino acid that did not absorb light at that wavelength. In F_1 hybrids from reciprocal crosses between *N. tabacum* and either *N. glauca* or *N. glutinosa,* the tyrosine-containing peptide was biparentally inherited, suggesting that the structure of this peptide and, therefore, the small subunit was controlled by a nuclear gene. Paper chromatography of tryptic peptides revealed that both *N. glutinosa* and *N. glauca* lacked a peptide present in *N. tabacum* (peptide I). In addition, two peptides (a, b) in *N. glauca* moved differently than in *N. tabacum*. Peptide I, characteristic of *N. tabacum*, was present in both reciprocal F_1 hybrids from the cross *N. glutinosa* × *N. tabacum*. The same was true in the *N. tabacum* × *N. glauca* reciprocal crosses. In addition the position of peptides a and b was similar to that seen for the *N. tabacum* parent in F_1 hybrids from this reciprocal cross.

Kawashima and Wildman (1972) concluded that this biparental inheritance of the *N. tabacum* traits indicates that a nuclear gene determines the primary structure of the small subunit of fraction I protein. Since peptides a and b were positioned in reciprocal F_1 hybrids between *N. glauca* and *N. tabacum* as they were in *N. tabacum* alone, the *N.*

tabacum genetic information was thought to suppress the *N. glauca* genetic information for the small subunit. Subsequently Sakano *et al.* (1974) found that the small subunit of *N. tabacum* could be resolved into two polypeptides by isoelectric focusing of carboxymethylated material, and these were present in approximately equal amounts (Fig. 2). In *N. glauca* only one of these peptides was present, but in double the amount of the same peptide in *N. tabacum*. The reciprocal interspecific hybrids contained both peptides, indicating that they received genetic information from *N. tabacum*. However, the staining of the peptide corresponding to that of the *N. glauca* parent was more intense, suggesting that this peptide was present in an amount greater than expected if only the *N. tabacum* gene had been expressed in the hybrids. Sakano *et al.* (1974) concluded that, contrary to the earlier finding of Kawashima and Wildman (1972), the *N. glauca* information for the small subunit was expressed in the F_1 hybrids. While this conclusion may be valid for the carboxymethylated peptides resolved by isoelectric focusing, it does not explain the earlier results for the tryptic peptides a and b. If these peptides were inherited biparentally, one would have expected to see both the *N. tabacum* and *N. glauca* a and b peptides in the F_1 hybrids instead of only the *N. tabacum* a and b peptides, as found.

Sakano *et al.* (1974) also found, using carboxymethylation and isoelec-

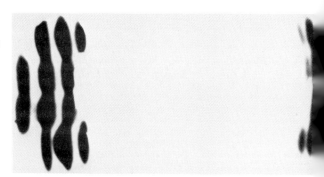

N. tabacum × *N. glauca* F_1

N. glauca × *N. tabacum* F_1

N. glauca

N. tabacum

Large subunit   Small subunit

FIG. 2. Inheritance of the polypeptides comprising the large and small subunits of fraction I protein in reciprocal F_1 hybrids from an interspecific cross in the genus *Nicotiana,* as resolved by isoelectric focusing of carboxymethylated material. From "Evolution of Ferredoxin and Fraction I Protein in the Genus *Nicotiana,*" by S. G. Wildman, K. Chen, J. C. Gray, S. D. Kung, P. Kwanyuen, and K. Sakano, in *Genetics and Biogenesis of Mitochondria and Chloroplasts,* edited by C. William Birky, Jr., Philip S. Perlman, and Thomas J. Byers. Ohio State University Biosciences Colloquia, no 1. Columbus: Ohio State University Press, 1975.

tric focusing, that *N. excelsior* small subunits contained four peptides and those of *N. gossei* contained three. Inheritance of the extra peptide was biparental, supporting the original conclusion of Kawashima and Wildman (1972) that the small subunit is coded by a nuclear gene.

Since no tryptic peptide differences were found between large subunits of fraction I protein obtained from these five American *Nicotiana* species, Chan and Wildman (1972) then examined the Australian species, which were considered to be far more distant from the American species in an evolutionary sense than the latter were from each other. Peptide maps of the large subunits of three Australian species, *N. gossei, N. excelsior,* and *N. suaveolens,* were found to be identical to one another and to the American species with the exception that the Australian species contained an extra tryptic peptide. Reciprocal F_1 hybrids between *N. gossei* and *N. tabacum* demonstrated that this extra peptide was maternally inherited. Accordingly, Chan and Wildman (1972) concluded that the large subunit of fraction I protein was a product of the chloroplast genome. Subsequently, Sakano *et al.* (1974) found that the large subunit could be separated into three polypeptides using carboxymethylation and isoelectric focusing, and the isoelectric points of these three polypeptides showed interspecific differences. In reciprocal F_1 hybrids of *N. glauca* × *N. tabacum, N. glutinosa* × *N. tabacum,* and *N. glauca* × *N. langsdorfii,* the isoelectric points of the three polypeptides always resembled those of the maternal parent (Fig. 2). These results further strengthened the original conclusions of Chan and Wildman (1972) that this subunit is coded by chloroplast DNA.

Singh and Wildman (1973) compared the RubPCase isolated from different *Nicotiana* species for K_m of ribulose 1,5-bisphosphate and found that the Australian species *N. gossei* had a lower K_m than any of the other species studied. In reciprocal F_1 hybrids employing *N. gossei* as one of the parents, the low K_m for ribulose 1,5-bisphosphate was found to be maternally inherited. Thus, the RubPCase catalytic site appears to be maternally inherited, a finding consistent with its localization in the large subunit of fraction I protein.

Although one can criticize the conclusions of Wildman and co-workers regarding the nuclear or chloroplast location of the genes coding for the large and small subunits of fraction I protein because they have not followed the inheritance of the interspecific differences beyond the F_1 generation, owing to the sterility of the *Nicotiana* hybrids, their evidence is the best presently available regarding the location of the genes coding for this enzyme. No point mutations have been identified that specifically affect the structure or catalytic activity of the large subunit of RubPCase. On the basis of the interspecific hybridization data in *Nicotiana* which suggest that the large subunit is coded by chloroplast DNA, mutants affecting this subunit would be expected to show a non-Mendelian pattern of inheritance, either maternal or biparental, depending on the plant species.

In contrast, two nuclear gene mutations have been identified that alter the carboxylase:oxygenase activity ratio of fraction I protein. Kung and Marsho (1976) described a dominant Mendelian mutant in tobacco (*Su*) which produces yellow plants when in the heterozygous (*Su su*) condition, and is lethal when homozygous. The mutant shows a 2- to 3-fold higher rate of photorespiration than the wild type, and reduced carboxylase and oxygenase activities. Although no differences in the isoelectric focusing pattern of either the large or small subunits could be detected between mutant and wild type, Kung and Marsho were able to demonstrate that the ratio of oxygenase to carboxylase activity was lower in purified fraction I protein from the mutant compared to wild type. Kung and Marsho (1976) concluded that, since the *su* mutant specifically affected the activity of fraction I protein, particularly its carboxylase:oxygenase activity ratio, and since the mutant was inherited in a Mendelian fashion, it must modify the small subunit of this enzyme in some manner that in turn affects the catalytic activity of the large subunit. Hence they suggested that the small subunit plays a regulatory role in the function of this enzyme aggregate. They also questioned whether the oxygenase activity of this enzyme can totally account for photorespiration, since the low oxygenase activity of purified fraction I protein from the mutant could not account for the observation that its photorespiration rate is 2- to 3-fold higher than that of wild type.

The Mendelian acetate-requiring mutant *ac-i72* of *Chlamydomonas reinhardtii* isolated by Nelson and Surzycki (1976a,b) is also reported to alter the carboxylase:oxygenase activity of fraction I protein. This mutant is similar to the yellow mutant of tobacco studied by Kung and Marsho (1976) in having subnormal levels of CO_2 fixation and RubPCase activity together with altered oxygenase activity and high glycolate production. The *ac-i72* mutant is thought to die under normal phototrophic growth conditions because of accumulated glycolate but will survive under an amber light regime, where total light intensity is presumably reduced, or in a low-oxygen environment. Purified RubPCase from *ac-i72* shows reduced specific activity, lower V_{max}, increased requirement for Mg^{2+} for maximal activity, greater sensitivity to inhibition by Cl^-, and slightly altered isoelectric point compared to the wild type when isolated in HEPES buffer, but not when isolated in Tris buffer of same molarity and pH. However, the RubPCase from the *ac-i72* mutant and wild type are identical in amount present, pH requirement, temperature sensitivity, molecular weight, subunit structure, and sedimentation coefficient as well as in K_m for the CO_2 and ribulose-1,5 bisphosphate substrates. Clearly this nuclear gene mutant appears to affect the functioning of fraction I protein, but this alteration, like that characterized in the yellow tobacco mutant by Kung and Marsho (1976), has not yet been shown to lead to alteration of a specific peptide in either subunit of the enzyme, or for that matter, to affect directly either subunit in any physical way. Until point mutants with better defined effects on the RubPCase molecule are isolated and characterized, one cannot use this direct genetic approach to determine where the genes coding for the peptides of the two subunits of Fraction I protein are localized or how many genes are involved.

A more direct molecular approach to this question has been undertaken recently in *Chlamydomonas* by Howell *et al.* (1976), and Gelvin *et al.* (1977). Labeled mRNA presumed to code for the large subunit of RubPCase was isolated from chloroplast polysomes using immunoprecipitation with antibodies specific to the large subunit. These antibodies recognize nascent chains of the large subunit and thereby select only those polysomes carrying mRNA for this subunit. The mRNA obtained was shown to have a sedimentation velocity of 14 S and to hybridize with chloroplast DNA, specifically to one of the 30 restriction fragments obtained upon *Eco*RI digestion. The 16 S and 23 S chloroplast rRNAs were found to hybridize to this same fragment, which has a molecular weight of 3.9×10^6 or one of similar size that comigrated with it in the gels. Gelvin *et al.* (1977) considered that there is sufficient evidence to rule out the possibility that the 14 S RNA that they isolate is a degradation product from one of the two chloroplast rRNA species, and they suggested that the genes coding for the large subunit of RubPCase and for the rRNAs may lie adjacent to each other on chloroplast DNA. These results represent the first attempt in any plant to localize the structural gene coding for a specific protein to its physical position on either nuclear or organelle DNA, using a direct molecular approach. However, they are consistent with results of Wildman and co-workers showing that in tobacco the large subunit of fraction I protein is coded by chloroplast DNA.

The interspecific differences found for the peptides comprising the large and small subunits of fraction I protein have also been useful in tracing the evolution of species within the genera *Nicotiana* (Chen *et al.*, 1976; Kung *et al.*, 1975b) and *Triticum* (Chen *et al.*, 1975) to reveal the parents of the allotetraploid species cultivated today, as well as migration of the species with continental drift. In addition, these protein differences have been utilized to follow the inheritance of chloroplast and nuclear genes in parasexual hybrids in *Nicotiana*. In hybrid plants regenerated from fused protoplasts, the small subunit polypeptides are biparentally inherited in all cases, but the large subunit polypeptides are uniparentally, although randomly, inherited in all except one of the plants studied (Kung *et al.*, 1975a; Chen *et al.*, 1977).

B. SITES OF TRANSLATION OF FRACTION I POLYPEPTIDES

In the case of RubPCase enzyme, the partitioning of genetic information between the chloroplast and the nucleus appears to parallel precisely the sites of translation of this information. Criddle *et al.* (1970) first suggested the involvement of the chloroplast and cytoplasmic protein synthesizing systems in the formation of the two subunits of fraction I protein in barley. In a double-labeling experiment they found that the organelle protein synthesis inhibitor chloramphenicol affected formation of the large subunit, whereas cycloheximide blocked synthesis of the small subunit. Subsequent inhibitor studies in other plants have yielded similar results (Givan and Criddle, 1972; Armstrong *et al.*, 1971; Margulies, 1971). Blair and Ellis (1973) then showed that isolated pea chloroplasts could synthesize the large, but not the small, subunit of RubPCase and that this was the only soluble chloroplast polypeptide made in this system. The *in vitro* labeled product was shown to be identical with the native large subunit in terms of tryptic peptide maps. Gray and Kekwick (1974) have described results complementary to those of Blair and Ellis, using an *in vitro* protein-synthesizing system in which 80 S polysomes from bean incorporate ^{14}C-labeled amino acids into a protein precipitated by antisera to the small subunit of RubPCase. Gooding *et al.* (1973) made antibodies to the large and small subunits of RubPCase from wheat. The 70 S chloroplast ribosomes and the 80 S cytoplasmic ribosomes were then separated by density gradient centrifugation, nascent peptides were released with [^{3}H]puromycin, and the puromycin-labeled peptides precipitated with antisera to the two subunits. The 70 S ribosomes were found to have only large subunit peptides attached to them, but the 80 S ribosomes had peptides derived from both subunits associated with them. Gooding *et al.* (1973) suggested that this association probably reflected the complexing of completed large subunits with nascent chains of small subunits being made on 80 S ribosomes and did not mean that large subunits were being made in both chloroplast and cytoplasm. Recently, Roy *et al.* (1976) have established that the small subunit of RubPCase is the product of a small proportion of cytoplasmic polyribosomes during greening of etiolated wheat seedlings.

Hartley *et al*. (1975) were the first to obtain *in vitro* translation of mRNA for the large subunit of fraction I protein. They translated total chloroplast RNA using a cell free extract from *E. coli* and found that two products, one of M_r 52,000 and the other of M_r 35,000, were made. The M_r 52,000 product was slightly smaller (ca. M_r 1500) than the native large subunit of RubPCase made *in vitro* by isolated pea chloroplasts, and contained seven of the nine chymotryptic peptides of these native large subunits. Wheeler and Hartley (1975) then separated the RNA from spinach chloroplasts into poly(A) and non-poly(A)-containing fractions and repeated the *in vitro* translation experiments using the *E. coli* cell-free system. They found that only the non-poly (A) containing RNA fraction programmed synthesis of the M_r 52,000 and 35,000 polypeptides and concluded that the mRNA coding for the large subunit of fraction I protein as well as the mRNA coding for the M_r 35,000 species were not polyadenylated. Sagher *et al*. (1976) prepared poly(A) and non-poly(A)-containing RNA from *Euglena* chloroplasts and assayed these RNAs using a cell-free protein synthesizing system from wheat germ. Like Wheeler and Hartley, they observed that the messenger for the large subunit of fraction I protein was present only in the non-poly(A) containing RNA fraction. Moreover, the product which they obtained was identical with native fraction I large subunit (M_r 59,000) as determined by a two-dimensional gel system employing isoelectric focusing followed by size filtration in SDS-polyacrylamide gels. Their results also revealed that this polypeptide was not synthesized by the non-poly(A) containing RNA fraction from the W_3BUL mutant lacking chloroplast DNA, and that the large subunit message from wild type sedimented in the 10–20 S fraction of a sucrose gradient.

Howell *et al*. (1977) have obtained similar evidence from *Chlamydomonas* that the large subunit mRNA is not polyadenylated. However they found that the *E. coli* translation system only yielded partial large subunit polypeptides. These were identified by a combination of immunoprecipitation and tryptic peptide analysis. The wheat germ system did not translate the large subunit mRNA to form an immunoprecipitable product. Howell *et al*. (1977) argue that the partial polypeptides obtained with the *E. coli* translation system probably did not result from mRNA degradation as the messenger fraction used was large enough to program synthesis of the entire polypeptide. Thus the *E. coli* system appears unable to translate large subunit mRNA in its entirety. Gelvin and Howell (1977) have also found that the large subunit mRNA of *C. reinhardtii* is translated mostly on small polyribosomes consisting of on the average 5 ribosomes.

Dobberstein *et al*. (1977) have performed essentially complementary experiments on the small subunit of fraction I protein from *Chlamydomonas*. They found that small subunit mRNA is polyadenylated and translated on cytoplasmic ribosomes. More important, however, Dobberstein *et al.* (1977) showed that the small subunit polypeptide is made as a precursor *in vitro*, and is then clipped by an endoprotease present in a postpolysomal supernatent of *C. reinhardtii*. Since there is little evidence that 80 S ribosomes are attached to the outer envelope of the chloroplast in *Chlamydomonas*, Dobberstein *et al.* (1977) speculate that the extra sequence may be required for recognition and possibly transfer of small subunits into the chloroplast following their completion and release. Finally, Iwanij *et al.* (1975) demonstrated a very tight control of synthesis of the two subunits of RubPCase using synchronous cells of *C. reinhardtii*. Both subunits are synthesized in synchrony during the light part of the light–dark cycle. In summary, there is overwhelming evidence that the large subunit of RubPCase is made on chloroplast ribosomes, and the small subunit on cytoplasmic ribosomes. This protein should be an ideal model for studying the mechanisms by which polypeptides are transported across the chloroplast envelope and assembled with partner polypeptides made inside the organelle.

III. Thylakoid Membrane Polypeptides

A. Location of Genes Coding for Thylakoid Membrane Polypeptides

Our understanding of the number and location of genes coding for specific structural and functional polypeptides of the thylakoid membranes is at a very primitive stage. As in the case of the genes coding for fraction I protein, the problem has been approached by characterizing photosynthetically defective mutations in hopes of identifying those whose primary defect involves a specific thylakoid structural or functional component, e.g., the two chlorophyll–protein complexes CP I and CP II (Kung and Thornber, 1971; Thornber *et al.*, 1967; Shiozawa *et al.*, 1974). Alternatively, workers have searched for differences in thylakoid peptides or the two chlorophyll–protein complexes in closely related species that can be hybridized to reveal their inheritance patterns.

Kung *et al.* (1972) isolated CP II from *N. glauca* and *N. tabacum* and found that the protein components of each had unique tryptic peptides designated G and T, respectively. In reciprocal interspecific hybrids the T peptide was inherited biparentally, indicating its coding by a nuclear gene. However, the G peptide was absent, an observation reminiscent of the a and b tryptic peptides of the small subunit of fraction I protein where the *N. tabacum* peptides were present in both hybrids and the *N. glauca* peptides were absent (Kawashima and Wildman, 1972). Thus, for two presumably unrelated proteins, only the *N. tabacum* peptides are present in the hybrids. One wonders if these F_1 hybrid plants really do have complete complements of *N. tabacum* and *N. glauca* chromosomes. However, the protein associated with the CP II complex may actually represent more than one polypeptide (see Section III,B on sites of translation of thylakoid polypeptides).

Isoelectric focusing of carbaminomethylated ferredoxins of different *Nicotiana* species revealed that the *S*-carbaminomethylated ferredoxin of *N. glutinosa* had a slightly more acid isoelectric point than those of nine other *Nicotiana* species, including *N. glauca* and *N. tabacum* (Kwanyuen and Wildman, 1975). Since the F_1 hybrid *N. glutinosa* ♀ × *N. glauca* ♂ had ferredoxins characteristic of both parents, Kwanyuen and Wildman (1975) concluded that the ferredoxin protein moiety was coded by a nuclear gene. The ferredoxins of *N. tabacum* and *N. glauca* contain methionine whereas that from *N. glutinosa* does not. The *N. glutinosa* ♀ × *N. glauca* ♂ hybrid contained half the methionine of the *N. glauca* parent, once again suggesting the presence of two ferredoxins in the hybrid, one of which contained methionine like the *N. glauca* parent whereas the other lacked methionine, like the *N. glutinosa* parent.

Mutants also have been described that lack certain thylakoid membrane polypeptides, and these have been used to associate particular polypeptides with specific photosynthetic functions in which the mutants are defective (Bennoun and Chua, 1976). Mutants affecting chlorophyll–protein complex I (CP I), which has PS I activity, have been studied in *Chlamydomonas* and higher plants. In *C. reinhardtii* these include both chloroplast and nuclear mutants. Chua *et al.* (1975) characterized two Mendelian mutants *F1* and *F14* lacking PS I and the ability to form CP I, using the high-resolution SDS-gradient gel system of Chua and Bennoun (1975) which separates 33 polypeptides ranging from M_r 10,000 to 68,000 (Figs. 3 and 4). These mutants specifically lacked only one thylakoid polypeptide which was designated as polypeptide 2. Chua *et al.* (1975) also isolated CP I from wild-type *C. reinhardtii* and showed that polypeptide 2 was the only polypeptide in the complex. Very recently, Bennoun *et al.* (1977) have described a chloroplast gene mutant of *C. reinhardtii*, C_1, which contains only 10–15% of the normal amount of CP I in its thylakoid membranes. This deficiency is correlated with a drastic reduction in the amount of polypeptide 2 in the membranes. Evidence to be presented in Section III,B indicates that polypeptide 2 is synthesized

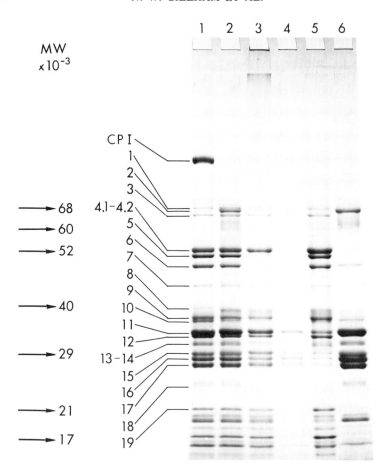

FIG. 3. Thylakoid membrane polypeptides of *Chlamydomonas reinhardtii* separated on a sodium dodecyl sulfate gel containing a 7.5 to 15% acrylamide concentration gradient. (1) Nonheated, 20 μg of chlorophyll; (2) heated, 20 μg of chlorophyll; (3) membranes extracted with 90% acetone, 30 μg of chlorophyll; (4) 90% acetone extract from (3); (5) membranes extracted with a 2:1 (v/v) mixture of chloroform:methanol, 30 μg of chloro- phyll; and (6) chloroform:methanol extract from (5). Redrawn from Chua *et al.* (1975).

on chloroplast ribosomes but that its insertion into these membranes requires one or more products of cytoplasmic protein synthesis. There- fore, the presence of polypeptide 2 in the thylakoid membranes is under the control of both chloroplast and nuclear genes in *Chlamydomonas*. In *Antirrhinum* a plastome mutant *en:alba-1* has been described (Herrmann, 1971a) which also lacks both CP I and the polypeptide associated with

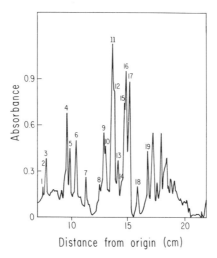

Distance from origin (cm)

FIG. 4. Densitometric tracings of thylakoid membrane polypeptides fractionated on a sodium dodecyl sulfate gel containing a 7.5 to 15% acrylamide concentration gradient. Sample was heated at 100° for 1 minute after solubilization. Load, 20 μg of chlorophyll. Redrawn from Chua and Bennoun (1975).

this complex, presumably equivalent to polypeptide 2 in *Chlamydomonas*.

Chua and Bennoun (1975) have investigated two Mendelian mutants in *Chlamydomonas* that are deranged in the reaction center complex of PS II. On the basis of fluorescence induction kinetics, the primary electron acceptor of PS II, Q, is either missing or inactive in both mutants. One of the mutants, *F34*, is nonconditional, but is suppressed by another nuclear mutation, *SU-1*. The second mutant, *T4*, is temperature-sensitive and has a wild-type phenotype at 25°C, but a mutant phenotype at 35°C. Analysis of the membrane polypeptides shows that *F34* lacks polypeptide 6 and polypeptide 5 is reduced in amount by 50% compared to wild type. When *F34* is coupled with *SU-1*, the activity of PS II is partially restored to about 60%, and this is paralleled by a partial recovery of polypeptide 6 to about 50%, whereas polypeptide 5 is restored completely to wild-type level. The stoichiometic recovery of polypeptide 6 in *F34 SU-1* strongly suggests that this polypeptide is a component of the PS II reaction-center complex. Growth of *T4* at the restrictive temperature causes the mutant to behave generally like *F34* except that polypeptides 4.1 and 4.2 become reduced in addition to 5 and 6 (Chua and Bennoun, 1975). These polypeptides are all products of chloroplast protein synthesis (see Section III,B). If polypeptides translated in the chloroplast are coded in the chloroplast, then the products of the *F34* and *T4* genes must be involved indirectly in the assembly or processing of polypeptides 5, 6, and possibly 4.1 and 4.2. The fact that each mutant is, to some extent, pleiotropic is consistent with such a hypothesis.

Chua (1976) has characterized a chloroplast gene mutant, *thm-u-1*, which synthesizes a variant polypeptide in place of lamellar polypeptide 5 (apparent M_r 50,000). The function of this polypeptide is unknown. The variant, designated 5′, is larger than the wild-type counterpart by about M_r 1000. The replacement of polypeptide 5 by 5′ does not have any apparent effect on the photosynthetic capacities of the membranes. The *thm-u-1* mutant is transmitted to all meiotic progeny when carried by the maternal parent, but is rarely transmitted by the paternal parent. When biparental transmission occurs, *thm-u-1* segregates somatically during mitosis of haploid zoospores and vegetative diploids with respect to its wild-type allele. A small percentage of the diploids remain heterozygous and continue to segregate the 5 and 5′ polypeptides for many cell generations. Since these biparental diploids synthesize both polypeptides 5 and 5′, these results strongly suggest that *thm-u-1* codes for polypeptide 5. Inhibitor studies have shown that polypeptide 5′ (Chua, unpublished results), like polypeptide 5 (Chua and Gillham, 1977), is made on chloroplast ribosomes (see Section III,B).

The membrane polypeptides of *C. reinhardtii* mutants deficient in PS II have also been studied by Levine *et al.* (1972), Levine and Duram (1973), and Anderson and Levine (1974b) using a different gel system devised by Hoober (1970). Polypeptides a, b, and c of Hoober were designated IIa, IIb, and IIc by Levine *et al.* (1972) because submembrane fragments enriched in PS II activity were also enriched in these polypeptides. Using membrane preparations that had been extracted with acetone, Levine *et al.* (1972) examined two Mendelian PS II mutants, *ac-141* and *1fd-27*, and found that polypeptide IIa was reduced in *ac-141*, but that all three PS II polypeptides were present in normal amounts in *1fd-27*. Since in wild-type *Chlamydomonas* polypeptide 5 is absent from gels of acetone-extracted material and polypeptide 6 is reduced in amount (Chua and Gillham, 1977), the possibility that these two polypeptides are deficient in these mutants cannot be ruled out.

In a later study with a pigment-deficient Mendelian mutant, *ac-5*, Levine and Duram (1973) reported that polypeptides IIa, IIb, and IIc were present in cells grown photosynthetically in reasonably normal amounts, but very deficient in cells grown with acetate as a carbon source. Nevertheless, the Hill reaction, a measure of PS II activity, is similar in both groups of cells. Since the Hill reaction rate on a chlorophyll basis in *ac-5* is much higher than it is in wild type, but is the same whether cells are grown with CO_2 or acetate as carbon sources (Goodenough and Staehelin, 1971) despite the great fluctuations in amount of polypeptides IIa, IIb, and IIc under the two sets of conditions one can only conclude that these polypeptides do not play a *direct* role in the PS II reaction. In light of these findings the experiments with *ac-141* cannot be regarded as establishing that polypeptide IIa is important in PS II activity, especially when acetone-extracted materials instead of unextracted membranes were analyzed. In any event polypeptide IIa of Levine *et al*. (1972) is equivalent to at least three polypeptides (8–10) in the system of Chua and Bennoun (1975) (see Section III,B).

Anderson and Levine (1974a) have examined the thylakoid membrane polypeptides of a chlorophyll *b*-less Mendelian mutant of barley, whose chlorophyll–protein complexes have also been studied by Thornber and Highkin (1974), and a chlorophyll-deficient pea mutant. The barley and pea mutants both contain diminished amounts of polypeptides IIb and IIc, but IIa is present in normal amounts. By analogy with *Chlamydomonas* the deficiency in IIb is expected, since polypeptide IIb is a constituent of the light-harvesting chlorophyll–protein complex (Anderson and Levine, 1974b). Herrmann (1971b) described a second plastome mutant of *Antirrhinum, en:viridis-1,* which is pale green and contains 37% of the chlorophyll of wild type. This mutant has an almost normal chlorophyll *a* to *b* ratio, but cannot perform the Hill reaction. In the mutant, two polypeptides, 2 and 13, associated with the two pigment-protein complexes are present in almost normal amounts, but a major polypeptide, 7, is missing, and polypeptides 8–12 are decreased in amount. Polypeptide 7 in *Antirrhinum* is very likely equivalent to polypeptide 6 in *C. reinhardtii*, which could explain why the mutant has no Hill reaction. Since this polypeptide is a product of chloroplast protein synthesis in *C. reinhardtii*, it would not be unreasonable if it were also coded by the chloroplast genome.

Experiments with the W_3BUL mutant of *Euglena* which lacks chloroplast DNA afford an alternative approach for ascertaining where specific thylakoid membrane polypeptides are coded and synthesized. By definition, any polypeptide present in the residual thylakoids formed by the undifferentiated plastid of the W_3BUL mutant must be coded by a nuclear gene and synthesized on cytoplasmic ribosomes. However, the converse does not hold, because certain polypeptides coded by nuclear DNA and translated in the cytoplasm might be missing from the residual thylakoid membranes of this mutant because they require products of the chloroplast genome and protein synthesizing system for integration. Bingham and Schiff (1976) have attempted to isolate the residual thylakoid membranes from the W_3BUL mutant using the unique sulfolipid of the chloroplast membranes as a marker and have examined the polypeptides of this membrane fraction with SDS-gel electrophoresis. Their results suggest that many of the thylakoid membrane polypeptides of *Euglena* must be coded by nuclear genes and the messages translated on cytoplasmic ribosomes since these polypeptides are present in the residual thylakoid membranes of the W_3BUL mutant lacking chloroplast DNA.

The foregoing discussion should indicate that at the moment we have only the faintest glimmerings of an idea concerning which thylakoid membrane polypeptides are coded by nuclear genes and which by chloroplast genes. Most of the mutants studied to date are Mendelian and have pleiotropic effects. It is clearly time for a thorough investigation employing many mutants, preferably in an organism such as *Chlamydomonas*, where both Mendelian and chloroplast genomes are amenable to systematic genetic analysis.

B. SITES OF TRANSLATION OF THYLAKOID MEMBRANE COMPONENTS

Like the mitochondrial inner membrane, the thylakoid membrane system of the chloroplast is a complex structure containing electron-transport components, an ATP generating apparatus, and a large number of polypeptides. This similarity also extends to the biosynthesis of the thylakoid membrane polypeptides, which is in part the province of the organelle and in part the responsibility of the protein-synthesizing machinery of the cytoplasm. The first indications that synthesis of thylakoid membrane polypeptides might depend on both the cytoplasmic and chloroplast protein-synthesizing systems came from experiments in which the increase in activities of certain membrane-bound enzymes were measured in the presence and in the absence of specific inhibitors. In synchronously growing cultures of *Chlamydomonas*, Armstrong *et al.* (1971) examined the effects of the chloroplast ribosome inhibitors chloramphenicol and spectinomycin as well as the inhibitor of cyto-plasmic ribosome function, cycloheximide, on the formation of various chloroplast components that normally increase 2- to 3-fold during the same period in the growth cycle. Increases in chlorophyll, ferredoxin, and ferredoxin–NADP reductase were blocked only by cycloheximide, indicating that these components require activity of cytoplasmic ribosomes. Increases in cytochrome 563 and cytochrome 553 were inhibited by cycloheximide, chloramphenicol, and spectinomycin, suggesting that these components require translational activity of both chloroplast and cytoplasmic ribosomes. A summary of the effects of these inhibitors on similar components associated with thylakoid membranes in *Euglena* can be found in Table 13.6 of the review by Schmidt and Lyman (1976). Rather than consider these experiments further, we will turn to more recent experiments that deal with the sites of synthesis of specific thylakoid membrane polypeptides. This knowledge is still far less complete than in the case of the inner mitochondrial membrane polypeptides, particularly with respect to the functional enzyme complexes to which the peptides belong (Schatz and Mason, 1974).

A notable exception is chloroplast coupling factor CF_1, the thylakoid-bound ATPase that catalyzes the terminal stage of photophosphorylation in the chloroplast (Penefsky, 1974). Biosynthesis of CF_1 appears to involve both chloroplast and cytoplasmic protein synthesis. The structure and properties of the CF_1 membrane-bound enzyme have been reviewed recently (Penefsky, 1974; Panet and Sanadi, 1976; see also the Chapter by Knaff and Malkin in Volume 7 of this series). Briefly, CF_1 can be solubilized by treatment of the thylakoid membranes with 1 mM EDTA, and the soluble form possesses latent ATPase activity.

Results obtained principally by Racker's group have shown that spinach CF_1 has an M_r of 325,000 and is made up of five nonidentical subunits: α, β, γ, δ, and ϵ. The α and β subunits are believed to be involved in the catalytic reaction, whereas the ϵ subunit is a potent inhibitor of the ATPase activity; the functions of the remaining two subunits, γ and δ, have not yet been defined (Penefsky, 1974; Panet and Sanadi, 1976). Early inhibitor studies by Horak and Hill (1971, 1972) indicated that the synthesis of CF_1 during greening of etiolated leaves of *Phaseolus vulgaris* requires the participation of both chloroplast and cytoplasmic ribosomes. In isolated pea chloroplasts, Eaglesham and Ellis (1974) found that the labeling pattern of thylakoid membranes is not altered by EDTA washing, and accordingly, they concluded that all CF_1 subunits are synthesized in the cytoplasm. However, more careful studies by Mendiola-Morgenthaler *et al.* (1976) revealed that the α, β, and ϵ subunits of CF_1 are labeled in intact spinach chloroplasts whereas the remaining two are presumably synthesized in the cytoplasm. These results are in contrast to those obtained with the mitochondrial coupling factor, F_1, in which all the five subunits are made in the cytoplasm (Tzagoloff *et al.*, 1973). The cooperation of chloroplast and cytoplasmic ribosomes in the biosynthesis of CF_1 is reminiscent of that in RubPCase, where synthesis of the two subunits is under stringent control and little or no free pool of subunits apparently exists in the cell (Iwanij *et al.*, 1975).

A hint as to regulation and control of the synthesis of CF_1 subunits also appears in recent experiments of Cashmore (1976), who examined the synthesis of thylakoid membrane polypeptides in short-term labeling experiments with green pea seedlings carried out in the presence of either chloramphenicol or cycloheximide. Two major polypeptide peaks were identified in his SDS gels of the thylakoid membrane fraction. One of these, P I, had an M_r of 58,000 and most probably represented both the α (59,000) and β (56,000) subunits, which are not resolved in his gels. Synthesis of P I was found to be inhibited completely by chloramphenicol, a result not unexpected if P I is indeed α and β. In the presence of cycloheximide P I continued to be labeled to the same extent, but 50% of the protein appeared in the soluble fraction rather than in the membranes. If P I is in fact the α and β subunits of CF_1, these results would mean that the two subunits are synthesized but 50% of them are not assembled onto the thylakoids because of a deficiency of γ and δ, which are presumably made on cytoplasmic ribosomes and therefore sensitive to cycloheximide inhibition.

Since the most complete picture of the biogenesis of individual thylakoid membrane polypeptides comes from experiments with *Chlamydomonas reinhardtii*, we will begin with a description of these experiments and then compare these results with those obtained in higher plants. Two approaches have been taken in studying the biogenesis of thylakoid membrane polypeptides in *Chlamydomonas*. The original studies of Hoober, Ohad, and their colleagues utilized a mutant called *y-1* which, unlike wild-type cells of *C. reinhardtii*, does not form chlorophyll in the dark when supplied with acetate as a carbon source, but instead becomes etiolated like a higher plant (Hudock and Levine, 1964; Hudock *et al.*, 1964; Ohad *et al.*, 1967a,b). The mutant stops making chlorophyll and thylakoid membranes when transferred to the dark and begins to accumulate protochlorophyllide. During successive cell divisions chlorophyll and preexisting lamellar structures are diluted out. If these "yellow" cells are returned to the light, the accumulated protochlorophyllide is photoconverted to chlorophyllide, and rapid synthesis of chlorophyll and thylakoid membranes commences (Ohad *et al.*, 1967b). Within 6–8 hours the regreening process is complete and the cells are indistinguishable from wild type in amount of chlorophyll and thylakoid membrane organization.

To study the sites of synthesis of the thylakoid membrane polypeptides, both Hoober and Ohad (Hoober *et al.*, 1969; Hoober, 1970, 1972, 1976; Hoober and Stegeman, 1973, 1975; Eytan and Ohad, 1972a,b; Bar-Nun and Ohad, 1975; Ohad, 1975) allowed dark-grown cells of *y-1* to regreen in the presence of inhibitors of chloroplast or cytoplasmic protein synthesis and labeled arginine. The polypeptides made in the presence of each inhibitor were radioactively labeled. This approach is similar to that used by Tzagoloff (Tzagoloff *et al.*, 1973) in studying the biosynthesis of the oligomycin-sensitive ATPase and cytochrome oxidase in yeast, in that polypeptide synthesis is followed in a derepressing organelle. The differentiation is triggered in the yeast experiments by glucose limitation and in the *Chlamydomonas* experiments with *y-1* by light. Compared to the yeast experiments, the *Chlamydomonas* experiments were generally rather long term.

A second, more recent, approach to studying the sites of synthesis of thylakoid membrane polypeptides was taken by Chua and Gillham (1977). This approach is similar to the one taken by Schatz and Bücher and their colleagues in defining the sites of synthesis of the cytochrome oxidase polypeptides in yeast and *Neurospora* (cf. Schatz and Mason, 1974; Sebald *et al.*, 1974). Light-grown wild-type cells of *C. reinhardtii* containing a fully developed chloroplast were pulse-labeled with radioactive acetate in the presence of inhibitors of cytoplasmic or chloroplast protein synthesis. In the absence of limiting pools of particular polypeptides, the labeling patterns would be expected to be complementary for the two sets of inhibitors; i.e., polypeptides whose synthesis is normally blocked by the cytoplasmic protein synthesis inhibitor should continue to be made in the presence of the inhibitor of chloroplast protein synthesis, and vice versa.

To determine the sites of synthesis of these lamellar polypeptides Chua and Gillham (1977) pulse-labeled the thylakoid membrane polypeptides of exponentially growing cells of *C. reinhardtii* with [^{14}C]acetate. Cytoplasmic protein synthesis was inhibited by anisomycin, and chloroplast protein synthesis by chloramphenicol or spectinomycin. The labeling patterns of isolated thylakoid membranes of wild-type cells were compared to those of a mutant, *spr-u-1-6-2,* known to have spectinomycin-resistant chloroplast ribosomes from both *in vivo* and *in vitro* experiments (Conde *et al.*, 1975). Polypeptides made on chloroplast ribosomes were predicted not to be labeled in the presence of either chloramphenicol or spectinomycin in wild type, but to be labeled in the *spr-u-1-6-2* mutant in the presence of spectinomycin. Thus, the results with *spr-u-1-6-2* would serve as positive confirmation for those polypeptides synthesized on chloroplast ribosomes. After labeling, the thylakoid membranes were isolated, the component polypeptides were subjected to SDS–gradient gel electrophoresis, and autoradiograms were made of the dried gels.

Chua and Gillham (1977) found that polypeptides 4.1, 4.2, 5, and 6 (see Fig. 3) and two low-molecular-weight polypeptides (LMW-1 and LMW-2) were made on chloroplast ribosomes. Polypeptide 2 appears to require both functioning chloroplast and cytoplasmic ribosomes for its formation and/or integration into the membrane. Synthesis of this polypeptide was inhibited completely by either chloramphenicol or spectinomycin in wild type and by chloramphenicol alone in the *spr-u-1-6-2* mutant, indicating that it was made in the chloroplast. However, partial and variable inhibition of incorporation of radioactivity into this polypeptide was also obtained with anisomycin. From these results, Chua and Gillham (1977) concluded that integration of polypeptide 2 was dependent on one or more cytoplasmically synthesized partner proteins whose pool size tended to be smaller than that of polypeptide 2. Inhibition of synthesis of this partner polypeptide by anisomycin would thus block integration of polypeptide 2. If this interpretation were correct, one should be able to minimize the apparent inhibition of polypeptide 2 synthesis by anisomycin by increasing the pool size of the putative partner protein. In fact, Chua and Gillham (1977) found that prior incubation of the cells with chloramphenicol relieved the inhibition of polypeptide 2 synthesis by anisomycin during subsequent treatment with the latter drug.

In addition to these polypeptides, two broad, diffuse bands of radioactivity (D-1 and D-2) were observed in the gels; the labeling was resistant to anisomycin but sensitive to the inhibitors of chloroplast protein synthesis. These bands of radioactivity correspond to polypeptides that stained faintly. All remaining polypeptides appeared to be products of cytoplasmic protein synthesis. Any polypeptide whose synthesis was blocked by chloramphenicol or spectinomycin in wild type was synthesized in the presence of the latter antibiotic in *spr-u-1-6-2*.

In the experiments of Hoober *et al.* (1969), Hoober (1970), and Hoober and Stegeman (1975), cells of the *y-1* mutant were transferred from darkness to light and allowed to regreen in the presence of chloramphenicol for 4 hours, and then [^{14}C]arginine was added to the medium for 4.5 more hours of greening. At the end of this time the cells were transferred to a medium containing cycloheximide plus [^{3}H]arginine for 2 additional hours of greening. Thus, only membrane polypeptides made in the cytoplasm should be labeled in the first part of the experiment, and only those polypeptides made in the chloroplast should be labeled in the second part. After labeling, the membranes were isolated and extracted with 90% acetone; the component polypeptides were separated by SDS-gel electrophoresis. The gels were then stained, sliced, and counted. Both chloramphenicol- and cycloheximide-resistant incorporation into thylakoid membrane polypeptides was observed.

A comparison between the gels of Hoober and co-workers and those of Chua and Gillham reveals a number of differences. The three major peaks in Hoober's gels are designated a, b, and c. Peak a appears to correspond to polypeptides 9, 10, possibly 8 and to diffuse band D-1; peak b, to polypeptides 11 and 12; and peak c, to polypeptides 13–17 in the system of Chua and Gillham. Polypeptides 2 and 5 are absent from gels of acetone-extracted material, and polypeptide 6 is greatly reduced

(Chua and Gillham, 1977; Chua et al., 1975). Therefore, these three polypeptides, all of which are products of chloroplast protein synthesis, are very likely absent in Hoober's gels. Near the top of Hoober's gel is a peak that is labeled in the presence of cycloheximide but not chloramphenicol. This peak probably corresponds to Chua and Gillham's polypeptides 4.1 and 4.2, which are products of chloroplast protein synthesis and are not affected by acetone extraction prior to gel electrophoresis.

Hoober (1970) reported both cycloheximide- and chloramphenicol-resistant incorporation into polypeptide a and suggested that either the synthesis of this polypeptide required the activity of both cytoplasmic and chloroplast protein-synthesizing systems or that this peak contained more than one polypeptide. The chloramphenicol-resistant incorporation in Hoober's peak a probably corresponds to polypeptides 9, 10, and possibly 8 of Chua and Gillham, whereas the cycloheximide-resistant incorporation corresponds to the broad, diffuse band of radioactivity D-1 obtained by Chua and Gillham with thylakoid membranes fractionated in chloroform–methanol. Hoober's polypeptide b is labeled in the presence of chloramphenicol but not of cycloheximide; this agrees with the finding of Chua and Gillham that polypeptides 11 and 12 are products of cytoplasmic protein synthesis. Incorporation into Hoober's polypeptide c is principally sensitive to cycloheximide, but a small peak of chloramphenicol-sensitive incorporation also occurs in this region. The cycloheximide-sensitive portion of Hoober's peak c probably corresponds to polypeptides 13–17 of Chua and Gillham, which are synthesized on cytoplasmic ribosomes, whereas the chloramphenicol-sensitive incorporation may be equivalent to the second band of radioactivity, D-2, seen in the gels of Chua and Gillham. The two peaks of chloramphenicol-sensitive incorporation seen in the low-molecular-weight region of Hoober's gels could be equivalent to the low-molecular-weight polypeptides (LMW-1 and LMW-2) reported by Chua and Gillham as being products of chloroplast protein synthesis.

Hoober (1972) and Hoober and Stegeman (1973, 1975) also have studied the regulation of thylakoid membrane polypeptide synthesis in y-1. They reasoned that the conversion of protochlorphyll(ide) might be the key step triggering thylakoid membrane assembly in Chlamydomonas as it seems to be in higher plants (Henningsen and Boynton, 1974), since the action spectrum for greening of the y-1 mutant in C. reinhardtii is identical to that of higher plants (McLeod et al., 1963). This proved to be the case, since cells regreened normally if exposed to light above 600 nm, but regreening stopped immediately if the cells were exposed to light above 675 nm (Hoober and Stegeman, 1973, 1975), which is above the 650 nm in vivo absorption maximum of the photoconvertible form of protochlorophyllide. Synthesis of polypeptides a, b, and c also declined over a period of 2–3 hours after transfer of cells to >675 nm light (Hoober and Stegeman, 1973, 1975). This finding was consistent with the hypothesis that the protochlorophyll–chlorophyll conversion and membrane biosynthesis go hand in hand. Hoober (1972) had previously reported that the major thylakoid membrane polypeptides, in particular polypeptide c, could be detected in gels of whole-cell protein. When dark-grown cells of y-1 were transferred to white light in the presence of high concentrations of chloramphenicol (200 μg/ml), synthesis of chlorophyll and thylakoid membranes was inhibited, but polypeptide c continued to be made and appeared in the soluble fraction instead of the membrane pellet. Subsequently, Hoober and Stegeman (1973, 1975) showed that in the presence of chloramphenicol and >675 nm white light, greening of y-1 was also inhibited, and again polypeptide c continued to be made, but not incorporated into thylakoid membranes. When greening cells of y-1 were transferred to darkness, the same result was obtained with respect to polypeptide c. Similar, but less definitive, results were reported for synthesis of polypeptide b.

On the basis of these findings, Hoober and Stegeman (1973, 1975) formulated a model for control of synthesis of polypeptides b and c in *C. reinhardtic*. This model hypothesizes that protochlorophyll controls the synthesis of polypeptides b and c. The model postulates that the mRNAs for these two polypeptides are transcribed from chloroplast DNA and exported to the cytoplasm, where they are translated on cytoplasmic ribosomes. Protochlorophyll, in combination with a polypeptide (P_R) synthesized on chloroplast ribosomes, which acts as a corepressor, blocks the transcription of the mRNAs for polypeptides b and c. Therefore, in y-1 growing in the dark, the protochlorophyllide accumulated blocks synthesis of these key thylakoid polypeptides, which in turn prevents formation of photosynthetic membranes. When the light is turned on, protochlorophyll is converted to chlorophyll, which cannot combine with P_R, and transcription of the mRNAs for polypeptides b and c occurs. These mRNAs are transported to the cytoplasm and translated, and thylakoid membrane assembly proceeds. If cycloheximide is added to the medium in either the light or dark, translation of the mRNAs for polypeptides b and c is blocked. If regreening cells are exposed to >675 nm light, the transcription of the mRNAs for polypeptides b and c is blocked by the accumulation of protochlorophyll which combines with P_R. If chloramphenicol is added to the medium, translation of the message for P_R on chloroplast ribosomes is blocked, there is no corepressor with which protochlorophyll can combine to block transcription of the mRNAs for polypeptides b and c, and synthesis of these polypeptides is not repressed.

This scheme now needs reevaluation in light of the finding by Chua and Gillham (1977) that polypeptides b and c of Hoober each consist of several polypeptides. It is interesting that synthesis of polypeptides in this region of the gel continues if regreening y-1 cells are transferred to the dark or to >675 nm light in the presence of chloramphenicol. However, polypeptides b and c are identified by gel electrophoresis of radioactively labeled total soluble protein and by comparison of the positions of these radioactive peaks with the protein staining peaks seen in the gels of membrane preparations. Thus, at least three interpretations can be made of these results. First, the synthesis of some or all of the polypeptides corresponding to peaks b and c is derepressed by chloramphenicol in y-1 under conditions where regreening is blocked. Second, the radioactive peaks that appear in the soluble fraction in the presence of chloramphenicol do not represent thylakoid membrane polypeptides at all, but, rather, are polypeptides of similar molecular weights whose synthesis is stimulated by an indirect effect of chloramphenicol. Third, during thylakoid membrane biogenesis there is tight coupling between the chloroplast and cytoplasmic protein-synthesizing systems involving insertion of chlorophyll into lamellae assembled from polypeptides made on both chloroplast and cytoplasmic ribosomes. Chloramphenicol uncouples the system by blocking chloroplast protein synthesis and the cytoplasmic system becomes "free running" so that thylakoid membrane polypeptides continue to be synthesized on cytoplasmic ribosomes. Clearly, the observations made by Hoober and Stegeman (1973, 1975) are of interest and the system needs to be analyzed further with a gel system giving better resolution and with the use of monospecific antibodies in order to prove that the polypeptides accumulated in the soluble fraction in the presence of chloramphenicol really are thylakoid membrane polypeptides.

Eytan and Ohad (1972a,b) also used cycloheximide and chloramphenicol to investigate the sites of synthesis of thylakoid membrane polypeptides in the regreening y-1 system of C. $reinhardtii$. However, their acetic acid–urea gel system had low resolution, and only one major polypeptide peak of staining and radioactivity, called the L protein, appeared to be synthesized during greening. They found that L protein synthesis was cycloheximide-sensitive and concluded that the L protein peak comprises one or more proteins associated with photosynthetic lamellae, which were made on cytoplasmic ribosomes. Several other poorly resolved protein peaks appeared to be labeled in the presence of cycloheximide, but not of chloramphenicol, and, thus, would be products of chloroplast protein synthesis. Meaningful comparisons of the gel patterns seen by Eytan and Ohad with those of Chua and Gillham or Hoober are difficult because of the poor resolution of the former gels.

However, the L protein of Eytan and Ohad may be equivalent to Hoober's peaks a, b, and c and polypeptides 8–17 in the system of Chua and Gillham.

On the basis of these results as well as more extensive physiological and ultrastructural findings, Eytan and Ohad (1972b) have also constructed a model to explain the regreening of *y-1*; the model supposes that membrane composition can be modulated through the use of inhibitors such as chloramphenicol and cycloheximide. Proteins of cytoplasmic origin (L proteins) assemble together with lipids to form a membrane framework, which then serves to accept proteins made on chloroplast ribosomes (activation proteins) and chlorophyll. Addition of the activation proteins and chlorophyll results in the formation of active photosynthetic lamellae. This is an interesting but difficult model to test, although a qualitative comparison of the polypeptide composition of the few remaining thylakoid membranes present in dark-grown cells of *y-1* with the membranes formed during regreening might give some idea of its validity.

More recently, Bar-Nun and Ohad (1975) have reexamined the regreening of dark-grown *y-1* cells, using SDS gels with somewhat increased resolution, and identified seven stained polypeptide peaks (I, II, IIb, III, IV, Va, and Vb). Peak I was not labeled in the presence of either chloramphenicol (18–19 hours of exposure) or cycloheximide (4 hours of exposure); peaks II and IIb were labeled in the presence of cycloheximide, but not of chloramphenicol; and incorporation into the remaining four peaks was inhibited by cycloheximide, but not by chloramphenicol. When CP I and CP II fractions (chlorophyll protein complexes) were prepared, the CP I fraction was found to be enriched in peak II and the CP II fraction in peaks III, IV, Va, and Vb. Comparison of these findings directly to those of Chua and Gillham (1977) is difficult because of differences in the two gel systems; however, a paper by Kretzer *et al.* (1976) permits certain tentative identifications. The latter authors divide their gels into five regions designated I (2 polypeptides), II (5 polypeptides), III (2 polypeptides), IV (several polypeptides), and V (2 polypeptides) of decreasing molecular weight. Both Chua and Bennoun (1975) and Kretzer *et al.* (1976) have studied the same temperature sensitive photosynthetic mutant, *T4*, which is deficient in certain thylakoid membrane polypeptides comprising the photosystem II reaction center when grown under restrictive conditions. Chua and Bennoun (1975) show that polypeptides 4.1, 4.2, and 5 are deficient and polypeptide 6 is missing, whereas Kretzer *et al.* (1976) report that polypeptides II-2 and II-4 are missing. Both groups find that chloramphenicol inhibits synthesis of these peptides in wild-type cells. Therefore, polypeptides 4.1–6 are probably equivalent to some of the polypeptides in region II of Kretzer *et al.* (1976). In addition at least one of the two region I polypeptides of Kretzer *et al.* (1976), whose synthesis and/or integration into the thylakoid membranes is blocked both by chloramphenicol and cycloheximide, very likely corresponds to polypeptide 2 of Chua and Gillham (1977) which shows the same pattern of inhibition and approximately the same gel position. Since both Chua and Bennoun (1975) and Kretzer *et al.* (1976) agree that the *T4* mutant is deficient in the photosystem II reaction center (CP II), the previous identification of peak II polypeptides as components of CP I by Bar-Nun and Ohad (1975) was probably in error. Certain of the polypeptides found by Bar-Nun and Ohad (1975) to be associated with CP II (regions III and IV), which are synthesized on cytoplasmic ribosomes, must be identical to some of the polypeptides (11–17) in the M_r 25,000–30,000 region of Chua and Gillham's (1977) gels.

Biosynthesis of thylakoid membrane polypeptides in higher plants has been studied both *in vivo* using inhibitors of chloroplast and cytoplasmic protein synthesis and *in vitro* in isolated chloroplasts. The *in vivo* studies of Machold and Aurich (1972), who exposed excised shoots of *Vicia faba* to radioactive amino acids in the presence of chloramphenicol or cycloheximide for 5 or 25 hours are the most comparable to those done with *Chlamydomonas*. The chloroplast lamellae were isolated, and polypeptides were subjected to electrophoresis in an SDS-gel system, which gave high resolution. Machold and Aurich (1972) were able to visualize 21 distinct polypeptides, and the synthesis of 12 of these was inhibited by chloramphenicol. The synthesis of the CP I polypeptide, designated B, was strongly inhibited by chloramphenicol and weakly inhibited by cycloheximide. Chua and Gillham (1977) subsequently observed exactly the same inhibition pattern for polypeptide 2 in *C. reinhardtii*. Synthesis of polypeptides E, F, and G, probably the equivalents of polypeptides 4–6 in *C. reinhardtii*, was inhibited by chloramphenicol, but not by cycloheximide, as would be expected from the *Chlamydomonas* results. Among the polypeptides synthesized on cytoplasmic ribosomes is one associated with CP II and equivalent to polypeptide 11 in *C. reinhardtii*. Thus, the results obtained by Machold and Aurich with *V. faba* parallel those of Chua and Gillham except that a higher proportion of the total polypeptides (ca. 60%) appears to be inhibited by chloramphenicol. In this regard one should note that the *V. faba* experiments are relatively long term, and some of the inhibitor effects might be indirect.

Ellis (1975a,b) analyzed thylakoid membrane polypeptide synthesis *in vivo* in etiolated pea shoots during regreening. Lincomycin and cycloheximide were used to inhibit chloroplast and cytoplasmic protein synthesis, respectively. Membrane polypeptides were labeled with

[^{35}S]methionine and membrane systems isolated for analysis by SDS-gel electrophoresis. Accumulation of a CP I polypeptide, which Ellis calls polypeptide 2, was completely inhibited by lincomycin and partially inhibited by cycloheximide (Ellis, 1975a,b). Thus, the inhibition patterns seen for this polypeptide are identical in *Chlamydomonas, Vicia,* and *Pisum.* However, Ellis (1975a) made the converse hypothesis to that of Chua and Gillham (1977) and supposed that polypeptide 2 was synthesized on cytoplasmic ribosomes and required the presence of a partner protein synthesized on chloroplast ribosomes for its integration. Three other cycloheximide-resistant peaks of incorporation were also observed (Ellis, 1975a). One of these, not discussed by Ellis, is of the correct molecular weight range to correspond to polypeptides 4–6 of Chua and Gillham (1977) in *C. reinhardtii* and E-F of Machold and Aurich (1972) in *Vicia.* The other two cycloheximide-resistant peaks are in the region of the gel that probably corresponds to polypeptides 8–17 of Chua and Gillham (1977). The major radioactive peak of Ellis, peak D, is in the same position as a stained band. In *C. reinhardtii* this region of the gel comprises polypeptides 8–10, which are made on cytoplasmic ribosomes, and D-1, which is the rapidly labeled component made on chloroplast ribosomes corresponding to a diffuse, poorly stained band. Whether peak D of Ellis corresponds to a stained band will be considered later, for this region of the gel may be as complex as it is in *C. reinhardtii.*

Both the gel positions and sites of synthesis of the CP II polypeptides in Ellis' experiments are the same as in *V. faba* and *C. reinhardtii.* The third cycloheximide-resistant peak of incorporation observed in Ellis' experiments is also in this part of the gel and may be equivalent to D-2 in *C. reinhardtii,* which also corresponds to a faintly stained polypeptide.

Most other studies on the biogenesis of thylakoid membrane polypeptides in higher plants have employed isolated chloroplasts. Eaglesham and Ellis (1974) examined light- or ATP-driven incorporation of [^{35}S]methionine into thylakoid membranes of isolated pea plastids, and observed five radioactive peaks, all of which were pronase digestible. However, only peak D (M_r 32,000) appeared to correspond to a stained polypeptide. Siddell and Ellis (1975) later performed a similar experiment with etioplasts isolated in various stages of the regreening process. Incorporation first occurred into the large subunit of RubPCase and later into peak D. No other incorporation of any significance was observed in the thylakoid membranes. Bottomley *et al.* (1974) examined the incorporation of [^{35}S]methionine into thylakoid membrane polypeptides of isolated spinach chloroplasts and reported incorporation of label into at least nine discrete products in the membrane fraction. One major peak of label corresponded to a stained band of M_r 72,000. One of the highly radioactive peaks, which did not correspond to a stained band, was equivalent to a polypeptide of M_r 36,000. Morgenthaler and Mendiola-Morgenthaler (1976), also using isolated spinach chloroplasts, observed five radioactive peaks in the thylakoid membrane fraction. The major peak of radioactivity corresponded to a stained band and had a molecular weight similar to that of peak D of Ellis and the nonstaining radioactive peak observed by Bottomley et al. (1974).

The experiments of Bottomley et al. (1974) and Morgenthaler and Mendiola-Morgenthaler (1976) differ only in the method of preparation of the spinach thylakoid membrane polypeptides for electrophoresis. Bottomley et al. (1974) extracted their membranes with acetone prior to electrophoresis, whereas Morgenthaler and Mendiola-Morgenthaler (1976) did not. As mentioned earlier, Chua and Gillham (1977) found in *Chlamydomonas* that acetone extraction leads to the selective loss of specific polypeptides from the gels. Conceivably peak D is not a homogeneous protein, but a group of two or more proteins of similar molecular weight. The staining protein or proteins would not run in the SDS gels after acetone extraction of the membranes whereas the rapidly labeling components would. The true nature of peak D will not be revealed until gels with better molecular weight resolution are employed and more sophisticated methods of membrane polypeptide fractionation, such as differential solubility in chloroform–methanol, are used. The stained polypeptides in the peak D region which do not label in isolated chloroplasts may well prove to be equivalent to polypeptides 8–10 in *Chlamydomonas,* which are made on cytoplasmic ribosomes, while the highly radioactive component may prove to be equivalent to peak D-1 in *Chlamydomonas,* which labels rapidly, does not stain well, and is a product of chloroplast protein synthesis.

As in the case of the mitochondrial inner membrane (Schatz and Mason, 1974) the most satisfactory picture of the sites of synthesis of individual polypeptides in the thylakoid membranes has come from *in vivo* experiments with specific inhibitors carried out over a short term rather than from *in vitro* experiments with isolated organelles. In order to establish rigorously whether the results obtained by Chua and Gillham (1977) for *Chlamydomonas* can be generalized satisfactorily to higher plants, one must establish the identity of specific polypeptides through immunological techniques. Until this is done, the comparisons made in this paper between the *Chlamydomonas* and higher-plant results must be regarded as suppositional.

Recently, Joy and Ellis (1975) have shown that 2 out of some 25 or more chloroplast envelope proteins become labeled in isolated pea chloroplasts. This labeling is inhibited by chloramphenicol, but not by cycloheximide. Morganthaler and Mendiola-Morganthaler (1976) have observed that envelope membranes from spinach chloroplasts labeled *in vitro* contain one major radioactive component and two minor components.

IV. Chloroplast Ribosomes and Aminoacyl-tRNA Synthetases

A. GENETIC CONTROL OF CHLOROPLAST RIBOSOME BIOGENESIS

The same two approaches discussed previously to determine the location of genes coding for polypeptides comprising the CO_2-fixing enzyme RubPCase and the thylakoid membranes have also been applied to localize the genes coding for chloroplast ribosomal components. Bourque and Wildman (1973) reported that the chloroplast large ribosomal subunits of *N. tabacum* and *N. glauca* differ in two proteins on one-dimensional gel electrophoresis. One protein present in *N. glauca* (a) was absent in a corresponding position in the gel of the *N. tabacum* proteins. The second protein (b) had a greater mobility in *N. tabacum* than in *N. glauca*. In reciprocal interspecific F_1 hybrids, protein a from the *N. glauca* parent was found in both hybrids, and both hybrids contained a protein with the same electrophoretic mobility as protein b of *N. tabacum*. The authors concluded that these two proteins were coded by nuclear genes; however, a careful repetition of these experiments using gel techniques capable of greater resolution is desirable before the conclusions are taken for granted.

The genetic control of chloroplast ribosome biogenesis is being actively investigated through the use of mutations in *C. reinhardtii* (cf. reviews by Chua and Luck, 1974; Harris *et al.*, 1976; Gillham *et al.*, 1976). Some 20 genes have now been identified through the use of appropriate assembly-defective or antibiotic-resistant and -dependent mutants (Table I). Before discussing these mutants and their defects in detail, the structure and properties of *Chlamydomonas* chloroplast ribosomes will be discussed briefly (cf. reviews by Harris *et al.*, 1976; Gillham *et al.*, 1976). Since chloroplasts are very difficult to isolate from *Chlamydomonas*, chloroplast and cytoplasmic ribosomes are generally separated from whole-cell supernatants on sucrose gradients. Bourque *et al.* (1971) identified five principal generic classes of ribosomes with sedimentation velocities of 83, 70, 66, 54, and 41 S as determined by linear extrapolation against the 83 S monomer peak. The 70 S ribosomes are derived from the chloroplast and contain large (54 S) and small (41 S) subunits. In certain mutants 66 S ribosomes virtually replace the 70 S chloroplast ribosomes. The chloroplast large ribosomal subunit contains 5 S (Chua, unpublished results) and 23 S rRNA, and the small subunit contains 16 S rRNA. The 16 S and 23 S rRNAs appear to be coded by chloroplast DNA, and transcriptional mapping studies suggest that each chloroplast DNA molecule of *C. reinhardtii* contains two to three copies of each rRNA cistron arranged in tandem repeats of a transcriptional unit coding for 16 S and 23 S rRNA. Electrophoretic and chromatographic analyses indicate that the large subunit contains 26–34 proteins, and the small subunit 19–25 proteins.

Harris *et al.* (1974, 1976) have identified seven or possibly eight nuclear genes involved in chloroplast ribosome biogenesis through the isolation of assembly-defective mutants (Table I). They are designated *cr-* with the exception of *ac-20*, the first mutant described as belonging to this class (Goodenough and Levine, 1970). The mutants so far isolated fall into two classes in terms of chloroplast ribosome phenotype. Mutants at three gene loci cannot synthesize normal amounts of either ribosomal subunit and are thought to be blocked in ribosomal assembly at a very early point (e.g., at the level of rRNA precursors). Mutants at five gene loci, which accumulate what appear to be 54 S subunits and are deficient to varying degrees in small subunits, may be blocked specifically at various points in small subunit assembly. No mutations have yet been found which are blocked in large subunit formation and accumulate small subunits.

The assembly-defective mutants grow normally when acetate is supplied as a carbon source, but very slowly under photosynthetic conditions. All mutants isolated to date have been leaky to various extents and no mutants completely deficient in chloroplast ribosomes have been found. The poor photosynthetic growth of these mutants presumably results because chloroplast protein synthesis cannot be carried out at normal rates. As a consequence the mutants develop a characteristic syndrome of photosynthetic defects including the inability to synthesize normal amounts of RubPCase and the inability to carry out the Hill reaction at wild-type rates. These specific defects are to be expected, since both the large subunit of RubPCase and polypeptide 6, the reaction-center polypeptide of PS II in *Chlamydomonas,* are products of chloroplast protein synthesis. Chloroplast protein synthesis mutants in *C. reinhardtii,* thus constitute a special class of photosynthetic mutants defective in both RubPCase and the Hill reaction and can be distinguished from other photosynthetic mutants by this double defect. Assembly-defective mutants make normal amounts of chlorophyll and thylakoid membranes, but these membranes are structurally defective and improperly assembled into grana.

With the possible exception of *cr-7* and *ac-20*, all of the assembly defective mutants are nonallelic by recombination analysis and all, including *ac-20* and *cr-7,* appear to complement in diploids. By comparing the photosynthetic capacity and chloroplast ribosome profiles of the individual mutants, of their double mutants, and of their diploids (with

TABLE I

SUMMARY OF NUCLEAR AND CHLOROPLAST GENES IN *Chlamydomonas reinhardtii* KNOWN TO AFFECT THE FUNCTION OR ASSEMBLY OF CHLOROPLAST RIBOSOMES[a]

Type of alteration	Reference	Location of gene	Locus designation	Number of alleles	Effect of Mutant on Chloroplast Ribosomes		
					Levels of antibiotic resistance		Other phenotypic modifications
					In vitro	In vivo	
Antibiotic resistance							
Streptomycin	c,d	N	sr-1	3	Low	—	—
	i	N	sr[1]	1	Low	—	Unnamed sr mutation reported to alter a ribosomal protein
	c,d	C	sr-u-sm-3a	13	Low to high	—	—
	c,d	C	sr-u-sm3	1	Intermediate	—	
	b–e	C	sr-u-2-60	2	High	High	sr-u-2-60 affects assembly of the small subunit
	c,d	C	sd-u-3-18[1]	1	High	—	sd-u-3-18 affects assembly of the small subunit
	b–f	C	sr-u-sm2	2	Very high (S)	—	sr-u-sm2 alters a protein of the small subunit?
	h	C	sr-u-35[1]	1	—	High	sr-u-35 reported to alter a protein of the large subunit?
	i	C	sr[1]	1	High	—	
Spectinomycin or neamine	b–e	C	spr-u-1-H-4[2]	6	Low to high (S)	Low to high	spr-u-1-27-3 alters a protein of the small subunit?
Erythromycin	g,j	N	ery-M-1	4	—	—	All 4 mutant alleles alter the same protein of the large subunit
	g,j	N	ery-M-2	4	—	—	ery-M-2d alters a protein of the large subunit?
	b–e,g	C	er-u-1a	2	High (L)	High	er-u-1a alters a protein of the large subunit?
	b–e	C	er-u-37	2	High	High	—

244

Mutant		Location	Gene		In vitro			Phenotype
Carbomycin	k	C	car¹	1	High (L)	—	—	—
Cleocin	k	C	cle¹	1	High (L)	—	—	—
Assembly defective								
Small subunit								
	b,d	N	cr-1	1	—	—	—	Accumulates large subunits, deficient in 70 S monomers
	b,d	N	cr-2	1	—	—	—	Accumulates large subunits, deficient in 70 S monomers
	b,d	N	cr-3	1	—	—	—	Accumulates large subunits, deficient in 70 S monomers
	b,d	N	cr-5	1	—	—	—	Accumulates large subunits, deficient in 70 S monomers
	b,d	N	cr-7	1	—	—	—	Accumulates large subunits, deficient in 70 S monomers
Both subunits	b,d	N	ac-20	1	—	—	—	Deficient in 70 S monomers
	b,d	N	cr-4	1	—	—	—	Deficient in 70 S monomers
	b,d	N	cr-6	1	—	—	—	Deficient in 70 S monomers

[a] N, gene located in the nucleus; C, gene located in the chloroplast; S, *in vitro* resistance localized to the small ribosomal subunit; L, *in vitro* resistance localized to the large ribosomal subunit; superscript 1, allelic relationship to other genes not determined; superscript 2, neamine- and spectinomycin-resistant chloroplast mutants appear to be allelic based on recombination analysis. Data are taken from the references cited in footnotes b–k.

[b] Harris et al. (1976).
[c] Harris et al. (1977).
[d] Gillham et al. (1976).
[e] Conde et al. (1975).
[f] Ohta et al. (1975).
[g] Mets and Bogorad (1972).
[h] Brügger and Boschetti (1975).
[i] Spiess and Arnold (1975).
[j] Davidson et al. (1974).
[k] Schlanger and Sager (1974).

each other and with wild type) a working model can be formulated for the assembly of chloroplast ribosomes analogous to the formal schemes used to deduce enzymic pathways from the behavior of auxotrophic mutants. Within a given portion of the pathway the mutants can be tentatively ordered, based on the phenotype seen in the double mutants. Thus *ac-20* and *cr-4* appear to affect sequentially the formation of components common to both the small and large ribosomal subunits, whereas *cr-1, cr-2,* and *cr-3* (all of which accumulate 54 S particles) are probably impaired in specific sequential steps in the pathway leading to the formation of the 41 S subunit. Since *ac-20* and *cr-4* are epistatic to *cr-1, cr-2,* and *cr-3,* and block the formation of both subunits, they are thought to affect an early part of the pathway common to both subunits, whereas the latter three mutants specifically affect only one branch of the pathway by which the small subunit is synthesized.

This model for ribosome assembly indicates possible biochemical roles for each of the individual genes in ribosome formation, and suggests experimental approaches to verify the molecular defects in each of the mutants. However, to date none of the *cr*-mutants have been shown to alter a specific chloroplast ribosomal protein.

Although cells of *C. reinhardtii* are sensitive to antibacterial antibiotics known to block organelle protein synthesis in other eukaryotic cells, one-step mutants resistant to these antibiotics are readily isolated. Mutations at three nuclear and seven to nine chloroplast loci confer antibiotic resistance or dependence on chloroplast ribosomes (Table I). In *E. coli* mutations that confer resistance to these antibiotics at the ribosomal level do so by altering specific ribosomal proteins by amino acid substitutions (see review by Jaskunas *et al.*, 1974). Davidson *et al.* (1974) have studied four mutant alleles at the nuclear gene locus *ery-M-1* which confer erythromycin resistance on chloroplast ribosomes. Three of these alleles alter the charge of a specific protein (LC 6) of the large chloroplast ribosomal subunit, and the fourth, *ery-M-1b,* leads to deletion of 30% of this protein. So far this is one of the most rigorous pieces of evidence to date that any gene, organellar or nuclear, in any eukaryotic organism is the structural gene for a specific organelle ribosomal protein. Bogorad *et al.* (1976) have shown that chloroplast ribosomes from vegetative diploids between the *ery-M-1b* mutant and wild type contain both the mutant and wild type forms of the LC 6 protein. These results further strengthen the hypothesis that the *ery-M-1* gene codes for the LC 6 protein. If the *ery-M-1* gene were instead responsible for enzymatically modifying the structure of the LC 6 protein, one would expect mutants in this gene to be recessive to wild type rather than being co-dominant as observed. In addition, Bogorad *et al.* (1976) have isolated and characterized three phenotypic revertants of the *ery-M-1b* mutant to antibiotic sensitivity, designated *es-M-1*, *es-M-2* and *es-M-3*. The *es-M-1* mutant is linked but not allelic with mutants at the *ery-M-1* locus, while the *es-M-2* mutant is unlinked to the *es-M-1* mutant, but may be allelic or linked to the *es-M-3* mutant. The observation that the *es-M-1 ery-M-1b* double mutant still contains the mutant form of protein LC 6 suggests that the *es-M-1* mutant may alter a second ribosomal protein which restores the wild-type antibiotic sensitive phenotype in combination with the mutant form of the LC 6 protein present in the double mutant. The *ery-M-2* locus, which is also nuclear, may specify a different protein of the chloroplast large ribosomal subunit (Mets and Bogorad, 1972), but here the evidence is not so compelling. Nuclear mutations to streptomycin resistance (*sr-1*) are all allelic and confer resistance on chloroplast ribosomes *in vitro* (Table I) (Gillham *et al.*, 1976; Harris *et al.*, 1977).

Chloroplast gene mutations confer resistance directly on chloroplast ribosomes to the antibiotics streptomycin, spectinomycin, neamine, erythromycin, carbomycin, and cleocin (Conde *et al.*, 1975; Harris *et al.*, 1976, 1977; Schlanger and Sager, 1974). In addition, mutants dependent on streptomycin and neamine are known (Sager, 1972; Adams *et al.*, 1976), and one streptomycin-dependent mutant has been shown to confer antibiotic resistance on chloroplast ribosomes *in vitro* (Table I) (Harris *et al.*, 1977).

By mapping and allele testing, Boynton *et al.* (1976) have identified four chloroplast loci for streptomycin resistance, two for erythromycin resistance, and one for spectinomycin/neamine resistance. Allele tests of the cleocin and carbomycin mutations with respect to each other and to other chloroplast mutations to antibiotic resistance remain to be done. The same is true of the streptomycin- and neamine-dependent mutants. Present evidence suggests that all these antibiotic-resistant and -dependent mutants map in a single linkage group in the chloroplast genome (Sager, 1972; Adams *et al.*, 1976; Singer *et al.*, 1976; Boynton *et al.*, 1976; Harris *et al.*, 1977).

Perhaps the simplest and most direct method for demonstrating that these antibiotic-resistant mutations directly affect chloroplast ribosomes is an *in vitro* assay in which isolated ribosomes are assayed for their ability to incorporate a particular radioactive amino acid in response to a given synthetic polynucleotide in the presence of the antibiotic. This method was first adapted from the *Escherichia coli* system as a polyuridylic acid/phenylalanine assay for streptomycin, spectinomycin, neamine, carbomycin, and cleocin resistance of chloroplast ribosomes by Schlanger and Sager (1974). Subsequently the method was modified to include the polyuridylic acid/isoleucine misreading reaction for streptomycin resistance and a polycytidilic acid/proline assay for erythromycin resistance by Conde *et al.* (1975). Schlanger and Sager (1974) have also done subunit exchange experiments between mutant and wild-type chloroplast ribosomes and then analyzed these "hybrid" chloroplast ribosomes in the polyuridylic acid/phenylalanine system to localize the site of antibiotic resistance to either the small or large ribosomal subunit. Table I summarizes results of Schlanger and Sager (1974), Conde *et al.* (1975), and Harris *et al.* (1977), who have now used this system to assess the drug resistance of chloroplast ribosomes from many nuclear and chloroplast gene mutants of *Chlamydomonas* that are resistant to antibiotics known to inhibit protein synthesis on bacterial ribosomes. Several significant conclusions can be drawn from these data. First, many, if not all, of the antibiotic-resistant mutants isolated in *Chlamydomonas* confer resistance directly on chloroplast ribosomes. Second, each mutant allele has a unique phenotype and, in cases where several different chloroplast gene loci confer resistance to the same antibiotic, e.g., to streptomycin, the general level of resistance tends to be locus-specific as well. Third, in cases where subunit exchange experiments have been done, resistance to a given antibiotic is the property of the same ribosomal subunit in the chloroplast of *Chlamydomonas* and in *E. coli*. Fourth, the chloroplast genes conferring antibiotic resistance in *Chlamydomonas* show certain similarities in their map order to similar genes in *E. coli* and *B. subtilis* (Fig. 5), although more genes appear to confer streptomycin resistance on the chloroplast ribosomes of *Chlamydomonas* than on the 70 S ribosomes of either bacterial species. Where the levels of resistance of chloroplast protein synthesis in individual mutants have been assessed *in vivo*, they are found to mirror the resistance levels measured *in vitro* in every case examined (Conde *et al.*, 1975).

In several cases (Boynton *et al.*, 1973; Mets and Bogorad, 1972; Ohta *et al.*, 1975; Brügger and Boschetti, 1975) specific chloroplast gene mutants have been reported to alter particular proteins of a given chloroplast ribosomal subunit, although each of these cases needs to be substantiated further. To prove unequivocally that a specific gene codes for a given ribosomal protein, a minimal requirement is the demonstration that a series of allelic mutants at the same gene locus all affect that same protein. So far this has been done only by Davidson *et al.* (1974) for the Mendelian mutants at the *ery-M-1* locus.

In conclusion, one can certainly say that a number of chloroplast and nuclear genes involved in the biogenesis of chloroplast ribosomes now have been identified in *Chlamydomonas,* and their functions are beginning to be understood. Chloroplast rRNA is known to be a chloroplast gene product (cf. Harris *et al.*, 1976). Although some of the nuclear and chloroplast genes identified by antibiotic resistance mutations could conceivably act directly on the rRNA or its processing, their phenotypes

C. reinhardtii chloroplast

E. coli

B. subtilis

FIG. 5. Comparison of the chloroplast gene map of *Chlamydomonas reinhardtii* with that for *Escherichia coli* and *Bacillus subtilis* in the region where antibiotic-resistant mutations that alter 70 S ribosomes fall. Antibiotic resistance is designated as follows: erythromycin = er, ery; neamine = nr, nea; spectinomycin = spr, spc; streptomycin = sr, str; ribosomal ambiguity = ram; neomycin/kanamycin = nek; lincomycin = lin; kanamycin = kan; fusidic acid = fus; micrococcin = mic. The specific ribosomal proteins affected by the mutations in *E. coli* are shown with the designations *rps* and S for small subunit and *rpl* and L for large subunit. Figure is taken from Harris *et al.* (1977); *E. coli* map, from Bachmann *et al.* (1976); *B. subtilis* map from Jaskunas *et al.* (1974).

are strikingly similar to bacterial mutations known to alter ribosomal proteins. In each case examined the same ribosomal subunit is involved in conferring resistance to a given antibiotic in both the *Chlamydomonas* chloroplast and in *E. coli* and *B. subtilis*. Assuming, therefore, that chloroplast ribosomal proteins may be products of both nuclear and chloroplast genes, there would appear to be no correlation between the location of the gene and either the ribosomal subunit affected or the type of antibiotic resistance conferred. To account for the ability of mutations at any one of four gene loci (two nuclear and two chloroplast) to confer erythromycin resistance on chloroplast ribosomes, and of mutations at any one of five gene loci (one nuclear and four chloroplast) to confer streptomycin resistance, ribosomal "domains" have been postulated to exist on the large and small subunits, respectively, where these two drugs bind and block protein synthesis (Conde *et al.*, 1975). Assuming that these ribosomal domains are composed of a number of proteins, each of which is coded by a separate gene, an appropriate alteration of any *one* of these proteins by gene mutation must be sufficient to change the conformation of the binding site and thus confer antibiotic resistance on the chloroplast ribosome. Alternatively, certain of the antibiotic resistance genes could code for enzymes which process ribosomal proteins coded by another gene. In this case, resistance would result from improper processing of a given ribosomal protein rather than from a direct change in its primary structure.

B.   Sites of Translation of Chloroplast Ribosomal Proteins

Nothing is known about the sites of synthesis of any individual chloroplast ribosomal protein. Honeycutt and Margulies (1973) have obtained the general impression that the bulk of chloroplast ribosomal proteins are made in the cytoplasm of *Chlamydomonas* using experiments analogous to those carried out by Küntzel (1969) and Neupert *et al.* (1969) for *Neurospora* mitochondrial ribosomes. Chloramphenicol and cycloheximide were added to exponentially growing cells of the *arg-1* mutant of *C. reinhardtii*, and the incorporation of [^{14}C]arginine into chloroplast and cytoplasmic ribosomes was followed. In the presence of chloramphenicol, many of the chloroplast ribosomes become bound to the thylakoid membranes by nascent polypeptide chains because chloramphenicol prevents translocation of chloroplast ribosomes along the message. Since the ribosomes were isolated from supernatants of whole-cell extracts in which the membrane fragments had been pelleted out by centrifugation, the supernatants from the chloramphenicol-treated cells were depleted of chloroplast ribosomes. To overcome this problem Honeycutt and Margulies (1973) included a 4-hour posttreatment incubation in which the cells labeled in the presence of chloramphenicol were subsequently incubated in the absence of chloramphenicol and labeled arginine. During this period chloroplast ribosomes "frozen" onto the thylakoid membranes by chloramphenicol were slowly released into the supernatant.

Honeycutt and Margulies (1973) detected no change in relative specific activities of chloroplast ribosomes labeled in the presence of chloramphenicol. On the other hand, cycloheximide appeared to block incorporation into chloroplast ribosomes virtually completely. Honeycutt and Margulies (1973) concluded that most chloroplast ribosomal proteins were probably made on cytoplasmic ribosomes. Although this conclusion may well be correct, these experiments in no way rule out the possibility that some chloroplast ribosomal proteins are made in the chloroplast, as Honeycutt and Margulies (1973) are careful to point out.

Conde *et al.* (1975) have also observed that chloroplast ribosomes continue to be made for several cell generations in the chloroplast mutant *spr-u-1-27-3* under conditions where chloroplast protein synthesis is almost totally blocked and the CO_2 fixing enzyme RubPCase dilutes out with each cell doubling. While these data support the conclusion of Honeycutt and Margulies (1973), they also do not rule out the possibility that certain proteins are made in the chloroplast, and that the chloroplast ribosomes may assemble, but not necessarily function, without these proteins.

In contrast, Ellis and Hartley (1971) have studied the accumulation of chloroplast ribosomes in greening pea apices in the presence of the chloroplast protein synthesis inhibitor lincomycin. Chloroplast rRNA continued to be made in the presence of lincomycin, but accumulation of chloroplast ribosomes was blocked under these conditions. They, therefore, suggested that some chloroplast ribosomal proteins were probably made on chloroplast ribosomes and that the newly synthesized chloroplast rRNA was degraded because it could not be assembled into mature chloroplast ribosomes owing to the absence of specific ribosomal proteins.

In the case of ribosomal proteins of yeast mitochondria, most if not all appear to be translated in the cytoplasm and then imported into the mitochondrion (Groot, 1974). In *Neurospora* this is true for all but one mitochondrial ribosomal protein (Lizardi and Luck, 1972; Lambowitz *et al.*, 1976). However, in neither organism is there any definitive evidence regarding the site of information coding for these proteins, although the general assumption appears to be that they are coded by nuclear genes.

Obviously for the sake of simplicity, one is tempted to conclude that organelle ribosomal proteins coded by nuclear DNA should be translated by the cytoplasmic protein synthesizing system whereas proteins coded by organelle DNA should be translated by the organelle protein-synthesizing system. Although the data on sites of synthesis of chloroplast ribosomal proteins in *Chlamydomonas* are still rather sketchy, they do suggest that, as in the case of the yeast and *Neurospora* mitochondrial ribosomes, most of the proteins are translated in the cytoplasm. Yet, the chloroplast mutants of *Chlamydomonas* resulting in drug resistance of chloroplast ribosomes, very likely fall in structural genes for ribosomal proteins. The notion that chloroplast ribosomal proteins coded by the chloroplast genome of *Chlamydomonas* are translated in the cytoplasm and imported back into the chloroplast becomes intellectually more palatable in light of the hypothesis of Boynton *et al.* (1973) and of Conde *et al.* (1975), which postulates that these proteins are shared between the chloroplast and the mitochondrial ribosomes. This hypothesis also accounts for how *Chlamydomonas* can produce one-step mutations that appear to confer antibiotic resistance on both the chloroplast and the mitochondrial protein-synthesizing systems. The chief objection to this hypothesis is that it requires an export of mRNA from the chloroplast to the cytoplasm, and no strong precedent exists in the literature for such transport of mRNA across organelle membranes.

C. Genetic Control and Sites of Synthesis of Chloroplast Aminoacyl-tRNA Synthetases

Barnett *et al*. (1969) found that when dark-grown cells of *Euglena* which contain proplastids were transferred to the light, new chromatographic species of isoleucyl, phenylalanyl, and glutamyl tRNAs were formed during the course of chloroplast development. None of these tRNA species were found in light-grown cells of the permanently bleached mutant W₃BUL, which lacks chloroplast DNA; so the authors concluded that the induction of these tRNAs depended on the cells' ability to grow photosynthetically. These tRNAs are now known to be coded by the plastid genome (Schwartzbach *et al*., 1976). Reger *et al*. (1970) then studied the formation of the aminoacyl-tRNA synthetases for these tRNAs during greening and found that there were two isoleucyl-tRNA synthetases. One was present constitutively in light- and dark-grown cells, and aminoacylated only the cytoplasmic species of isoleucyl-tRNA. The second enzyme was light inducible, localized in the chloroplast, and acylated only the light-inducible chloroplast isoleucyl-tRNA. In contrast, all three of the phenylalanyl-tRNA synthetases were constitutive in dark-grown *Euglena* cells. However, one of these (synthetase I) could aminoacylate the light-inducible phenylalanyl-tRNA and was found in the chloroplast. Hecker *et al*. (1974) then demonstrated that synthetase I was really a mixture of the mitochondrial phenylalanyl-tRNA synthetase and the chloroplast phenylalanyl-tRNA synthetase. Although the chloroplast enzyme was present constitutively in dark-grown cells, it clearly increased in amount when these cells were transferred to the light. The chloroplast phenylalanyl-tRNA synthetase and also a valyl-tRNA synthetase specific for the chloroplast were found to be present in the bleached mutant W₃BUL, which lacks chloroplast DNA, but the isoleucyl-tRNA synthetase could not be found in the mutant. This suggests that the former two tRNA synthetases are coded by nuclear genes and translated in the cytoplasm. Further evidence that the phenylalanyl- and valyl-tRNA synthetases were synthesized on cytoplasmic ribosomes was provided by Hecker *et al*. (1974), who showed that the synthesis of these enzymes was inhibited by cycloheximide, but not by streptomycin, which is a chloroplast protein-synthesis inhibitor in *Euglena*.

Parthier (1973) has also studied the sites of synthesis of the chloroplast aminoacyl-tRNA synthetases of *Euglena* using a somewhat different approach. He employed the regreening *Euglena* system to detect light-inducible synthetases, but instead of assaying these synthetases with chloroplast tRNA from *Euglena*, he used in the assay tRNA extracted from the blue-green alga *Anacystis nidulans*. The obvious assumption is that prokaryotic tRNA obtained from this alga should be more similar to chloroplast tRNA than would be *Euglena* cytoplasmic tRNA. To assay for increases in cytoplasmic synthetases during regreening, Parthier used tRNA extracted from a permanently bleached mutant lacking chloroplast DNA. While low concentrations of cycloheximide were found to stimulate both chlorophyll accumulation and synthesis of the light-inducible leucyl-tRNA synthetase, higher cycloheximide concentrations were found to inhibit both. The levels of plastid leucyl-tRNA synthetases were similar in cells grown in the presence or in the absence of chloramphenicol, but addition of an inhibitory concentration of cycloheximide in either case blocked the enzyme increase. Parthier also obtained the same results with the other light-induced synthetases (i.e., for aspartyl, lysyl, phenylalanyl, and valyl tRNAs) as well as for the synthetases not induced by light (i.e., for arginyl, isoleucyl, seryl, and threonyl tRNAs). Thus all the aforementioned experiments point to most light-stimulated aminoacyl-tRNA synthetases of *Euglena* being localized in the chloroplast and being made on cytoplasmic ribosomes. In at least two cases, these enzymes are also known to be coded by nuclear genes. Thus during chloroplast development in *Euglena*, synthesis of chloroplast specific tRNAs coded by the chloroplast and their aminoacyl synthetases coded by the nuclear genome occurs coordinately. The translation of the mRNAs for these synthetases occurs on cytoplasmic ribosomes, and the synthetases are then imported into the chloroplast.

Nothing is presently known about how this light-triggered regulation of chloroplast tRNAs and their synthetases is achieved in *Euglena*. Goins *et al.* (1973) have observed that, in a golden mutant of *Euglena* (G_1BU) with reduced levels of chlorophyll, the light-inducible isoleucyl and methionyl tRNAs of the plastid are formed constitutively in the dark. In some direct or indirect way, the G_1BU mutation appears to block the normal regulation of these two light-induced tRNAs in the dark, but there is no information as to whether this is true of their synthetases as well.

V. Concluding Remarks

Evidence has been presented in this review that both chloroplast and nuclear genes code for the large and small subunits comprising fraction I protein, for certain of the thirty or more thylakoid membrane polypeptides, and for a few of the many chloroplast ribosomal proteins. In the case of fraction I protein, the inheritance patterns of peptide differences observed in interspecific F_1 hybrids of *Nicotiana* strongly suggest that the structural gene(s) for the large subunit is in the chloroplast and the structural gene(s) for the small subunit is in the nucleus. In addition, Howell and his colleagues have demonstrated directly that large subunit mRNA is coded by a specific restriction fragment of chloroplast DNA in *Chlamydomonas*. In the case of the genetic control of thylakoid membrane polypeptides, the involvement of two chloroplast genes, *thm-u-1* and C_1, has been demonstrated in *Chlamydomonas* and another two, *en:alba-1* and *en:viridis-1*, in *Antirrhinum*. Experiments with bleached mutants of *Euglena* suggest that the large majority of thylakoid membrane polypeptides are likely to be coded by nuclear genes whose messages are translated in the cytoplasm. However, a systematic approach to the problem in any one organism where mutations in the structural genes coding for specific thylakoid membrane polypeptides can be identified in both nucleus and chloroplast remains to be undertaken. The coding of specific chloroplast ribosomal proteins is only slightly better established. Unequivocal evidence exists that one nuclear gene codes for the structure of one protein in the large subunit of the chloroplast ribosome. Much indirect evidence supports the tentative conclusion that the genes coding for the remaining chloroplast ribosomal proteins are distributed among the nuclear and chloroplast genomes. The same shared genetic control of thylakoid membrane polypeptides is highly likely also, but here the evidence is far more indirect than in the case of the chloroplast ribosomal proteins.

The same dichotomy seen in the location of structural genes for the aforementioned polypeptides is also reflected in their sites of translation. Only in the case of fraction I protein does reasonably convincing evidence exist regarding both the location of the structural genes for the component polypeptides and their sites of translation. In the case of thylakoid membrane polypeptides, a fairly complete picture of their sites of synthesis exists, but very little is known regarding the location of their structural genes. Nevertheless, we have the technology presently at hand to determine both the location of many of these structural genes and the sites of translation of their mRNAs. In addition to the more traditional methods of genetic and biochemical analysis of specific mutants presently being used to unravel the problem in *Chlamydomonas*, one can now attack the question in both *Chlamydomonas* and higher plants by isolating mRNAs for specific chloroplast components and hybridizing them to individual restriction fragments of chloroplast DNA. Very rapid progress is to be expected in this area over the next few years. In fact, the first chloroplast gene has already been successfully cloned by Gelvin *et al.* (1977) in a chimeric bacterial plasmid, although the authors have yet to demonstrate expression of the cloned gene *in vivo*. Before long it should be known whether the "orthodox" transcription/translation solution which the cell seems to have evolved for synthesis of fraction I protein and for mitochondrial proteins as well (see Bücher *et al.*, 1976) also applies to all other protein complexes in the chloroplast where there is a division of labor between the organelle, the nucleus, and the cytoplasm. Whether all plants have evolved the same solution with respect to the location of structural genes for specific chloroplast components and the sites of translation of these components remains to be seen.

ACKNOWLEDGMENTS

This review was supported in part by NIH Grant GM-19427 to J. E. B. and N. W. G., by NIH Grant GM-21060 to N.-H. C., and by NIH RCDA Awards GM-70437 to N. W. G., GM-70453 to J. E. B., and GM-00223 to N.-H. C.

REFERENCES

Adams, G. M. W., Van Winkle-Swift, K. P., Gillham, N. W., and Boynton, J. E. (1976). *In* "The Genetics of Algae" (R. A. Lewin, ed.), pp. 69–118. Univ. of California Press, Berkeley.

Alberte, R. S., Hesketh, J. D., and Kirby, J. S. (1976). *Z. Pflanzenphysiol.* **77,** 152–159.

Allen, M. B. (1971). *In* "Photosynthesis and Nitrogen Fixation," Part A (A. San Pietro, ed.), Methods in Enzymology, Vol. 23, pp. 168–171. Academic Press, New York.

Anderson, J. M., and Levine, R. P. (1974a). *Biochim. Biophys. Acta* **333,** 378–387.

Anderson, J. M., and Levine, R. P. (1974b). *Biochim. Biophys. Acta* **357,** 118–126.

Armstrong, J. J., Surzycki, S. J., Moll, B., and Levine, R. P. (1971). *Biochemistry* **10,** 692–701.

Bachmann, B. J., Low, K. B., and Taylor, A. L. (1976). *Bacteriol. Rev.* **40,** 116–167.

Bachmann, M. D., Robertson, D. S., and Bowen, C. C. (1969). *J. Ultrastruct. Res.* **28,** 435–451.

Bachmann, M. D., Robertson, D. S., Bowen, C. C., and Anderson, I. C. (1973). *J. Ultrastruct. Res.* **45,** 384–406.

Badger, M. R., and Andrews, T. J. (1974). *Biochem. Biophys. Res. Commun.* **60,** 204–210.

Barnett, W. E., Pennington, C. J., and Fairfield, S. A. (1969). *Proc. Natl. Acad. Sci. U.S.A.* **63,** 1261–1268.

Bar-Nun, S., and Ohad, I. (1975). *Proc. Int. Congr. Photosynth., 3rd, Rehovot, 1974* Vol. 3, 1627–1638.

Baur, E. (1909). *Z. Vererbungsl.* **1,** 330–351.

Bedbrook, J. R., and Bogorad, L. (1976a). *Proc. Natl. Acad. Sci. U.S.A.* **73,** 4309–4313.

Bedbrook, J. R., and Bogorad, L. (1976b). *In* "Genetics and Biogenesis of Chloroplasts and Mitochondria" (T. Bücher, W. Neupert, W. Sebald, and S. Werner, eds.), pp. 369–373. Elsevier/North-Holland Biomed. Press, Amsterdam.

Behn, W., and Herrmann, R. G. (1977). *Mol. Gen. Genet.* **157,** 25–30.

Benedict, C. R., and Kohel, R. J. (1968). *Plant Physiol.* **43,** 1611–1616.

Bennoun, P., and Chua, N.-H. (1976). *In* "Genetics and Biogenesis of Chloroplasts and Mitochondria" (T. Bücher, W. Neupert, W. Sebald, and S. Werner, eds.), pp. 33–39. Elsevier/North-Holland Biomed. Press, Amsterdam.

Bennoun, P., Girard, J., and Chua, N.-H. (1977). *Mol. Gen. Genet.* **153,** 343–348.

Bingham, S., and Schiff, J. A. (1976). *In* "Genetics and Biogenesis of Chloroplasts and Mitochondria" (T. Bücher, W. Neupert, W. Sebald, and S. Werner, eds.), pp. 79–86. Elsevier/North-Holland Biomed. Press, Amsterdam.

Birky, C. W., Jr., Perlman, P. S., and Byers, T. J., eds. (1975). "Genetics and Biogenesis of Mitochondria and Chloroplasts," 361 pp. Ohio State Univ. Press, Columbus.

Bishop, N. I. (1971a). *Annu. Rev. Biochem.* **40**, 197–226.

Bishop, N. I. (1971b). *In* "Photosynthesis and Nitrogen Fixation," Part A (A. San Pietro, ed.), Methods in Enzymology, Vol 23, pp. 130–143. Academic Press, New York.

Blair, G. E., and Ellis, R. J. (1973). *Biochim. Biophys. Acta* **319**, 223–234.

Bottomley, W., Spencer, D., and Whitfeld, P. R. (1974). Protein synthesis in isolated spinach chloroplasts: comparison of light-driven and ATP-driven synthesis. *Arch. Biochem. Biophys.* **164**, 106–117.

Bourque, D. P., and Wildman, S. G. (1973). *Biochim. Biophys. Res. Commun.* **50**, 532–537.

Bourque, D. P., Boynton, J. E., and Gillham, N. W. (1971). *J. Cell Sci.* **8**, 153–183.

Boynton, J. E. (1966). *Hereditas* **56**, 238–254.

Boynton, J. E., Gillham, N. W., and Chabot, J. F. (1972). *J. Cell Sci.* **10**, 267–305.

Boynton, J. E., Burton, W., Gillham, N. W., and Harris, E. H. (1973). *Proc. Natl. Acad. Sci. U.S.A.* **70**, 3463–3467.

Boynton, J. E., Gillham, N. W., Harris, E. H., Tingle, C. L., Van Winkle-Swift, K., and Adams, G. M. W. (1976). *In* "Genetics and Biogenesis of Chloroplasts and Mitochondria" (T. Bücher, W. Neupert, W. Sebald, and S. Werner, eds.), pp. 313–322. Elsevier/North-Holland Biomed. Press, Amsterdam.

Brügger, M., and Boschetti, A. (1975). *Eur. J. Biochem.* **58**, 603–610.

Bücher, T., Neupert, W., Sebald, W., and Werner, S., eds. (1976). "Genetics and Biogenesis of Chloroplasts and Mitochondria." 895 pp. Elsevier/North-Holland Biomed. Press, Amsterdam.

Cashmore, A. R. (1976). *J. Biol. Chem.* **251**, 2848–2853.

Chan, P.-H., and Wildman, S. G. (1972). *Biochim. Biophys. Acta* **277**, 677–680.

Chen, K., Gray, J. C., and Wildman, S. G. (1975). *Science* **190**, 1304–1306.

Chen, K., Johal, S., and Wildman, S. G. (1977). *In* "Nucleic Acids and Protein Synthesis in Plants" (J. H. Weil and L. Bogorad, eds.), pp. 183–194. Plenum, New York.

Chen, K., Sarjit, J., and Wildman, S. G. (1976). *In* "Genetics and Biogenesis of Chloroplasts and Mitochondria" (T. Bücher, W. Neupert, W. Sebald, and S. Werner, eds.), pp. 3–11. Elsevier/North Holland Biomed. Press, Amsterdam.

Chua, N.-H. (1976). *In* "Genetics and Biogenesis of Chloroplasts and Mitochondria" (T. Bücher, W. Neupert, W. Sebald, and S. Werner, eds.), pp. 323–330. Elsevier/North-Holland Biomed. Press, Amsterdam.

Chua, N.-H., and Bennoun, P. (1975). *Proc. Natl. Acad. Sci. U.S.A.* **72**, 2175–2179.

Chua, N.-H., and Gillham, N. W. (1977). *J. Cell Biol.* **74**, 441–452.

Chua, N.-H., and Luck, D. J. L. (1974). *In* "Ribosomes" (M. Nomura, A. Tissières, and P. Lengyel, eds.), pp. 519–539. Cold Spring Harbor Lab., New York.

Chua, N.-H., Matlin, K., and Bennoun, P. (1975). *J. Cell Biol.* **67**, 361–377.

Conde, M. F., Boynton, J. E., Gillham, N. W., Harris, E. H., Tingle, C. L., and Wang, W. L. (1975). *Mol. Gen. Genet.* **140**, 183–220.

Correns, C. (1909). *Z. Vererbungsl.* **1**, 291–329.

Criddle, R. S., Dau, B., Kleinkopf, G. E., and Huffaker, R. C. (1970). *Biochem. Biophys. Res. Commun.* **41**, 621–627.

Davidson, J. N., Hanson, M. R., and Bogorad, L. (1974). *Mol. Gen. Genet.* **132**, 119–129.

Dobberstein, B., Blobel, G., and Chua, N.-H. (1977). *Proc. Natl. Acad. Sci. U.S.A.* **74**, 1082–1085.

Eaglesham, A. R. J., and Ellis, R. J. (1974). *Biochim. Biophys. Acta* **335**, 396–407.

Ellis. R. J. (1975a). *Phytochemistry* **14**, 89–93.

Ellis, R. J. (1975b). *In* "Membrane Biogenesis: Mitochondria, Chloroplasts and Bacteria" (A. Tzagoloff, ed.), pp. 247–276. Plenum, New York.

Ellis, R. J., and Hartley, M. R. (1971). *Nature (London), New Biol.* **233**, 193–196.

Eytan, G., and Ohad, I. (1972a). *J. Biol. Chem.* **247**, 112–121.

Eytan, G., and Ohad, I. (1972b). *J. Biol. Chem.* **247**, 122–129.

Gelvin, S., and Howell, S. H. (1977). *Plant Physiol.* **59**, 471–477.

Gelvin, S., Heizmann, P., and Howell, S. H. (1977). *Proc. Natl. Acad. Sci. U.S.A.* **74**, 3193–3197.

Gibor, A., and Herron, H. A. (1968). *In* "The Biology of *Euglena*" (D. E. Buetow, ed.), Vol. II, pp. 335–349. Academic Press, New York.

Gillham, N. W., Boynton, J. E., Harris, E. H., Fox, S. B., and Bolen, P. L. (1976). *In* "Genetics and Biogenesis of Chloroplasts and Mitochondria" (T. Bücher, W. Neupert, W. Sebald, and S. Werner, eds.), pp. 69–76. Elsevier/North-Holland Biomed. Press, Amsterdam.

Givan, A. L., and Criddle, R. S. (1972). *Arch. Biochem. Biophys.* **149**, 153–163.

Goins, D. J., Reynolds, R. J., Schiff, J. A., and Barnett, W. E. (1973). *Proc. Natl. Acad. Sci. U.S.A.* **70**, 1749–1752.

Goodenough, U. W., and Levine, R. P. (1969). *Plant Physiol.* **44**, 990–1000.

Goodenough, U. W., and Levine, R. P. (1970). *J. Cell Biol.* **44**, 547–562.

Goodenough, U. W., and Staehelin, L. A. (1971). *J. Cell Biol.* **48**, 594–619.

Gooding, L. R., Roy, H., and Jagendorf, A. T. (1973). *Arch. Biochem. Biophys.* **159**, 324–335.

Gough, S. (1972). *Biochim. Biophys. Acta* **286**, 36–54.

Granick, S. (1971). *In* "Photosynthesis and Nitrogen Fixation," Part A (A. San Pietro, ed.), Methods in Enzymology, Vol. 23, pp. 162–168. Academic Press, New York.

Gray, J. C., and Kekwick, R. G. O. (1974). *Eur. J. Biochem.* **44**, 491–500.

Groot, G. S. P. (1974). *In* "The Biogenesis of Mitochondria" (A. M. Kroon and C. Saccone, eds.), pp. 443–452. Academic Press, New York.

Harris, E. H., Boynton, J. E., and Gillham, N. W. (1974). *J. Cell Biol.* **63**, 160–179.

Harris, E. H., Boynton, J. E., and Gillham, N. W. (1976). *In* "The Genetics of Algae" (R. A. Lewin, ed.), pp. 119–144. Univ. of California Press, Berkeley.

Harris, E. H., Boynton, J. E., Gillham, N. W., Tingle, C. L., and Fox, S. B. (1977). *Mol. Gen. Genet.* **155**, 249–265.

Hartley, M. R., Wheeler, A., and Ellis, R. J. (1975). *J. Mol. Biol.* **91**, 67–77.

Hecker, L. I., Egan, J., Reynolds, R. J., Nix, C. E., Schiff, J. A., and Barnett, W. E. (1974). *Proc. Natl. Acad Sci. U.S.A.* **71**, 1910–1914.

Henningsen, K. W., and Boynton, J. E. (1974). *J. Cell Sci.* **15**, 31–55.

Henningsen, K. W. Boynton, J. E., von Wettstein, D., and Boardman, N. K. (1973). *In* "The Biochemistry of Gene Expression in Higher Organisms" (J. K. Pollak and J. W. Lee, eds.), pp. 457–478. Aust. N. Z. Book Co., Sydney.

Herrmann, F. (1971a). *FEBS Lett.* **19**, 267–269.

Herrmann, F. (1971b). *Exp. Cell Res.* **70**, 452–453.

Herrmann, R. G., Bohnert, H.-J., Driesel, A., and Hobom, G. (1976). *In* "Genetics and Biogenesis of Chloroplasts and Mitochondria" (T. Bücher, W. Neupert, W. Sebald, and S. Werner, eds.), pp. 351–359. Elsevier/North Holland Biomed. Press, Amsterdam.

Herrmann, R. G., Bohnert, H.-J., Kowallik, K., and Schmitt, J. M. (1975). *Biochim. Biophys. Acta* **378**, 305–317.

Highkin, H. R., Boardman, N. K., and Goodchild, D. J. (1969). *Plant Physiol.* **44**, 1310–1320.

Honeycutt, R. C., and Margulies, M. M. (1973). *J. Biol. Chem.* **248**, 6145–6153.

Hoober, J. K. (1970). *J. Biol. Chem.* **245**, 4327–4334.

Hoober, J. K. (1972). *J. Cell Biol.* **52**, 84–96.

Hoober, J. K. (1976). *In* "Genetics and Biogenesis of Chloroplasts and Mitochondria" (T. Bücher, W. Neupert, W. Sebald, and S. Werner, eds.), pp. 87–94. Elsevier/North-Holland Biomed. Press, Amsterdam.

Hoober, J. K., and Stegeman, W. J. (1973). *J. Cell Biol.* **56**, 1–12.

Hoober, J. K., and Stegeman, W. J. (1975). *In* "Genetics and Biogenesis of Mitochondria and Chloroplasts" (C. W. Birky, Jr., P. S. Perlman, and T. J. Byers, eds.), pp. 225–251. Ohio State Univ. Press, Columbus.

Hoober, J. K., Siekevitz, P., and Palade, G. E. (1969). *J. Biol. Chem.* **244**, 2621–2631.

Horak, A., and Hill, R. D. (1971). *Can. J. Biochem.* **49**, 207–209.

Horak, A., and Hill, R. D. (1972). *Plant Physiol.* **49**, 365–370.

Howell, S. H., Heizmann, P., and Gelvin, S. (1976). *In* "Genetics and Biogenesis of Chloroplasts and Mitochondria" (T. Bücher, W. Neupert, W. Sebald, and S. Werner, eds.), pp. 625–628. Elsevier/North-Holland Biomed. Press, Amsterdam.

Howell, S. H., Heizmann, P., Gelvin, S., and Walker, L. (1977). *Plant Physiol.* **59**, 464–470.

Hudock, G. A., and Levine, R. P. (1964). *Plant Physiol.* **39**, 889–897.

Hudock, G. A., McLeod, G. C., Moravkova-Kiely, J., and Levine, R. P. (1964). *Plant Physiol.* **39**, 898–906.

Iwanij, V., Chua, N.-H., and Siekevitz, P. (1975). *J. Cell Biol.* **64**, 572–585.

Jaskunas, S. R., Nomura, M., and Davies, J. (1974). *In* "Ribosomes" (M. Nomura, A. Tissières, and P. Lengyel, eds.), pp. 333–368. Cold Spring Harbor Lab., New York.

Joy, K. W., and Ellis, R. J. (1975). *Biochim. Biophys. Acta* **378**, 143–151.

Kawashima, N., and Wildman, S. G. (1970). *Annu. Rev. Plant Physiol.* **21**, 325–358.

Kawashima, N., and Wildman, S. G. (1972). *Biochim. Biophys. Acta.* **262**, 42–49.

Kawashima, N. Kwok, S.-H., and Wildman, S. G. (1971). *Biochim. Biophys. Acta* **236**, 578–586.

Kirk, J. T. O. (1972). *Sub-Cell. Biochem.* **1**, 333–361.

Kirk, J. T. O., and Tilney-Bassett, R. A. E. (1967). "The Plastids," 607 pp. Freeman, San Francisco, California.

Kolodner, R., and Tewari, K. K. (1975). *Biochim. Biophys. Acta* **402**, 372–390.

Kretzer, F., Ohad, I., and Bennoun, P. (1976). *In* "Genetics and Biogenesis of Chloroplasts and Mitochondria" (T. Bücher, W. Neupert, W. Sebald, and S. Werner, eds.), pp. 25–32. Elsevier/North-Holland Biomed. Press, Amsterdam.

Küntzel, H. (1969). *Nature (London)* **222**, 142–146.

Kung, S.-D. (1976). *Science* **191**, 429–434.

Kung, S.-D., and Marsho, T. V. (1976). *Nature (London)* **259**, 325–326.

Kung, S.-D., and Thornber, J. P. (1971). *Biochim. Biophys. Acta* **253**, 285–289.

Kung, S.-D., Thornber, J. P., and Wildman, S. G. (1972). *FEBS Lett.* **24**, 185–188.

Kung, S.-D., Sakano, K., and Wildman, S. G. (1974). *Biochim. Biophys. Acta* **365**, 138–147.

Kung, S.-D., Gray, J. C. Wildman, S. G., and Carlson, P. S. (1975a) *Science* **187**, 353–355.

Kung, S.-D., Sakano, K., Gray, J. C., and Wildman, S. G. (1975b). *J. Mol. Evol.* **7**, 59–64.

Kwanyuen, P., and Wildman, S. G. (1975). *Biochim. Biophys. Acta* **405**, 167–174.

Lambowitz, A. M., Chua, N.-H., and Luck, D. J. L. (1976). *J. Mol. Biol.* **107**, 223–253.

Lambowitz, A. M., Merril, C. R., Wurtz, E. A., Boynton, J. E., and Gillham, N. W. (1976). *J. Cell. Biol.* **70**, 217a.

Lefort, M. (1959). *Rev. Cytol. Biol. Veg.* **20**, 1–160.

Levine, R. P. (1968). *Science* **162**, 768–771.

Levine, R. P. (1969). *Annu. Rev. Plant Physiol.* **20**, 523–540.

Levine, R. P. (1971). *In* "Photosynthesis and Nitrogen Fixation," Part A (A. San Pietro, ed.), Methods in Enzymology, Vol. 23, pp. 119–129. Academic Press, New York.

Levine, R. P., and Duram, H. A. (1973). *Biochim. Biophys. Acta* **325**, 565–572.

Levine, R. P., and Goodenough, U. W. (1970). *Annu. Rev. Genet.* **4**, 397–408.

Levine, R. P., Burton, W. G., and Duram, H. A. (1972). *Nature (London), New Biol.* **237**, 176–177.

Lizardi, P. M., and Luck, D. J. L. (1972). *J. Cell Biol.* **54**, 56–74.

Lorimer, G. H., Badger, M. R., and Andrews, T. J. (1976). *Biochemistry* **15**, 529–536.

Machold, O., and Aurich, O. (1972). *Biochim. Biophys. Acta* **281**, 103–112.

McLeod, G. C., Hudock, G. A., and Levine, R. P. (1963). *In* "Photosynthetic Mechanisms of Green Plants," Publ. No. 1145, pp. 400–408. Nat. Acad. Sci.—Nat. Res. Counc. Washington, D.C.

Margulies, M. (1971). *Biochem. Biophys. Res. Commun.* **44**, 539–545.

Mendiola-Morgenthaler, L. R., Morgenthaler, T. T., and Price, C. A. (1976). *FEBS Lett.* **62**, 96–100.

Mets, L., and Bogorad, L. (1972). *Proc. Natl. Acad. Sci. U.S.A.* **69**, 3779–3783.

Miles, C. D., and Daniel, D. J. (1974). *Plant Physiol.* **53**, 589–595.

Morgenthaler, J. J., and Mendiola-Morgenthaler, L. (1976). *Arch. Biochem. Biophys.* **172**, 51–58.

Nasyrov, Y. S., and Šesták, Z., eds. (1975). "Genetic Aspects of Photosynthesis," 392 pp. Junk, The Hague.

Nelson, P. E., and Surzycki, S. J. (1976a). *Eur. J. Biochem.* **61**, 465–474.

Nelson, P. E., and Surzycki, S. J. (1976b). *Eur. J. Biochem.* **61**, 475–480.

Neupert, W., Sebald, W., Schwab, A. J., Massinger, P., and Bücher, T. (1969). *Eur. J. Biochem.* **10**, 589–591.

Nielsen, O. F. (1974). *Hereditas* **76**, 269–304.

Ohad, I. (1975). *In* "Membrane Biogenesis: Mitochondria, Chloroplasts and Bacteria" (A. Tzagoloff, ed.), pp. 279–347. Plenum, New York.

Ohad, I., Siekevitz, P., and Palade, G. E. (1967a). *J. Cell Biol.* **35**, 521–552.

Ohad, I., Siekevitz, P., and Palade, G. E. (1967b). *J. Cell Biol.* **35**, 553–584.

Ohta, N., Inouye, M., and Sager, R. (1975). *J. Biol. Chem.* **250**, 3655–3659.

Panet, R., and Sanadi, D. R. (1976). *Curr. Top. Membr. Transp.* **8**, 99–160.

Parthier, B. (1973). *FEBS Lett.* **38**, 70–74.

Penefsky, H. S. (1974) *In* "The Enzymes" (P. D. Boyer, ed.), 3rd Ed., Vol. 10, pp. 375–394. Academic Press, New York.

Poulsen, C., Strøbaek, S., and Haslett, B. G. (1976). *In* "Genetics and Biogenesis of Chloroplasts and Mitochondria" (T. Bücher, W. Neupert, W. Sebald, and S. Werner, eds.), pp. 17–24. Elsevier/North-Holland Biomed. Press, Amsterdam.

Rédei, G. P. (1973). *Mutat. Res.* **18**, 149–162.

Rédei, G. P., and Plurad, S. B. (1973). *Protoplasma* **77**, 361–380.

Reger, B. J., Fairfield, S. A., Epler, J. L., and Barnett, W. E. (1970). *Proc. Natl. Acad. Sci. U.S.A.* **67**, 1207–1213.

Röbbelen, G. (1959). *Z. Vererbungsl.* **90**, 503–506.

Rochaix, J. D. (1976). *In* "Genetics and Biogenesis of Chloroplasts and Mitochondria" (T. Bücher, W. Neupert, W. Sebald, and S. Werner, eds.), pp. 375–378. Elsevier/North-Holland Biomed. Press, Amsterdam.

Roy, H., Patterson, R., and Jagendorf, A. T. (1976). *Arch. Biochem. Biophys.* **172**, 64–73.

Sager, R. (1972). "Cytoplasmic Genes and Organelles," 405 pp. Academic Press, New York.

Sagher, D., Grosfeld, H., and Edelman, M. (1976). *Proc. Natl. Acad. Sci. U.S.A.* **73**, 722–726.

Sakano, K., Kung, S.-D., and Wildman, S. G. (1974). *Mol. Gen. Genet.* **130**, 91–97.

Schatz, G., and Mason, T. L. (1974). *Annu. Rev. Biochem.* **43**, 51–87.

Schiff, J. A., Lyman, H., and Russell, G. K. (1971). *In* "Photosynthesis and Nitrogen Fixation," Part A (A. San Pietro, ed.), Methods in Enzymology, Vol. 23, pp. 143–162. Academic Press, New York.

Schlanger, G., and Sager, R. (1974). *Proc. Natl. Acad. Sci. U.S.A.* **71**, 1715–1719.

Schmid, G. (1967). *J. Microsc. (Paris)* **6**, 485–498.

Schmid, G. (1971). *In* "Photosynthesis and Nitrogen Fixation," Part A (A. San Pietro, ed.), Methods in Enzymology, Vol. 23, pp. 171–194. Academic Press, New York.

Schmid, G., Price, J. M., and Gaffron, H. (1966). *J. Microsc. (Paris)* **5**, 205–212.

Schmidt, G. W., and Lyman, H. (1976). *In* "The Genetics of Algae" (R. A. Lewin, ed.), pp. 257–299. Univ. of California Press, Berkeley.

Schwartzbach, S. D., Hecker, L. I., and Barnett, W. E. (1976). *Proc. Natl. Acad. Sci. U.S.A.* **73**, 1984–1988.

Sebald, W., Machleidt, W., and Otto, J. (1974). *In* "The Biogenesis of Mitochondria: Transcriptional, Translational and Genetic Aspects" (A. M. Kroon and C. Saccone, eds.), pp. 453–463. Academic Press, New York.

Senger, H., and Bishop, N. I. (1972a). *Plant Cell Physiol.* **13**, 633–649.

Senger, H., and Bishop, N. I. (1972b) *Plant Cell Physiol.* **13**, 937–953.

Shiozawa, J. A., Alberte, R. S., and Thornber, J. P. (1974). *Arch. Biochem. Biophys.* **165**, 388–397.

Shumway, L. K., and Kleinhofs, A. (1973). *Biochem. Genet.* **8**, 271–280.

Shumway, L. K., and Weier, T. E. (1967). *Am. J. Bot.* **54**, 773–780.

Siddell, S. G., and Ellis, R. J. (1975). *Biochem. J.* **146**, 675–685.

Singer, B., Sager, R., and Ramanis, Z. (1976). *Genetics* **83**, 341–354.

Singh, S., and Wildman, S. G. (1973). *Mol. Gen. Genet.* **124**, 187–196.

Spiess, H., and Arnold, C. G. (1975). *Ber. Dtsch. Bot. Ges.* **88**, 391–398.

Strøbaek, S., and Gibbons, G. C. (1976). *Carlsberg Res. Commun.* **41**, 57–72.

Thornber, J. P., and Highkin, H. R. (1974). *Eur. J. Biochem.* **41**, 109–116.

Thornber, J. P., Gregory, R. P. F., Smith, C. A., and Bailey, J. L. (1967). *Biochemistry* **6**, 391–396.

Tilney-Bassett, R. A. E. (1975). *In* "Genetics and Biogenesis of Mitochondria and Chloroplasts" (C. W. Birky, Jr., P. S. Perlman, and T. J. Byers, eds.), pp. 268–308. Ohio State Univ. Press, Columbus.

Tzagoloff, A., Rubin, M. S., and Sierra, M. F. (1973). *Biochim. Biophys. Acta* **301**, 71–104.

Veleminský, J., and Röbbelen, G. (1966). *Planta* **68**, 15–35.

von Wettstein, D., Henningsen, K. W., Boynton, J. E., Kannangara, G. C., and Nielsen, O. F. (1971). *In* "Autonomy and Biogenesis of Mitochondria and Chloroplasts" (N. K. Boardman, A. Linnane, and R. M. Smillie, eds.), pp. 205–223. North-Holland Publ. Amsterdam.

von Wettstein, D., Kahn, A., Nielsen, O. F., and Gough, S. (1974). *Science* **184**, 800–802.

Walles, B. (1965). *Hereditas* **53**, 247–256.

Wang, W.-Y., Wang, W. L., Boynton, J. E., and Gillham, N. W. (1974). *J. Cell Biol.* **63**, 806–823.

Wang, W.-Y., Boynton, J. E., and Gillham, N. W. (1975). *Cell* **6**, 75–84.

Wheeler, A. M., and Hartley, M. R. (1975). *Nature (London)* **257**, 66–67.

Wildman, S. G., and Bonner, J. (1947). *Arch. Biochem.* **14**, 381–413.

Wildman, S. G., Lu-Liao, C., and Wong-Staal, F. (1973). *Planta* **113**, 293–312.

Wildman, S. G., Chen, K., Gray, J. C., Kung, S. D., Kwanyuen, P., and Sakano, K. (1975). *In* "Genetics and Biogenesis of Mitochondria and Chloroplasts" (C. W. Birky, Jr., P. S. Perlman, and T. J. Byers, eds.), pp. 309–329. Ohio State Univ. Press, Columbus.

Wong-Staal, F., and Wildman, S. G. (1973). *Planta* **113**, 313–326.

Mutations and Genetic Manipulations as Probes of Bacterial Photosynthesis

BARRY L. MARRS*

Edward A. Doisy Department of Biochemistry
Saint Louis University School of Medicine
St. Louis, Missouri

I. Introduction

There are, by and large, two communities of scientists who are uniquely interested in the photosynthetic bacteria: those who view the intracellular photosynthetic membrane system as a fertile field for the study of membrane biosynthesis and its regulation, and those who

* Supported by Public Health Service Research Career Development Award GM-00098, U. S. Public Health Service Grant GM21073, and National Science Foundation Grant PCM-19843 A01.

recognize bacterial photosynthesis as a process particularly well suited for probing many basic problems in bioenergetics. Many workers in each of these groups are engaged in biochemical research that could undoubtedly be facilitated by the use of mutants and attendant genetic techniques; however, the biochemical genetics of photosynthetic bacteria remains a remarkably underdeveloped area. Although several notable exceptions can and will be cited, the lack of exploitation of mutants in the area of bioenergetics may be attributable to a lack of familiarity with genetic techniques on the part of researchers with heavily physical training. On the other hand, although the biologists pursuing the regulation of membrane biosynthesis are presumably familiar with the genetic analyses necessary for defining and describing regulatory interactions, systems suited to these somewhat more sophisticated genetic analyses have not been available. Thus the dual goals of this article are to describe, through examples where possible, the ways in which mutants may be advantageously applied to bioenergetic problems, and to report on certain new developments in the genetics of photosynthetic bacteria which extend their usefulness in the study of membrane development. Because one goal of this article is to promote an understanding of the potentials and limitations of genetics as a tool for aiding bioenergetic researchers, the work reviewed in this area will be organized according to the techniques employed rather than the results obtained.

The excellent review by Gibson and Cox (1973) addresses many of the same topics to be considered in this chapter, using, however, the study of oxidative phosphorylation in *Escherichia coli* as the focal point for discussion. The recent review by Haddock and Jones (1977) also includes several examples and a discussion of the uses of mutants defective in energy transduction.

II. General Considerations

A. RATIONALE

"Why do you study crippled strains? It's hard enough to figure out how the normal ones work." Although most scientists today do realize that mutants can be useful, it is instructive to try to answer this (actual) question explicitly. It is simple to assert that genetics can catalyze and facilitate biochemical research, and indeed this has been pointed out to the bioenergetics community periodically for many years (Kováč et al., 1967; Bishop, 1969; Levine, 1969, 1973; Cox and Gibson, 1974; Miles, 1975), but still the tool is seldom used. The possible reasons for this situation would seem to be either that genetic techniques have been tried and found to be less useful than was anticipated, or they have not been tried. I assume that the latter is more nearly true, and further assume that one remedy for this disuse is to try to organize a logical framework within which the applications of the mutant approach can be classified and thereby be made more accessible to the nongeneticist.

Table I presents a scheme for classifying applications of mutants to biochemical and biophysical problems. The initial subdivision is into (I) applications using mutants in ways that require little or no genetic analysis beyond the initial construction and characterization of the mutants, and (II) applications in which genetic analyses play a more central role. This division is convenient because the former class of uses is compatible with a wide range of experimental approaches and is currently available, whereas the latter requires more expertise in genetics, and systems to realize these potential uses are currently only under development in photosynthetic bacteria. It must be emphasized at the outset that while mutants may be used without genetic analyses, it is usually possible to draw more information from a system when genetic studies accompany the other experimental designs.

Table I is thus a partial answer to the question: Why do you study crippled strains? i.e., mutants provide experimental material that may have a variety of advantages over wild type, and mutations can be used to probe a system in a fashion analogous to studies involving inhibitors. Mutations, however, have two major advantages over chemical inhibitors as experimental probes, namely, specificity and range. A reaction

TABLE I

APPLICATION OF MUTANTS TO BIOCHEMICAL PROBLEMS

I. Uses of mutants not relying on genetic analyses
  A. As a source of material more amenable to experimental analysis than wild type (for the following reasons)
    1. Lacking interfering substances or reactions similar to those studied
    2. Lacking interfering substances or reactions not related to those studied
    3. Showing improved incorporation of isotopically labeled compounds
    4. Containing an increased amount of a substance or system under study
    5. As reagents for *in vitro* complementation or reconstitution
  B. To assign a biological role to a protein or enzyme product
  C. To resolve and define a sequence of reactions
II. Uses of mutants requiring genetic analyses
  A. To resolve mutations contributing to a phenotype
    1. The number of mutations involved
    2. The number of genes involved
  B. Strain construction
  C. To establish which genes code for which proteins
  D. To order a sequence of reactions (epistasis)
  E. To study the regulation of gene expression

can be usefully probed by either an inhibitor or a mutation only if it can be *affected* with some degree of *specificity*. The range of reactions that can be perturbed by mutations clearly approaches the limit of the entire set of cellular processes, since almost all reactions require that certain protein structures be "active," whether these structures are enzymes, parts of cytochromes, membrane proteins, etc.; and because mutations often have their primary effect on a single protein, they have the potential for great specificity.

To provide a more complete answer to the preceding question, it is necessary to consider the relationship between a normal system and mutants derived from it. If there were no logical way of relating the two, then indeed very little insight could be gained from the study of mutants. In general, mutations may affect either the primary structure of a protein or the pattern of synthesis of a protein. The majority of mutations thus have a direct effect on a single protein. Ignoring for the moment those classes of mutations that directly affect the structures of several proteins, the simplest situation to analyze is one in which a single protein is altered in such a way as to lose all its biological activity. A mutant strain containing a mutation causing a straightforward lesion can be examined biochemically to determine which activities are lost and what are their biological consequences. If the activity lost is essential for normal growth, a way must be found to propagate the mutant strain in order to have material to study biochemically. Several solutions to this problem are possible, but, in the area of bioenergetics of photosynthetic bacteria, the simplest answer is to grow the mutant by an alternative energy metabolism mode, e.g., nonphotosynthetic mutants may be propagated by respiration or fermentation (Sistrom and Clayton, 1964; Yen and Marrs, 1977).

Even with the simplest conceivable primary lesion, great care must be taken in comparing the parental or wild-type strain to a mutant derivative. Biological systems are usually regulated at several levels, and blocking one system often results in compensatory changes in allied systems. Ideally, such compensatory changes can be minimized by propagating mutants under conditions in which the affected system is gratuitous, i.e., it is made but not required. A good example of this is the propagation by photosynthetic growth of mutants with altered respiratory oxidases. In *Rhodopseudomonas capsulata* there are two alternative terminal oxidases, and mutants lacking one or the other can be grown by respiration. However, since the oxidases are thought to function during aerobic growth, the absence of one or the other could induce the synthesis of atypical amounts of the residual system. The comparison of mutant and wild-type cells propagated in this way may thus involve complications. Cells grown in the absence of oxygen synthesize an active, branched respiratory chain. The comparison of mutant and wild-type respiratory chains from cells grown photosynthetically is more likely to be valid, because neither terminal oxidase functions in the absence of oxygen (as far as we know).

Drews *et al.* (1976) have analyzed by SDS–polyacrylamide gel electrophoresis the membrane proteins of various mutants and recombinants of *R. capsulata* with alterations in the photosynthetic apparatus. This study shows that the synthesis of many proteins can be affected by a single mutation. In Fig. 1, for example, a mutation blocking bacteriochlorophyll (BChl) synthesis results in the loss of several proteins that are thought to bind light-harvesting BChl. The genetic replacement of the single block in the BChl pathway restores the synthesis of the proteins in question. This phenomenon is almost certainly due to regulatory interactions in the cell, as was proposed by Takemoto and Lascelles (1973) as a

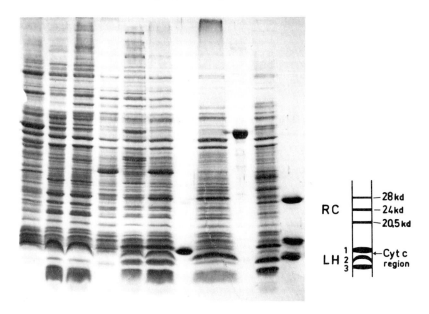

FIG. 1. Proteins from the membranes of various strains of *Rhodopseudomonas capsulata* after sodium dodecyl sulfate–polyacrylamide gel electrophoresis. The membrane proteins fractionated in the fourth slot from the left are derived from a mutant (YS) blocked in bacteriochlorophyll synthesis. Slots five and six were loaded with membranes from two independently isolated photosynthetically competent recombinants that were produced by treating YS cells with gene transfer agents from wild-type strains of *R. capsulata*. Note that many protein bands reappear in the photosynthetically competent recombinants. The recombinants' protein patterns are indistinguishable from wild type. The diagram on the right indicates the protein pattern of the photosynthetic apparatus of wild-type strains. RC, reaction center components; LH, light-harvesting components; kd = kilodalton. From Drews *et al.* (1976).

result of their studies of the membranes of *R. sphaeroides* mutants blocked in BChl synthesis. Although they could not unambiguously establish that only one genetic lesion is involved, the reversion rates of the mutants suggest that this is the case. The fact that several mutants examined all show the same absence of specific membrane proteins, even though they represent blocks at different points in the BChl pathway, also supports the assumption that synthesis of mature BChl is necessary for incorporation of several specific proteins into chromatophore membranes.

B. Isolation of Mutants

How are useful mutants isolated? It is not my goal to describe procedures for mutant isolation here, especially since it has been treated elsewhere (Marrs *et al.*, 1978; Haddock, 1977). However, since the task of isolating mutants of interest is one of the most challenging and most rewarding aspects of microbial genetics, I shall outline the essence of the process for those potential users who may be unfamiliar with mutant hunts. The first step is usually mutagenesis to increase the frequency of mutants so that smaller numbers of cells need be examined. For mutant phenotypes that may be caused by a single mutation, one usually thinks of a typical frequency of mutant per wild type in the range between 10^{-5} and 10^{-7}, and strong mutagenesis may increase the frequency by about 10^2. Some phenotypes are much rarer than others because only very few of the many possible genetic changes will result in the desired properties, whereas some mutants can be created in many alternative ways, and they represent the more frequent end of range. There are three classes of procedures that enable one to isolate a particular mutant from a large population. These are *screening, enrichment,* and *selection.* I think most microbial geneticists will share my subjective ranking of these techniques, which places selection as the most elegant (involves the least busy work), enrichment next, and screening at the bottom of the totem pole, although a certain bravado is felt in the synonymous term "brute force."

Selection implies that the only cells capable of growth are the desired mutants, all others being killed or inhibited. An example is the selection of mutants resistant to the antibiotic neomycin. 10^8 or 10^9 individuals can be "examined" for neomycin resistance by simply spreading a suspension of cells on an agar plate containing suitable growth medium and the antibiotic, and only those which are resistant will multiply and form visible colonies. Selections seldom give only the mutants of interest, and is is usually necessary to *screen* the survivors for the clones that actually show desired properties. In a screening procedure, each colony must be examined in some way. For example, some neomycin-resistant mutants of *Escherichia coli* are deficient in heme synthesis (Sasarman *et al.*, 1968). These heme-deficient colonies could be located among the neomycin-resistant population by picking each colony and assaying spectrally for cytochromes, but staining colonies on plates is generally a superior technique because it is quicker, and more colonies can thus be checked. The Nadi reaction, for example, could be used to screen for the presence of the cytochrome (Cyt) *c* oxidase system (Marrs and Gest, 1973).

When the mutant phenotype sought limits the growth capacity of cells rather than extending it, an alternative strategy is usually employed as typified by penicillin selection. Penicillin kills only growing cells. Thus, for example, nonphotosynthetic mutants may be selected for by incubating a mutagenized population under photosynthetic conditions with penicillin. After allowing sufficient time for wild-type cells to die, the penicillin is removed (by washing or penicillinase) and the cells are plated under respiratory growth conditions. The colony-forming survivors are then screened for photosynthetic growth.

Enrichment procedures give a growth advantage to the desired mutant, but do not totally inhibit the rest of the population; thus they are usually coupled with more extensive screening to recognize the mutants of interest. For example, if one wanted to isolate an organism producing a nitrogenase less sensitive to molecular oxygen, one might grow cells with N_2 as the sole nitrogen source under a controlled atmosphere in which the O_2 concentration was just high enough to cause about 90% inhibition in growth rate of wild type. A mutant that contained a slightly more resistant form of the enzyme would grow faster than other cells and would eventually become the predominant organism in the culture. Of course a variety of phenotypic changes could lead to a growth advantage under a particular stress condition, and it is possible that mutants with entirely different properties from those initially sought would be predominant in the culture of the enrichment, so an effective screening to follow the enrichment is essential. This type of selection often results in the isolation of mutants in which a series of "fitter" mutations have occurred, each one conferring a slight growth advantage under the particular culture conditions employed.

It should be clear that in order to obtain a clone of cells with a particular mutant phenotype, conditions must be available to allow the mutant to grow. If the desired mutant is to be affected in an essential property, *conditional lethal* mutations must be sought. These mutations may change the structures of proteins so that they are functional at one temperature but not at another, so-called *temperature-sensitive* mutations, or they may be *nonsense* mutations, which cause premature termination of polypeptide synthesis. Nonsense mutations can be suppressed by the presence of certain genetic elements, and strains bearing these mutations may be propagated only when the suppressor element is present. In practice nonsense mutations are not convenient conditional lethals for bacterial mutations because of the difficulty of removing the suppressor when the mutant phenotype is to be expressed, and this system finds most application among bacterial viruses, or as a useful type of mutation in a nonessential function. Temperature-sensitive mutants of *Rhodospirillum rubrum* which are unable to grow photosynthetically at the nonpermissive temperature have been isolated by Weaver (1971).

III. Uses of Mutants Not Relying on Genetic Analyses

A. Mutants as Sources of Material

1. Mutants Lacking Interfering Substances or Reactions Similar to Those Studied

It is not unusual to find that a particular biological function can be performed by more than one series of reactions in a given organism. For example: ammonia can be assimilated by two or three alternative routes in most cells; some bacteria have multiple systems for concentrating iron from the environment; cells have many ribonucleases; most cells have two or three distinct DNA repair mechanisms; and the list could be extended greatly. While it is probably true that cells possess each of these redundant systems for a particular advantage it confers compared to the others, an experimenter contemplating analysis of one such process is nonetheless faced with "interfering" activities that may confuse analyses. Mutants lacking all but one of a multiple set of pathways make very convenient tools for studying the residual system. Often the multiple nature of the systems involved is first revealed by mutational analysis. An example illustrating many of these points is the analysis of the respiratory electron-transport pathway of *Rhodopseudomonas capsulata*. It was observed that a mutant that had lost all terminal oxidase activity appeared to have two independent lesions, both of which were required to block respiration. This was interpreted to indicate that a branched electron-transport pathway was functioning (Marrs and Gest, 1973). By isolating revertants for each of the two lesions independently, strains were developed that used exclusively one or the other oxidase for respiration. These strains made possible the identification and characterization of both oxidases, an accomplishment that had eluded investigators prior to the availability of the mutants (Klemme and Schlegel, 1969). The loss of Cyt c-oxidase activity in one strain (M7) was shown to correlate with the loss of one Cyt *b* with a 410 mV midpoint potential, strongly suggesting that the missing cytochrome was part of the oxidase (Zannoni *et al.*, 1974). The residual respiratory oxidase activity in M7 was carbon monoxide sensitive, in contrast to oxidase activity in wild-type organisms. When membranes from M7 are exposed to CO, another *b*-type cytochrome shows a shift in midpoint potential from 270 to 355 mV. In a second mutant strain (M6), which lacks the CO-sensitive oxidase, there is no midpoint shift in the presence of CO, suggesting that the Cyt b_{270} is the CO-sensitive oxidase (Zannoni *et al.*, 1976). The model for electron transport in *R. capsulata* shown in Fig. 2 is a result of these studies.

2. Mutants Lacking Interfering Substances or Reactions Not Related to Those Studied

Sometimes a cell product interferes with a procedure or assay technique, and if the appropriate mutant is available the problem may be circumvented. For example, absorbancy changes related to the redox state of various endogenous cytochromes are sensitive and versatile signals related to events in electron transport, but the photosynthetic membranes of wild-type photosynthetic bacteria usually contain an enormous excess of strongly absorbing carotenoid pigments, which obscure the cytochrome signals. "Carotenoidless" mutants are available, but for several reasons "green" mutants present a more elegant solution to the problem. Carotenoidless mutants, which usually are

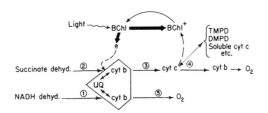

Fig. 2. A model of electron-transport pathways present in membranes of photosynthetically grown *Rhodopseudomonas capsulata*. The numbered arrows represent steps in respiratory electron–transport pathways; some steps may involve more than one electron carrier. The box enclosing ubiquinone and cytochrome *b* (cyt b) represents the branching point and symbolizes the interchange of reducing equivalents from the different dehydrogenases (dehyd.) to either branch. The order of carriers within the box is not specified. Strain M6 is blocked in step 5, and this block is associated with an altered *b*-type cytochrome with a midpoint potential of 270 mV; strain M7 is blocked in step 4 and lacks a *b*-type cytochrome with a 410 mV midpoint potential; cyt c is thought to be a *c*-type cytochrome with a 340 mV midpoint potential; BChl, bacteriochlorophyll; UQ, ubiquinone; e, electron; DMPD and TMPD, *N,N*-dimethyl-and *N,N,N',N'*-tetramethyl-*p*-phenylenediamine. From La Monica and Marrs (1976).

strains blocked early in the carotenoid biosynthetic pathway that accumulate colorless carotenoids, have the drawback of additional membrane changes. The full extent and significance of these changes cannot yet be assessed, but one clear signal of the changes is the loss of the absorbance peaks corresponding to a variety of light-harvesting chlorophyll (LH II), which usually absorbs at approximately 800 and 855 nm, and both of these peaks are greatly reduced or absent in all carotenoidless mutants described to date (e.g., Weaver *et al.*, 1975). Green mutants have blocks later in the carotenoid biosynthetic pathway and usually accumulate one or more neurosporene derivatives. Green mutants normally have wild-type BChl absorption spectra, and thus their membranes represent a less perturbed source of material. Furthermore, the carotenoids accumulated by green mutants undergo a characteristic change in absorbancy, the so-called carotenoid shift, which is thought to be associated with the development of an electrical potential across the membrane. Of course this signal is unavailable from carotenoidless mutants.

On the other hand, carotenoidless mutants have proved to be very useful as starting material for the preparation of purified photosynthetic reaction centers (Clayton and Wang, 1971; Wang and Clayton, 1973; Prince and Crofts, 1973). For reasons not understood, it is much easier to separate the light-harvesting and reaction-center components of the photosynthetic apparatus when blue-green (carotenoidless) strains are used as membrane sources (Clayton and Haselkorn, 1972).

Mutants that have lost any of a wide variety of inconvenient or annoying traits can be selected or constructed. *Rhodopseudomonas palustris* cells tend to clump during growth, resulting in inhomogeneous suspensions, but mutants that do not clump may be selected by transferring samples of culture supernatant through a serial series of growth cycles (Yen, personal communication). Examples of this approach are very numerous in fields outside of bacterial photosynthesis, the most noteworthy recent example being the construction of a strain of *Escherichia coli* by Curtiss and his colleagues (Maturin and Curtiss, 1977) which has been enfeebled so that it is unable to survive passage through intestinal tracts (of rats), and this is a safer host for experiments involving recombinant DNA. This strain (χ^{1776}) was constructed by bringing together a constellation of known mutations causing sensitivity to bile and requirements for diaminopimelic acid and thymine.

3. Mutants Facilitating the Incorporation of Isotopes

Isotopic labeling can be put to a variety of uses, and mutants can be used to increase incorporation and to limit the incorporation into a restricted class of compounds. These ends are accomplished by inactivating normal biosynthetic pathways that either dilute an exogenous (labeled) compound or lead that compound into undesired products. For example, mutants that require δ-aminolevulinic acid (ALA) might be used to promote the incorporation of ALA into BChl at high, defined specific activity. If a wild-type organism is fed labeled ALA, the label will be diluted, usually to an unknown extent, by ALA synthesized by the organism. If there is only one possible biosynthetic route by which the tetrapyrrole portion of BChl can be synthesized, the specific activity of BChl synthesized by an ALA auxotroph growing in the presence of labeled ALA can be calculated. This is clearly a great advantage for quantitative studies.

4. Mutants Producing More of a Substance or System under Study

Purification of any biological entity is simplified by having as starting material organisms rich in whatever is to be prepared. It is often possible to isolate strains that produce ten or even one hundred times more of a product than does wild type. Such strains usually are the result of inactivation of the normal homeostatic controls of the organism, but the product must be carefully compared to the normal to ascertain if it is truly the same and simply "overproduced." I know of no examples of this technique applied to bioenergetics problems, but it is commonplace in molecular biology (and industrial microbiology). My colleagues and I have successfully used this approach to isolate quantities of a genetic exchange vector, the gene-transfer agent (GTA), from *R. capsulata* (see below). Wild-type strains produce very little GTA even under the most favorable culture conditions. We developed a technique to recognize strains which could produce more GTA (by measuring the genetic exchange which GTA mediates), and we were rewarded with a strain that produced about 1000 times as much material. This made possible detailed analysis of the structural and chemical composition, because it made purification quick and simple. The amount of GTA recoverable from 1 liter of culture was increased even more than one thousandfold, apparently a result of avoiding nonspecific losses that often accompany the purification of minute amounts of materials.

5. The Use of Mutants in Reconstitution Studies

The classic approach of resolution and reconstitution is widely applicable to problems of bioenergetics. The essence of the technique involves physically separating the components of an active system into simple, easily characterizable elements. Usually the resolved elements carry out only partial reactions, which may be analyzed, and the role of the elements in the overall process then may be verified by recombining the components necessary to restore the original activity. Mutant strains have been used to facilitate the analysis of bacterial photosynthesis by this approach, and they hold promise for aiding in the isolation of previously uncharacterized electron-transport components.

Jones and Plewis (1974) reconstituted light-dependent electron transport in cytoplasmic membrane fragments by mixing purified reaction centers from wild-type *R. sphaeroides* with membranes obtained from a BChl-less mutant of the same species. The reaction-center complex preparation alone showed light-induced bleaching of the reaction-center BChl peak at 598 nm, but only slight absorbancy changes attributable to light-induced Cyt *c* oxidation or *b* reduction were detected. Membranes from the BChl-less mutant showed no light-induced redox changes. Reconstitution was effected by mixing a suspension of membranes prepared from the nonphotosynthetic mutant with a reaction-center preparation in detergent (2.0% lauryldimethylamine oxide), and then destroying the detergent by reduction. The reconstituted membranes showed light-induced absorbancy changes indicative of oxidation of Cyt *c* and reduction of Cyt *b*. The mutant membranes were chosen for this study because they had no endogenous light-induced activities. The authors might have used cytoplasmic membranes from wild-type cells in which BChl synthesis was inhibited by O_2, but aerobic membrane fragments prepared from respiring cells usually retain traces of active BChl and thus higher background activities would have been experienced. The hypothetical alternative of complete removal of only the reaction centers from photosynthetic membranes was not technically feasible.

Garcia *et al.* (1974, 1975) performed analogous reconstitution studies with membranes from a phototrophic-negative *R. capsulata* mutant, and also provided evidence for reassembly of active structures by demonstrating photophosphorylation and NAD^+ photoreduction. These studies did suffer somewhat from high background activities (of light-induced Cyt *b* reduction and Cyt *c* oxidation) attributed to residual cytochromes in the reaction-center preparation, and, had mutants defective in these activities been available, they would have served as a superior source of reaction centers. It should be mentioned that both Jones and Plewis (1974) and Garcia *et al.* (1974, 1975) made use of carotenoidless mutants in preparing reaction-center complexes, because active reaction centers can be more easily extracted from their photosynthetic membranes.

A general scheme for purification of components of complex systems using mutant strains has been applied in research outside the field of bioenergetics. Restoration of an *in vitro* activity to a mutant system may be used as an assay for purification. The fractionated wild-type system can serve as a source of active components that replace those missing in the mutant. A refinement on this general concept is the procedure, sometimes called *in vitro* complementation, in which two mutants are used as sources of material, neither of which is active alone, but which can be recombined to give an active system. The advantage gained by using two complementary mutations is that since neither possesses activity alone, screening for conditions that permit activity to be reconstituted can be done without fractionating either input material, thus minimizing the risk of inactivating essential components.

Schekman *et al.* (1974) give an interesting general discussion of the advantages and pitfalls of using complementation assays to resolve multienzyme systems, using their studies on DNA replication to illustrate their points. The difficulties encountered included the following: (i) The mutant protein may behave differently when purified, and in some cases may actually be stabilized by a variety of nonspecific compounds. (ii) If the mutant produces an inactive form of a protein that binds normally in a complex with other components, it must be physically displaced for complementation to occur. (iii) Simple mutations may have pleiotropic effects that complicate the situation. Inactive gene product A may interfere with the activity of another component, B. If B determines a reaction that is rate limiting in the reaction assayed, then B will be purified rather than A. (iv) A mutation for each component of the system may not be available, in which case the system will not be completely resolved by this approach.

B. Use of Mutants to Assign a Biological Role to a Protein or an Enzyme Product

Several examples of this type of application can be given, and particular attention should be paid to possible overinterpretation of information obtained from mutants when used in this context.

Blue-green mutants lack the polyunsaturated carotenoids that give wild type their characteristic colors. Blue-green mutants are much more sensitive than wild type to photooxidative killing; i.e., they die when simultaneously exposed to light and oxygen (Sistrom *et al.*, 1956). This correlation suggests that the colored carotenoids may function *in vivo* to protect against photooxidative killing. The difficulty with this logic is the implicit assumption that the *only* difference between mutant and wild type is the carotenoid content. In this particular case there are several reasons to suspect that other changes accompany the loss of colored carotenoids. The *in situ* BChl absorption spectra of blue-green mutants of *R. sphaeroides* and *R. capsulata* are not like wild type, and certain polypeptides found in wild-type membranes are missing from blue-green mutants (Fig. 1). These changes have been described as a loss of a particular form of light-harvesting BChl (LH II). How is one to decide whether the sensitivity to photooxidative killing is a consequence of the absence of carotenoids or the absence of LH II? The examination of many independently isolated mutants is often helpful. If the correlation between carotenoids and sensitivity were coincidental, one might expect to find a mutant that lacks carotenoids but retains LH II, and this strain might show resistance to photooxidative killing. However, carotenoid-minus LH II-plus strains have never been found. This last observation suggests that there is a connection between the appearance of colored carotenoids and LH II BChl. It has been suggested that the LH II absorbance spectrum is caused by a BChl–carotenoid conjugate, and thus the loss of carotenoids must always lead to a loss of LH II (Weaver *et al.*, 1975). Another possible connection is more insidious. Suppose that carotenoidless mutants were unable to grow photosynthetically unless they lacked LH II. Whenever viable, blue-green, photosynthetically competent mutants were isolated, they would necessarily contain a second mutation blocking LH II formation. This type of selection can lead to the isolation of complex mutants that require genetic resolution before they can be used with confidence. Indeed, attempting to correlate the loss of a compound or reaction with a biological effect can often be misleading unless genetic analyses can demonstrate that only one mutation is involved, and even then one must be cautious, since one mutation may have more than one effect. In the case of blue-green mutants, genetic analysis in *R. capsulata* suggests that two separate mutations are indeed involved in the blue-green phenotype (see Section V).

Another example of this type of analysis is found in the study by Zannoni *et al.* (1974), concerning the identity of the terminal oxidase of *R. capsulata*. The mutant strain M7 lacks Cyt *c* oxidase activity, and it also lacks a high-potential cytochrome with a spectrum of the *b* type. The correlation of the oxidase activity with the cytochrome (Cyt b_{410}) content is compelling, especially because of the $+410$ mV midpoint potential of the cytochrome. However, the logical possibility remains that the absence of Cyt b_{410} is coincidental. Several possible explanations for a coincidence of this type are known. Some mutagens used in the creation of mutant strains characteristically cause a cluster of genetic changes in one region of the bacterial chromosome. This is very well documented for the mutagen N-methyl-N'-nitro-N-nitrosoguanidine (Guerola *et al.*, 1971), which was in fact used to produce strain M7 (Marrs and Gest, 1973). Because genes with related functions are often clustered on the bacterial chromosome, it is likely that several related proteins will be altered in a mutant selected after nitrosoguanidine treatment. Thus there is a possibility that Cyt b_{410} is not the terminal oxidase, but rather a protein which serves some other function, and the gene that codes for Cyt b_{410} was damaged when the true (unknown) cytochrome oxidase gene was inactivated.

This scenario can be tested by genetic analysis and further biochemical testing. It is also possible that Cyt b_{410} requires that another protein (the terminal oxidase) be bound to the membrane before it can bind. If this were true the loss of Cyt b_{410} would correlate with the loss of the terminal oxidase, but Cyt b_{410} need not function as the terminal oxidase. This type of artifact may be revealed by either further genetic analysis (see Section V) or detailed biochemical studies. Examples of this type of interaction among membrane proteins have been encountered in the study of mitochondrial biogenesis (Schatz and Mason, 1974).

The studies of Lascelles and Altshuler (1969) on δ-aminolevulinic acid (ALA) auxotrophic mutants of *R. sphaeroides* are a good example of the use of mutants to assign or define the role of an enzyme. The enzyme ALA synthetase, which catalyzes the reaction between glycine and succinyl-CoA to form ALA, is the first committed step on the tetrapyrrole biosynthetic pathway. As such, ALA synthetase is expected to be a site of regulation of the tetrapyrrole pathway, and indeed this enzyme has been shown to be sensitive to feedback inhibition by heme. In the photosynthetic bacteria, however, the main quantitative function of ALA synthetase is to form BChl, and BChl formation is inhibited by O_2, clearly a different type of regulatory response than that heme synthesis must show. It had been proposed that this dilemma might be resolved if there were two distinct ALA synthetases, one which provided ALA for heme synthesis and another for BChl, but Lascelles and Altshuler isolated mutants lacking ALA synthetase activity and showed that both heme and BChl synthesis were simultaneously blocked, thus suggesting that isozymes did not exist. In order to rule out the possibility that two mutations had actually been selected for, one blocking each of the two putative enzymes, they isolated revertants and showed that both biosynthetic activities were recovered together. The stratagem of isolating revertants that have regained an activity lost by the original mutant is often useful in helping to decide whether a constellation of differences between mutant and wild-type phenotypes are all due to one underlying cause. There are, however, hazards in this logic. If multiple mutations were involved in the mutant phenotype, but were not independent of each other, selection for reversion to wild type might lead to the isolation of mutants in which several reverting changes occurred. Once again, genetic analysis is necessary for complete confidence in the number of mutations involved.

With respect to the particular example of ALA synthetase mutants, the possibility also exists that a common catalytic subunit of ALA synthetase could combine with two different regulatory subunits, resulting in two enzymes that could each respond differently to regulatory signals, but could both be inactivated by a single mutation in a gene coding for the catalytic subunit.

Another complication in analyzing revertants is that occasionally a new level of complexity is introduced by mutations resulting in a third "state" rather than a return from the mutant state to the original wild-type "state." One transparent example is seen when blue-green mutants of photosynthetic bacteria are treated with photooxidative killing conditions to select for cells that have regained the ability to synthesize carotenoids. Although some revertants that can synthesize carotenoids survive this selection, another major class of survivors are "pseudorevertants," which can no longer synthesize BChl (in addition to the original block of carotenoid biosynthesis) (Yen and Marrs, 1976). These latter mutants are selected for because mature BChl is the mediator of photooxidative killing, and they are easily distinguished because they cannot grow photosynthetically. Not all "pseudorevertants" are thus easily deciphered, and caution must be exercised when one selects for revertants lest one overinterpret the results.

C. Use of Mutants to Resolve and Define a Sequence of Reactions

A great deal of information about photosynthetic electron transport and photophosphorylation can be gathered by studying a collection of mutants in which each is unable to grow by photosynthesis. Mutants may be used to resolve complex systems only when techniques are available to distinguish or categorize blocks at different points. There might be, for example, a series of electron carriers between Cyt b and Cyt c, and a variety of mutant strains could exist, each blocked at a different point, but if the only signal available was whether Cyt c could be reduced by electrons from Cyt b, the mutants would all appear to be of one type. Fortunately, there are many signals associated with photosynthetic electron-transport carriers that may be differentiated by their various midpoints and the kinetics of their appearance, EPR signals, relationships to proton uptake, carotenoid absorbancy changes, and more. If differences can be detected among the various mutants blocked between Cyt b and Cyt c, each distinct type of mutant will define a step. This approach becomes even more useful when coupled with genetic techniques that allow the mutants to be classified as to how many genes, and thus how many polypeptides, are involved (see Section V).

An analysis of this sort is in progress in a collaborative effort between P. L. Dutton and his colleagues and the author's group. A nonphotosynthetic mutant of *R. capsulata* was isolated which grew normally by respiration and appeared to have all the normal photopigments and cytochromes. Biochemical studies revealed that although NADH and succinate could drive O_2 uptake in membrane fragments from this mutant, neither substrate could drive the reduction of horse heart Cyt c added to the suspension (an activity easily demonstrable in an isogenic strain that had regained the ability to photosynthesize). This seemed to localize the block between Cyt b and Cyt c, according to current models for electron transport in this organism (see Fig. 2). When membranes prepared from mutant cells grown under low partial pressures of O_2 were tested for light-induced reactions, it was found that a rapid oxidation of Cyt c_{340} occurred, but rereduction did not occur (see Fig. 3). This result was consistent with a block between Cyt b and Cyt c, but further analysis revealed that Cyt b was not reduced following a singal flash of light, in contrast to the control strain (W. van den Berg, personal communication). This might indicate that one particular Cyt b was not able to properly accept electrons from the reaction center, and not able therefore to rereduce the Cyt c_{340} after it was oxidized by the reaction center. Other studies, however, indicate that the normal reductant for

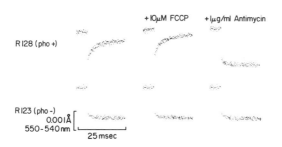

FIG. 3. The flash-induced kinetics of cytochrome c oxidation and rereduction in membranes isolated from a nonphotosynthetic mutant (R123) and a photosynthetically competent isogenic control strain (R128) of *Rhodopseudomonas capsulata* (20 μM bacteriochlorophyll (BChl), pH 7.0, E_h 30 mV). Membranes were isolated from cells grown under a constant low O_2 tension (2.5% v/v O_2, 5% CO_2, 92.5% N_2) in the dark. At the indicated ambient potential, essentially all the Cyt c_{340} is reduced before a single flash a few nanoseconds in duration is administered in each trial. The flash is of sufficient intensity to saturate the BChl present, and about one half of all Cyt c_{340} present is oxidized by each flash. R128 shows rapid oxidation of Cyt c_{340} followed by a multiphasic, slower rereduction, which is somewhat accelerated by uncoupler, and blocked by antimycin. The nonphotosynthetic mutant shows the same initial oxidation of Cyt c_{340}, indicating that its reaction centers are active and properly associated with Cyt c_{340}, but no rereduction is observed on this time scale under any of the conditions examined. By courtesy of P. L. Dutton.

Cyt c_{340} is a two-electron carrier with a midpoint of about 150 mV, which does not fit with any of the known b cytochromes involved in photosynthesis (Prince and Dutton, personal communication).

Another possible interpretation is that a special form of ubiquinone may be needed both for the reduction of Cyt b and Cyt c. If the model shown in Fig. 4 is correct, ZH_2 must be formed to reduce both cytochromes. If a protein responsible for this activity were missing, be it an enzyme for ZH_2 formation, a protein to stabilize the product, or a catalyst to facilitate the reaction between ZH_2 and Cyt c, the observed results would be obtained. Other findings from the study of this mutant that support this model are that only one proton is bound per electron by mutant chromatophores (as opposed to almost two protons per electron in wild-type chromatophores), and the third phase of the carotenoid shift

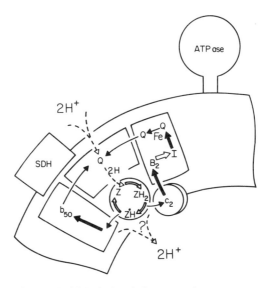

Fig. 4. A schematic model of light-induced electron and proton transport in membranes of *Rhodopseudomonas capsulata*. Succinic dehydrogenase (SDH) is on the outside of the membrane vessicle (chromatophore) and Cyt c_{340} (c_2) is on the inside. Electron transport is represented by arrows with filled heads. Filled fat arrows indicate those electron transport steps thought to contribute to the three phases of the carotenoid shift. B_2, I, FeQ, and Q are components of the reaction center. Electrons leaving the reaction-center complex are donated to an "outer Q pool" and in turn are passed to a cycling inner carrier, Z, that is the immediate reductant of Cyt c_{340}. Following a flash of light, Cyt c_{340} is rapidly oxidized and rereduced, creating Z·H. Z·H donates electrons to an unspecified carrier on the inner side of the membrane, perhaps Cyt b_{-90}, producing Z, which can then accept electrons from the outer Q pool. Cyt b_{50} passes electrons from the inside of the membrane to the outer Q pool, which ultimately cycle back to reduce Z. The protons transported across the membrane during this process may return to the outside with concomitant synthesis of ATP. By courtesy of R. Prince and P. L. Dutton.

(thought to be a result of electron transport from Cyt b_{-90} to Cyt b_{50} in wild type) does not occur in the mutant. Further studies are required to test these hypotheses, but it is clear that the study of this mutant helps to suggest and test hypotheses.

A recent example of how the study of a mutant may lead to a new understanding of a sequence of reactions is seen in an interesting series of papers by del Valle-Tascón, Giménez-Gallego, and Ramírez. These authors initially reported the very curious observation of a mutant of *Rhodospirillum rubrum* that could not grow photosynthetically (del Valle-Tascón *et al.*, 1975), but could carry out photophosphorylation. They identified a defect in the ability of this strain to carry out light-driven oxygen uptake and reasoned that the defect must be on a side branch, off the main cyclic electron flow pathway, but somehow essential for adjusting the levels of oxidation of the carriers in the cyclic pathway. They later confirmed this hitherto unexpected relationship by showing that a low level of O_2 stimulated the initiation of photosynthetic growth in wild-type strains, but failed to do so in their mutant strain Fll (Giménez-Gallego *et al.*, 1976; del Valle-Tascón *et al.*, 1977). Their hypothesis can be described using the model of electron transport developed for *Rhodopseudomonas capsulata* and shown in Fig. 2. They would suggest that the cyanide-insensitive oxidase portion of the branched respiratory pathway is necessary for oxidizing carriers on the low-potential side of the cyclic pathway (i.e., in the region of Cyt b and ubiquinone) to allow cyclic electron flow to begin. They found a cyanide-sensitive pathway in *Rhodospirillum rubrum*, which is analogous to the Cyt c oxidase of *Rhodopseudomonas capsulata*, that was capable of serving the same function under higher O_2 tension. Their observations with *Rhodospirillum rubrum* suggest an analogous function for the alternate oxidase of *Rhodopseudomonas capsulata*, and that their Fll mutant of *Rhodospirillum rubrum* is analogous to the M6 mutant of *Rhodopseudomonas capsulata*. If this is true, it predicts that *Rhodospirillum rubrum* has a branched oxidase pathway like that in *Rhodopseudomonas capsulata*, and Fll may help in the sorting out of these two paths, just as M6 served in the study of *R. capsulata* (Marrs and Gest, 1973; Zannoni *et al.*, 1976).

IV. Genetic Systems Available in Photosynthetic Bacteria

There is currently only one genetic exchange system operational for a photosynthetic bacterium, the gene transfer system of *R. capsulata* (Marrs, 1974), although as of this writing several laboratories are making rapid progress toward increasing the number and types of genetic tools available to us. Since most bacterial genetic exchange systems are merozygotic, that is they transfer only a portion of the donor's genome to the recipient, the most important single characteristic of each system is the size of the piece of DNA that is incorporated into the recipient cell. To determine the linkage relationships necessary to map genes by genetic crosses, pairs of genes must be cotransferred. If the piece of DNA transferred is shorter than the distance between the genes, linkage cannot be established. If, at the other extreme, very large tracts of DNA are transferred, markers that are close together will only rarely be separated, and thus fine-structure mapping is impractical. It would be ideal if two exchange systems were known for some photosynthetic bacterium, one for establishing gross map structure, the other for fine structure. The gene transfer system of *R. capsulata* is suitable for fine-structure mapping, and a map of the photopigment region of the chromosome has been constructed (Yen and Marrs, 1976). Promising techniques for mapping large portions of the chromosome will be discussed in Section VI.

The genetic exchange discovered in *R. capsulata* is mediated by a vector termed the gene transfer agent (GTA). GTAs are produced by the majority of strains of *R. capsulata* (Wall *et al.*, 1975a) and can transfer genes to most strains. GTA particles resemble tailed phages in general morphology, possessing a head 300 Å in diameter with short spikes and a tail of variable length (300–500 Å) with a 60 Å diameter. GTA is thus smaller than any morphologically similar bacterial virus. Tail fibers and a collar are also visible. Sodium dodecyl sulfate–polyacrylamide gel electrophoresis of the proteins of purified GTA reveals polypeptides of the following apparent molecular weights (M_r): 27, 40.2, 19, 14, 22.6, 39.4, 15, and 13.6 (given in order of decreasing staining intensity). DNA from GTA has been analyzed by agarose gel electrophoresis of restriction endonuclease digests and by C_0T determinations. The results of both analyses show no detectable phagelike DNA, but rather suggest that GTA particles contain only chromosomal DNA from the donor bacterium. The M_r 3 × 10^6 GTA DNA (Solioz and Marrs, 1977) is taken up by recipient cells and appears to recombine with the chromosome resident in the recipient, thereby replacing some of the recipient's genetic information with a portion of the donor's genome. Any one particle carries only about 5 or 6 genes or about 0.1% of the donor genome. One recipient cell may take up several GTA particles, and their DNA may or may not be successful in replacing the corresponding resident genes (Marrs, unpublished observations). GTA-borne DNA does not seem to be able to exist and replicate autonomously in the recipient; therefore it is not now possible to construct strains that are diploid for particular regions, and cis-trans genetic complementation tests are not yet available.

The GTA system is excellent for constructing strains, one marker at a time, because strains are self-fertile. They can receive GTA from any producer strain, and the frequency of recombinant per recipient can be as high as 0.1%; thus detection of the desired recombinants need not involve selective techniques if a screening procedure is available. Because GTA carries a relatively small fragment of the donor genome, it is well suited for the resolution of mutant phenotypes that involve more than one mutation, since even markers that lie fairly close to one another are frequently separated upon gene transfer. An example of some genetic manipulations that are possible with the GTA system is shown in Table II. The objective of this exercise was to obtain two strains that carried the same mutation, one having received the marker via GTA, so that the stability of transferred markers could be measured. Rifampicin resistance markers were introduced so that control crosses between the two tryptophan-requiring strains could be performed to

TABLE II

PEDIGREE OF STRAINS CONSTRUCTED TO TEST THE STABILITY OF MARKERS TRANSFERRED VIA GENE TRANSFER AGENT (GTA)[a]

Strain	Selection Procedure	Properties
B10	Isolated from nature	Wild-type *Rhodopseudomonas capsulata*
BB101	Plated B10 cells with rifampicin, selected spontaneous resistance mutant	Rifampicin resistant
BB103	Plated B10 cells with streptomycin, selected spontaneous resistance mutant	Streptomycin resistant
YB1021	Penicillin selection for tryptophan auxotrophy following nitrosoguanidine mutagenesis of BB103	Streptomycin-resistant tryptophan auxotroph, respiration-deficient (unexpected)
YB1020	YB1021 cells were treated with GTA from B10; selected for cells capable of aerobic dark growth	Streptomycin-resistant tryptophan auxotroph
YB1022	YB1020 cells were treated with GTA from BB101; plated with rifampicin; selected resistant clone	Streptomycin-resistant tryptophan auxotroph, rifampicin resistant
YB1023	BB103 cells were treated with GTA from YB1020; penicillin selection for tryptophan auxotrophy	Streptomycin-resistant tryptophan auxotroph
YB1024	YB1023 cells were treated with GTA from BB101; plated with rifampicin; selected resistant clone	Streptomycin-resistant tryptophan auxotroph, rifampicin resistant

[a] From Yen and Marrs (unpublished data).

assure that the same tryptophan marker (*trpA20*) was present in each. This check was necessary because the transfer of the marker that causes tryptophan auxotrophy was detected by penicillin selection, and a small but finite possibility existed that a spontaneous tryptophan requirer was picked instead of a product of gene transfer. The control crosses (YB1020 cells × YB1024 GTA and YB1023 cells × YB1022 GTA) gave no tryptophan-independent colonies per several thousand rifampicin-resistant colonies, thus indicating that the same mutation was present in both donor and recipient. When YB1020 or YB1023 cells were treated with BB101 GTA, approximately equal numbers of tryptophan-independent and rifampicin-resistant colony-forming units were recovered. A new spontaneous *trp* mutation, which lay so close to *trpA20* that they are separated only rarely, could not be entirely ruled out, but the odds are strongly against it. Similarly, a deletion covering the *trpA20* site could not be ruled out by this test. However, the goal of this construc-tion exercise was to test the reversion rate of the *trp* marker in each strain, and since the results showed that both markers indeed had the same rate of reversion, the existence of a deletion can be ruled out. The identical stability of the marker in each strain argues that genes incorporated via the GTA are as stably integrated as markers arising in a strain, and thus GTA-borne markers are probably integrated into the host chromosome.

Several points of technical interest are demonstrated in this construc-tion: (1) there is no immunity to gene transfer among various clones all derived from the same parental strain; (2) strains are active as both donors and recipients, e.g., YB1020; (3) strains may repeatedly take up genes from GTA; e.g., both YB1022 and YB1024 resulted from two sequential gene transfers. These characteristics make the *R. capsulata* GTA system convenient for simple strain construction tasks.

V. Uses of Mutants Requiring Genetic Analyses

The objective of this section is to catalog the basic types of informa-tion to be gained from genetic analyses that can be useful in biochemical studies, and only brief mention will be made of how the analyses are actually performed. The text by Hayes (1968) is recommended as an excellent general reference for microbial genetics principles and applica-tions.

A. RESOLUTION OF MULTIPLE MUTATIONS

1. Number of Mutations Involved

A mutant phenotype may be caused by one or more mutations. It is important to understand how many mutations contribute to the phenotype, and what changes each mutation contributes, if the wild-type system is to be successfully probed.

Revertants are often studied when no genetic exchange system is available. If it is found that the entire constellation of mutant properties simultaneously return to normal in cells selected for the return of one property, it is suggestive of one underlying mutation. However, many exceptions are known, one example being the pleiotropic suppression of nonsense mutations by suppressor mutations. If different revertants have different properties, it may be because the reverting mutations succeed in restoring activity to different extents. Most reverting mutations do not restore the original base sequence or even amino acid sequence, but rather result in a new sequence that can function more like the original than the mutant.

When genetic exchange can be used, the analysis of the number and types of mutations contributing to a phenotype is simplified. The mutant is genetically crossed by wild type, and the progeny are examined. The phenotypes of recombinants are closely examined to see whether contributing mutations conferring intermediate properties can be identified. For example, when blue-green (carotenoidless) cells of *R. capsulata* are crossed with GTA from wild-type cells, many cells that regain the ability to synthesize carotenoids still have an altered BChl infrared absorption spectrum, indicating that a second mutation was present in the blue-green mutant which caused the BChl spectral alteration (Marrs, 1978). A second example may be found in the analysis of the mutations causing the respiratory-deficient phenotype in *R. capsulata* strain M5 (Marrs and Gest, 1973). When M5 cells are treated with wild-type GTA, respiration-competent cells of two types are isolated. Some lack Cyt *c* oxidase, and some lack the cyanide-resistant oxidase. This indicates that two mutations were necessary to cause the respiratory-deficient phenotype, one in each oxidase (Marrs, 1974). This finding leads directly to the conclusion that the respiratory electron-transport pathway is branched in *R. capsulata*.

The *number* of wild-type recombinants can be analyzed to determine whether two mutations are necessary to cause a particular phenotype, even if the two individual mutations do not reveal themselves during a phenotypic examination after backcrossing. If two mutations must both be transferred to achieve a particular phenotype, the frequency of the transfer of the phenotype will be a function of the distance between the mutations on the chromosome. The depression in the frequency of transfer can be detected by reference to a standard marker in the same cross, hence the term "ratio test" for this type of analysis (Yen and Marrs, 1976).

2. Number of Genes Involved

The determination of the number of genes involved in a process provides useful information related to the number of distinct polypeptides required to carry out the process. The preferred method for determining the number of genes involved is to analyze by complementation testing a collection of mutants that are affected in the process to be studied. Complementation testing requires the construction of diploid or partially diploid organisms that bring together two mutant copies of the genetic region involved. If both mutations are in the same gene, neither copy will provide information to produce the required gene product, and the phenotype of the diploid will remain mutant. If the two mutations being tested lie in different genes, the diploid will have the wild-type phenotype, because undamaged genes from one genome will complement the lesion-bearing gene from the other genome. Complications may arise if the products of the genes being tested interact in a way such that their activities are not independent. For example, products of genes A and B may both be part of a macromolecular complex, and some mutations in the product of gene A may inactivate the product of gene B, a phenomenon usually called *negative complementation*. Another anomaly encountered in complementation analysis occurs when the products of a single gene interact to function as a multimer. This sometimes gives rise to so-called *intracistronic complementation* in which two different mutant forms of a polypeptide may interact so as to correct a structural defect in one or the other, thus complementing and restoring activity in the diploid state, even though both mutations are in the same gene.

Mapping of mutations is useful in determining whether two mutations that cause similar phenotypes are actually in different genes, but if there are two adjacent genes specifying two polypeptides, either of which can be altered to confer the same phenotype, then even fine-structure mapping cannot determine where one gene ends and the other begins, and thus cannot reveal the presence of two genes without additional evidence. If, on the other hand, the same two genes specifying products involved in the same phenotypic properties are separated by a third distinguishable gene, then mapping alone can separate and define the two genes. Mapping thus provides a determination of the minimum number of genes.

B. Strain Construction

It is often desirable to bring together two or more well described mutations into the same strain, and the motives for this may range from needing a genetic marker for test crosses to the construction of a strain for an industrial application. Constructing strains by genetically introducing known mutations is far superior to the alternative of isolating desired combinations of mutations by repeated rounds of mutagenesis and selection, because a well-studied mutation need not be completely recharacterized following transfer, and multiple rounds of mutagenesis usually introduce unwanted mutations in unknown places (see example in Table II).

A simple example of the utility of strain construction is the introduction of a mutation blocking carotenoid desaturation into a strain with altered electron transport, described in Section III,A,2. A nonphotosynthetic strain, Yll, was crossed with GTA from a green mutant. Progeny were inspected for pigmentation, and a green recombinant was picked. It had all the electron-transport characteristics of its normally pigmented parent. A second genetic exchange was used to construct an isogenic control strain by selecting for a recombinant that had regained the ability to photosynthesize.

The concept of constructing an isogenic control strain when using mutations to probe a pathway is very valuable. Mutations are usually first introduced into strains by mutagenesis, and it is possible, and sometimes likely, that more than one genetic difference exists between a mutant strain and the parental strain that gave rise to it. When one examines the characteristics of the mutant strain, one would like to have some assurance that any differences discovered are related to the mutant phenotype being studied, rather than some random change. As an illustration, suppose the above-mentioned nonphotosynthetic mutant were discovered to lack an EPR signal that was present in the wild type. Is it safe to conclude that whatever gives rise to that signal in wild type is an essential component of the photosynthetic apparatus? No. It could be due to the loss of an iron–sulfur center that is normally part of an entirely separate system. If, however, it can be shown that a control strain, which is isogenic with the mutant strain, has the signal lacking in the mutant, then the correlation of the signal with the function is greatly strengthened. Two strains are said to be isogenic if they have chromosomes that differ only in the mutation identified for study. In practice, such strains are constructed either by moving the mutation of interest into a standard strain or by replacing the mutation by wild-type material in the mutant background. The properties of the individual mutation dictate which path is chosen; in either case, however, the degree of isogenicity actually achieved depends upon the method used for transferring the genetic material. If a system is employed that is known to transfer only small pieces of DNA, then only genetic markers quite near the mutation in question will be cotransferred. Thus the chances of being misled into assigning an unrelated property to a mutation diminish as the size of the genetic region transferred diminishes. In this respect the gene transfer system of *R. capsulata* is quite useful, since the maximum length of material transferable is only enough to carry about five genes, and the average amount transferred is half of that (Yen and Marrs, 1976).

A well-established genetic map often makes strain construction easier. If it is not possible to easily select or screen for the introduction of a particular mutation, it is sometimes possible to do so for a closely linked marker. The closer the second marker, the higher the fraction of recombinants receiving the unselected marker. Knowing where genes are located relative to each other is thus useful for strain construction.

The construction of strains can also be used to test models. The model for electron transport in *R. capsulata* predicts that if the mutation in strain M6 were combined with that in Yll, a respiration-deficient photosynthesis-deficient strain should be produced. This construction is currently being undertaken.

C. CORRELATION OF GENES WITH GENE PRODUCTS

The most straightforward method for establishing which gene codes for which polypeptide is to determine the amino acid sequence of a polypeptide coded for by a mutated gene, and show a change compared to wild type. This is not usually necessary for genes whose products have an easily assayed activity, since it may then be possible to demonstrate a temperature-sensitive gene product associated with a particular mutation that causes a temperature-sensitive phenotype. This method can be applied only if the gene product activity can be purified and assayed.

Fillingame (1975) has recently identified the DCCD binding protein in *E. coli* coupling factor by determining which protein no longer binds radiolabeled DCCD after a mutation to DCCD-resistance.

Some mutations affect the activity of a particular protein by causing an amino acid substitution in that protein's primary structure, and if the change involves charged amino acids, the mutant protein may have a new isoelectric point, or a new mobility in a charge-dependent electrophoretic system. Such a correlation is strongly suggestive of a gene–gene product relationship, but alternative interpretations are possible. Many proteins are known to require posttranslational modifications for activity, and several of these can involve charge changes (Stenflo, 1972). This raises the possibility that the mutation actually affects a modifying enzyme, and the protein showing the altered mobility was not in fact coded for by the gene bearing the mutation. If several mutations that map together and cause the same phenotype can be shown to have different effects on mobility (corresponding to different amino acid substitutions), then the alternative of a modifying enzyme becomes unlikely.

Nonsense mutations, and some frameshift mutations, cause premature chain termination during protein synthesis, resulting in a truncated polypeptide representing the amino-terminal end of the wild-type protein. The disappearance of a polypeptide band from an SDS–polyacrylamide gel is not, however, a good criterion for identifying that polypeptide as the gene product of the cistron harboring the nonsense mutation, since examples are known in membrane biogenesis where the product of one gene is necessary to the proper synthesis and/or incorporation of other polypeptides into an organelle. The identification of polypeptides encoded by genes harboring nonsense mutations is more secure if nonsense suppressors have been described and characterized in the organism under investigation. Nonsense suppressors restore activity by causing the insertion of a characteristic amino acid at termination codons. Thus in the presence of a specific suppressor, a full-length polypeptide should again be produced, but it should now have an altered amino acid composition, lacking the amino acid whose codon mutated to give rise to the aberrant chain termination codon, and possessing a new amino acid specified by the suppressor. The demonstration of a new amino acid composition is important for a rigorous proof, since the simple suppressor-induced reappearance of a wild-type polypeptide pattern on SDS–polyacrylamide gels would be expected whether the mutation occurred in a required enzyme (which might never be visualized on a gel because of low concentration) or in the polypeptide which is observed to be missing.

D. SEQUENCING REACTION STEPS

The use of mutants to order a sequence of reactions in a biochemical pathway, flowing linearly from a substrate through a series of intermediates to a product, is a classical approach that need not be elaborated here. Can a similar logic be applied to elucidation of the sequence of events in electron transport or other bioenergetic processes? In one sense this problem has already been discussed in earlier sections (I,A and III,A,3) describing the analogy between inhibitor studies and mutant studies; however, let us now pursue the comparison between classical biosynthetic pathways and electron-transport systems. Imagine a collection of mutants with blocks at a variety of points in a biosynthetic pathway. These mutants may be studied to see what compounds (precursors or intermediates) are accumulated in the presence of each block, and which compounds may be added to bypass the block by entering the path past the block. Analogous information in electron-transport pathways involves determining the reduction state of carriers (crossover point), and determining which carrier-specific electron donors are capable of bypassing the block.

There is another technique that may be applied to the ordering of biosynthetic pathways, and it involves the construction of double mutants by genetic recombination. Each of two mutants is analyzed to see what compounds are accumulated, and then the double mutant is analyzed. It usually accumulates one or the other of the compounds found in the single mutants, and the two blocks represented by the mutants may thus be ordered, the earlier block being the one with the same accumulation product as the double mutant. This is termed epistasis. This type of analysis could be of use in electron-transport systems when a clear-cut crossover point is not found. The construction of double mutants is in most respects analogous to the use of two inhibitors simultaneously, a method that can sometimes provide a test of the relationship proposed between the two elements inhibited (Thore *et al.*, 1969).

E. IDENTIFYING REGULATORY MUTATIONS

The rationale for studying the regulation of genes expressed in connection with the synthesis of bacterial photosynthetic membranes is that they show a level of organization comparable to a eukaryotic organelle, yet they seem simple enough to be amenable to the types of analyses that have led to our highly developed understanding of bacterial operons. What types of regulation are involved in coordinating the expression of a diverse set of genes that must function together to produce an organelle?

The elements that regulate the expression of a group of genes are almost invariably defined first genetically, and only rarely has it ever been possible to analyze operators, repressors, inducers, promoters, and other transcriptional control elements biochemically. Control elements are often recognized in mutants with pleiotropic changes in the levels of a group of related gene products, or mutants that no longer respond normally to regulatory signals. In the case of the chromatophore membranes of Rhodospirillaceae, mutants are known that no longer respond to the normal suppressor of photosynthetic membrane biogenesis, O_2 (Lascelles and Wertlieb, 1971). It would be informative to learn whether the mutation(s) causing this phenotype map(s) in the structural gene for an enzyme in the BChl biosynthetic pathway, suggesting an enzyme made insensitive to allosteric inhibition, or whether it maps in a region adjacent or removed from a cluster of genes involved in some aspect of photosynthetic membrane synthesis, thus resembling an operon control element.

The map of some of the genes for photopigment synthesis in *R. capsulata* shows that genes for chlorophyll biosynthesis and genes for carotenoid biosynthesis are adjacent on the chromosome (Yen and Marrs, 1976). This suggests coordinate transcription as a possible control mechanism for ensuring synchronized production of these pigments. Are other integral parts of the photosynthetic apparatus similarly clustered? Is the observed cluster truly a functional unit, a type of superoperon, containing genes belonging to biochemically diverse but functionally related pathways? These are questions toward which genetic analysis of mutant photosynthetic bacteria may be directed in the future.

VI.   Prospects for the Future

A.   NEW GENETIC SYSTEMS

A great deal of productive research can be envisioned resulting from diligent cross fertilization between microbial genetics and bioenergetics using only the tools currently at hand. There can be no doubt, however, that more versatile genetic systems for the photosynthetic bacteria would enhance the utility of the mutant approach. There are at least two additional genetic tools that will soon be available in this family: conjugation and transduction. Conjugation has been achieved in *R. sphaeroides* by W. R. Sistrom (personal communication) by introducing into *R. sphaeroides* a promiscuous sex factor that appears to be capable of mobilizing the chromosome. This system offers hope that large blocks of genes may be readily mapped. Transduction has been achieved at low levels in *R. sphaeroides* by both Sistrom and S. Kaplan, and recently Yen and Kaplan have succeeded in isolating a phage that gives transduction at a frequency high enough to be useful for strain construction and simple mapping (personal communication). The addition of these tools will have the effect of extending genetics to a second species of Rhodospirillaceae and increasing the range of genetic manipulations possible by making available a system that conserves genetic linkage between markers located far apart on the chromosome. It should be possible to establish a conjugative system in *R. capsulata* by the same methods that served for *R. sphaeroides*. This would leave without genetics only *Rhodospirillum rubrum* among the species of nonsulfur purple photosynthetic bacteria that have been widely used in studies of photosynthesis. It is this author's opinion that for the present this field will benefit by having its options limited to a few species, each being more intensely studied, and having comparative biology deemphasized.

This would allow more direct transfer of information from one investigation to the next, an efficiency that history shows stimulated research with *Escherichia coli* and the T-even phages in their early stages of development.

Recombinant DNA technology should enable one to obtain a large supply of cloned genes for various components of the photosynthetic apparatus. These DNA molecules could then serve as probes for determining the levels of transcription of genes during transitions in conditions that affect the production of photosynthetic membranes, thus enabling a complete study of regulation at the transcriptional level. The NIH guidelines concerning the safety precautions required to carry out such a project raise the question of whether the possibility exists of extending the ecological range of the host as one outcome of cloning. Since the answer is unquestionably affirmative, a photosynthetic *E. coli* being a distinct possibility, the containment required might prove to be prohibitive. However, since promiscuous conjugative plasmids can pass between these species, less stringent precautions might suffice, according to the principle that if the recombinant strain could arise in nature, it does not represent an extremely successful ecological variant—otherwise it would already exist.

B. New Areas of Research

Coupling factors have not yet been probed by the mutant approach among the photosynthetic bacteria, although many interesting findings stemmed from the study of such mutants of *E. coli* (Haddock and Jones, 1977). The difficulty with the photosynthetics has been that they were known to ferment only poorly, or not at all. Thus a mutation inactivating the coupling factor would be lethal, since the same coupling factor is used for both photosynthesis and respiration (Lien and Gest, 1973). Methods have recently been developed to allow anaerobic dark growth of Rhodospirillaceae (Uffen and Wolfe, 1970; Yen and Marrs, 1977; Madigan and Gest, personal communication); thus it should be possible to isolate, propagate, and study coupling factor mutants, assuming that at least some of these anaerobic dark-growth methods involve substrate level phosphorylation.

Wall *et al.* (1975b) have begun analyses of mutants unable to grow by fixing nitrogen gas for their sole source of nitrogen. Presumably some mutants of this type are affected in the nitrogenase itself, and thus will serve as interesting probes of the nitrogen-reducing pathway. The potential of light as a most easily manipulated energy supply has also attracted workers to study active transport (Weckesser and Magnuson, 1976; Jasper and Silver, 1977), motility, and chemotaxis in photosynthetic bacteria (Ordal, personal communication), and examples of the usefulness of mutants in similar studies using nonphotosynthetic organisms should provide ample testimony for their potential contributions among the photosynthetics.

C. APPLIED RESEARCH

Photosynthetic bacteria are not currently considered to be "industrial microorganisms," although a variety of potential uses have been demonstrated (Kobayashi et al., 1971; Crofts, 1971; Shipman et al., 1977). The rhodopseudomonads display a good deal of the metabolic versatility of the pseudomonads, and it should be possible to isolate, or create by plasmid transfer, strains that can degrade a wide range of carbon compounds unassailable by other bacteria. The photosynthetics have the advantage that they do not need to obtain energy from the compounds they degrade, but can afford to spend ATP to obtain carbon from unusual sources. An ecological role of photosynthetic bacteria in "detoxifying" rice paddies has been suggested (Kobayashi and Hague, 1971), and their abundance in most soils suggests a similar role elsewhere. In a sense they serve as a natural sewage treatment plant, and pilot studies show that this is a role they can play in captivity as well as in the wild. A variety of industrial wastes have been treated with a photosynthetic bacteria-based sewage treatment system resulting in purified water and a crop of photosynthetic bacteria, which were then utilized as fertilizer and feed supplements (Kobayashi and Tchan, 1973). Other studies have probed the potential of these bacteria to convert solar energy into H_2, which can then be used by other bacteria to make methane, or to use solar energy in fixing commercial amounts of N_2. In each of these potential applications the prospect exists of using genetic engineering to alter the characteristics of wild-type strains so as to enhance their utility. The processes involved in tailoring strains to fit industrial demands would not be envisioned to be different from those genetic manipulations described elsewhere in this article. At an interdisciplinary meeting convened in part to speculate on the feasibility of such undertakings, two problems were perceived as obstacles by members of the engineering-oriented community. To improve on the characteristics of an enzyme or process, it was proposed that one must first understand the molecular details of the reactions involved, then design proteins with improved properties, then synthesize DNA sequences coding for the improved proteins, and finally graft these new genes into the organism. Since this sequence of tasks seemed formidable, the hope of improving upon wild-type properties was deemed futile. The fallacy in this line of reasoning is that in reality one does not approach the problem in this manner at all. It is only necessary to decide what new property is to be selected for, and to devise a reasonable set of conditions that will enable a mutant with the desired properties to be isolated. The combination of mutagenesis and the vast numbers of microorganisms that can conveniently be examined provides a vast wealth of variants from which to select a desired set of properties. In other words, the improved strain is generated at random, without knowledge of why it possesses new properties, and the challenge for the experimenter is simply to find it among the vast excess of uninteresting cells.

The second so-called obstacle perceived was that nature has already been experimenting for eons to optimize the systems of living organisms. How can we hope to improve on perfection? The answer, of course, is that we would not be attempting to make a more effective competitive organism (although this too should be possible, since I doubt that evolution has ended), but rather to create an organism that, when husbanded by man, can perform a certain biochemical task better than the wild type. This experience dates to when man first began selecting and raising certain individual plants or animals, thus eventually creating distinctive breeds. The techniques of modern microbial genetics that enable us to speak of genetic engineering are only extensions of the practices of selective breeding.

VII. General Conclusions

It is this author's admittedly biased impression that a relatively small investment in research effort has already demonstrated the utility of the mutant approach to bioenergetic problems studied in photosynthetic bacteria. Biochemical genetics is not expected to provide bioenergetic insights that could not be achieved in its absence, but it seems reasonable to expect it to accelerate progress in this field as it has in others. It can be envisioned that a slight increase in the awareness of applications of mutants to bioenergetic problems will stimulate collaborative endeavors between members of the bioenergetics community and microbial geneticists, and it has been with that goal in mind that this paper was written.

REFERENCES

Bishop, N. I. (1969). *Biophys. J.* **9**, 118.
Clayton, R. K., and Haselkorn, R. (1972). *J. Mol. Biol.* **68**, 97.
Clayton, R. K., and Wang R. T. (1971). *In* "Photosynthesis and Nitrogen Fixation," Part A (A. San Pietro, ed.), Methods in Enzymology, Vol. 23, p. 696. Academic Press, New York.
Cox, G. B., and Gibson, F. (1974). *Biochim. Biophys. Acta* **346**, 1.

Crofts, A. R. (1971). *Proc. R. Soc., Ser. B* **179**, 209.
del Valle-Tascón, S., Giménez-Gallego, G., and Ramírez, J. M. (1975). *Biochem. Biophys. Res. Commun.* **66**, 514.
del Valle-Tascón, S., Giménez-Gallego, G., and Ramírez, J. M. (1977). *Biochim. Biophys. Acta* **459**, 76.
Drews, G., Dierstein, R., and Schumacher, A. (1976). *FEBS Lett.* **68**, 132.
Fillingame, R. H. (1975). *J. Bacteriol.* **124**, 870.
Garcia, A. F., Drews, G., and Kamen, M. D. (1974). *Proc. Natl. Acad. Sci. U.S.A.* **71**, 4213.
Garcia, A. F., Drews, G., and Kamen, M. D. (1975). *Biochim. Biophys. Acta* **387**, 129.
Gibson, F., and Cox, G. B. (1973). *Essays Biochem.* **9**, 1.
Giménez-Gallego, G., del Valle-Tascón, S., and Ramírez, J. M. (1976). *Arch. Microbiol.* **109**, 119.
Guerola, N., Ingraham, J. L., and Cerdá-Olmedo, E. (1971). *Nature (London)* **230**, 122.
Haddock, B. A. (1977). *In* "Microbial Energetics" (B. A. Haddock and W. A. Hamilton, eds.), p. 178–200. Cambridge Univ. Press, London and New York.
Haddock, B. A., and Jones, C. W. (1977). *Bacteriol. Rev.* **41**, 47.
Hayes, W. (1968). "The Genetics of Bacteria and Their Viruses," 2nd Ed. Wiley, New York.
Jasper, P., and Silver, S. (1977). *In* "Microorganisms and Minerals" (E. D. Weinburg, ed.), pp. 7–47. Dekker, New York.
Jones, O. T. G., and Plewis, K. M. (1974). *Biochim. Biophys. Acta* **357**, 204.
Klemme, J.-H., and Schlegel, H. G. (1969). *Arch. Mikrobiol.* **68**, 326.
Kobayashi, M., and Haque, M. Z. (1971). *Plant Soil Spec. Vol.* **1971**, p. 443.
Kobayashi, M., and Tchan, Y. T. (1973). *Water Res.* **7**, 1219.
Kobayashi, M., Kobayashi, M., and Nakanishi, H. (1971). *Hakko Kogaku Zasshi* **49**, 817.
Kováč, L., Lachowicz, T. M., and Slonimski, P. P. (1967). *Science* **158**, 1564.
La Monica, R. F., and Marrs, B. (1976). *Biochim. Biophys. Acta* **423**, 431.
Lascelles, J., and Altshuler, T. (1969). *J. Bacteriol.* **98**, 721.
Lascelles, J., and Wertlieb, D. (1971). *Biochim. Biophys. Acta* **226**, 328.
Levine, R. P. (1969). *Annu. Rev. Plant Physiol.* **20**, 523.
Levine, R. P. (1973). *Stadler Genet. Sympt.* **5**, 61.
Lien, S., and Gest, H. (1973). *Arch. Biochem. Biophys.* **159**, 730.
Marrs, B. (1974). *Proc. Natl. Acad. Sci. U.S.A.* **71**, 971.
Marrs, B. (1978). *In* "The Photosynthetic Bacteria" (R. K. Clayton and W. R. Sistrom, eds.). Plenum, New York, in press.
Marrs, B., and Gest, H. (1973). *J. Bacteriol.* **114**, 1045.
Marrs, B., Kaplan, S., and Shepherd, W. (1978). *In* "Photosynthesis and Nitrogen Fixation," Part C (A. San Pietro, ed.), Methods in Enzymology, Academic Press New York, in press.
Maturin, L., and Curtiss, R. (1977). *Science* **196**, 216.
Miles, D. (1975). *Stadler Genet. Symp.* **7**, 135.
Prince, R. C., and Crofts, A. R. (1973). *FEBS Lett.* **35**, 213.
Sasarman, A., Surdeanu, M., Szegli, G., Horodniceanu, T., Greceanu, V., and Dumitrescu, A. (1968). *J. Bacteriol.* **96**, 570.
Schatz, G., and Mason, T. L. (1974). *Annu. Rev. Biochem.* **43**, 51.
Schekman, R., Weiner, A., and Kornberg, A. (1974). *Science* **186**, 987.
Shipman, R. H., Fan, L. T., and Kao, I. C. (1977). *Adv. Appl. Microbiol.* **21**, 161.
Sistrom, W. R., and Clayton, R. K. (1964). *Biochim. Biophys. Acta* **88**, 61.
Sistrom, W. R., Griffiths, M., and Stanier, R. Y. (1956). *J. Cell. Comp. Physiol.* **48**, 473.
Solioz, M., and Marrs, B. (1977). *Arch. Biochem. Biophys.* **181**, 300.
Stenflo, J. (1972). *J. Biol. Chem.* **247**, 8167.

Takemoto, J., and Lascelles, J. (1973). *Proc. Natl. Acad. Sci. U.S.A.* **70**, 799.
Thore, A., Keister, D. L., and San Pietro, A. (1969). *Arch. Mikrobiol.* **67**, 378.
Uffen, R. L., and Wolfe, R. S. (1970). *J. Bacteriol.* **104**, 462.
Wall, J. D., Weaver, P. F., and Gest, H. (1975a). *Arch. Microbiol.* **105**, 217.
Wall, J. D., Weaver, P. F., and Gest, H. (1975b). *Nature (London)* **258**, 630.
Wang, R. T., and Clayton, R. K. (1973). *Photochem. Photobiol.* **17**, 57.
Weaver, P. (1971). *Proc. Natl. Acad. Sci. U.S.A.* **68**, 136.
Weaver, P. F., Wall, J. D., and Gest, H. (1975). *Arch. Microbiol.* **105**, 207.
Weckesser, J., and Magnuson, J. (1976). *J. Supramol. Struct.* **4**, 515.
Yen, H. C., and Marrs, B. (1976). *J. Bacteriol.* **126**, 619.
Yen, H. C., and Marrs, B. (1977). *Arch. Biochem. Biophys.* **181**, 411.
Zannoni, D., Baccarini-Melandri, A., Melandri, B. A., Evans, E. H., Prince, R. C., and Crofts, A. R. (1974). *FEBS Lett.* **48**, 152.
Zannoni, D., Melandri, B. A., and Baccarini-Melandri, A. (1976). *Biochim. Biophys. Acta* **449**, 386.

Subject Index